BETSI WELLS

SYSTEMS ANALYSIS AND PROJECT MANAGEMENT

McGraw-Hill Series in Management

Keith Davis and Fred Luthans, *Consulting Editors*

Allen: Management and Organization
Allen: The Management Profession
Argyris: Management and Organizational Development: The Path from XA to YB
Beckett: Management Dynamics: The New Synthesis
Benton: Supervision and Management
Brown: Judgment in Administration
Buchele: The Management of Business and Public Organizations
Campbell, Dunnette, Lawler, and Weick: Managerial Behavior, Performance, and Effectiveness
Cleland and King: Management: A Systems Approach
Cleland and King: Systems Analysis and Project Management
Cleland and King: Systems, Organizations, Analysis, Management: A Book of Readings
Dale: Management: Theory and Practice
Dale: Readings in Management: Landmarks and New Frontiers
Davis: Human Behavior at Work: Organizational Behavior
Davis and Newstrom: Organizational Behavior: Readings and Exercises
Davis, Frederick, and Blomstrom: Business and Society: Concepts and Policy Issues
DeGreene: Systems Psychology
Dunn and Rachel: Wage and Salary Administration: Total Compensation Systems
Edmunds and Letey: Environmental Administration
Feldman and Arnold: Managing Individual and Group Behavior in Organizations
Fiedler: A Theory of Leadership Effectiveness
Finch, Jones, and Litterer: Managing for Organizational Effectiveness: An Experiential Approach
Flippo: Personnel Management
Glueck: Business Policy and Strategic Management
Glueck: Readings in Business Policy and Strategy from *Business Week*
Glueck: Strategic Management and Business Policy
Hampton: Contemporary Management
Hicks and Gullett: Management
Hicks and Gullett: Modern Business Management: A Systems and Environmental Approach
Hicks and Gullett: Organizations: Theory and Behavior
Johnson, Kast, and Rosenzweig: The Theory and Management of Systems
Karlins: The Human Use of Human Resources
Kast and Rosenzweig: Experiential Exercises and Cases in Management
Kast and Rosenzweig: Organization and Management: A Systems and Contingency Approach
Knudson, Woodworth, and Bell: Management: An Experiential Approach
Koontz: Toward a Unified Theory of Management
Koontz, O'Donnell, and Weihrich: Essentials of Management
Koontz, O'Donnell, and Weihrich: Management

Koontz, O'Donnell, and Weihrich: Management: A Book of Readings
Lee and Dobler: Purchasing and Materials Management: Text and Cases
Levin, McLaughlin, Lamone, and Kottas: Production/Operations Management: Contemporary Policy for Managing Operating Systems
Luthans: Introduction to Management: A Contingency Approach
Luthans: Organizational Behavior
Luthans and Thompson: Contemporary Readings in Organizational Behavior
McNichols: Policymaking and Executive Action
Maier: Problem-Solving Discussions and Conferences: Leadership Methods and Skills
Margulies and Raia: Conceptual Foundations of Organizational Development
Margulies and Raia: Organizational Development: Values, Process, and Technology
Mayer: Production and Operations Management
Miles: Theories of Management: Implications for Organizational Behavior and Development
Miles and Snow: Organizational Strategy, Structure, and Process
Mills: Labor-Management Relations
Mitchell: People in Organizations: An Introduction to Organizational Behavior
Molander: Responsive Capitalism: Case Studies in Corporate Social Conduct
Monks: Operations Management: Theory and Problems
Newstrom, Reif, and Monczka: A Contingency Approach to Management: Readings
Petit: The Moral Crisis in Management
Petrof, Carusone, and McDavid: Small Business Management: Concepts and Techniques for Improving Decisions
Porter, Lawler, and Hackman: Behavior in Organizations
Prasow and Peters: Arbitration and Collective Bargaining: Conflict Resolution in Labor Relations
Reddin: Managerial Effectiveness
Sartain and Baker: The Supervisor and the Job
Sayles: Leadership: What Effective Managers Really Do . . . and How They Do It
Schlesinger, Eccles, and Gabarro: Managing Behavior in Organizations: Text, Cases, Readings
Schroeder: Operations Management: Decision Making in the Operations Function
Shore: Operations Management
Shull, Delbecq, and Cummings: Organizational Decision Making
Steers and Porter: Motivation and Work Behavior
Steinhoff: Small Business Management Fundamentals
Sutermeister: People and Productivity
Tannenbaum, Weschler, and Massarik: Leadership and Organization
Vance: Corporate Leadership: Boards, Directors, and Strategy
Walker: Human Resource Planning
Werther and Davis: Personnel Management and Human Resources
Wofford, Gerloff, and Cummins: Organizational Communications: The Keystone to Managerial Effectiveness

SYSTEMS ANALYSIS AND PROJECT MANAGEMENT

THIRD EDITION

David I. Cleland

Professor of Engineering Management
School of Engineering
University of Pittsburgh

William R. King

Professor of Business Administration
Graduate School of Business
University of Pittsburgh

McGRAW-HILL BOOK COMPANY

New York St. Louis San Francisco Auckland Bogotá
Hamburg London Madrid Mexico Montreal New Delhi
Panama Paris São Paulo Singapore Sydney Tokyo Toronto

This book was set in Times Roman by Black Dot, Inc. (ECU).
The editors were Kathi A. Benson and Gail Gavert;
the production supervisor was John Mancia.
New drawings were done by Allyn-Mason, Incorporated.
Halliday Lithograph Corporation was printer and binder.

SYSTEMS ANALYSIS AND PROJECT MANAGEMENT

7890HALHAL89

ISBN 0-07-011311-4

Library of Congress Cataloging in Publication Data

Cleland, David I.
Systems analysis and project management.

(McGraw-Hill series in management)
Includes bibliographies and index.
1. Industrial project management. 2. System analysis. I. King, William Richard, date
II. Title. III. Series
HD69.P75C53 1983 658.4'032 82-14817
ISBN 0-07-011311-4

TO OUR FAMILIES

CONTENTS

PREFACE

After the first edition of this book was honored with the McKinsey Foundation Award, we were pleased to find that many people took the occasion to contact us to tell us how they were using the book and to give us their opinions on its strong and weak points. We were told of uses ranging from formal classes in universities and management development programs to an individual's use of the ideas in the book as a basis for successfully analyzing and operating a new business venture. Both the varied uses and specific ideas inevitably suggested changes to us that would improve the book. The situation was very much the same with the second edition. Now this third edition offered us an opportunity to make further improvements.

The systems applications that were innovative when the first edition was published are now so much a part of the everyday life of organizations that they sometimes go unrecognized. Thus, the third edition gave us a chance to go well beyond mere refinements and updating.

Nonetheless, the basic objective of the book remains the same—to present the ideas of systems analysis and project management in a manner which demonstrates their essential unity and their applicability in a wide variety of business, industrial, and public management situations.

We accomplished this by demonstrating how systems ideas are essential to management in complex organizations and a complex society. We did this by relating systems ideas to the two generic phases of management—*deciding* and *doing*—which were here termed *planning* and *implementation*. Planning and implementation are made operational through the *strategic planning system* and the *project management system*.

Within this framework the book deals with systems analysis and project management as the methodologies through which planning and implementation are achieved.

We hope that the book will continue to have wide appeal to those involved in the management of complex systems programs and organizations. Executive-level managers who are concerned with strategic planning for the integration of complex systems and programs into an overall organizational framework should also find it useful.

For students of business, engineering, or public management, the book helps to integrate traditional management thought with modern systems concepts. We have attempted to present operational systems approaches as complementary, rather than contradictory, to much of that thought.

The level of quantitative or mathematical sophistication of each of these audiences should be adequate for an understanding of the book. The emphasis is on the underlying concepts and methodology and not on the specialized mathematics, and no special mathematical competence is required. The text material itself is suitable for an upper-level undergraduate and for a graduate course in management. It is presumed that the student or practitioner using this book has been exposed to the rudiments of management theory, but no detailed knowledge is necessary.

The growing body of literature reflecting the awareness of the dynamics of management practices by contemporary academicians and practitioners has been drawn on extensively. There are sufficient footnotes and recommended readings so that instructors can structure presentations around the background of their students.

The format of the book is simple and adaptable to many different courses and the instructor may choose to emphasize various facets of the overall topic. Briefly, the material covered includes the following:

Part One: Basic Systems Concepts consists of a single chapter that introduces systems ideas and their applicability to management of complex systems, programs, and organizations.

Part Two: Systems Approaches in Planning and Implementation is made up of two chapters that show how these systems ideas are specifically related to the planning and implementation phases of management. Chapter 2 develops the basis for applying systems ideas to planning and describes an operational process for doing so and a management system—a *strategic planning system*—that permits this to be done on a continuing basis in an organization. Chapter 3 deals similarly with the implementation phase of management, ending with the description of a *project management system* for achieving the same kind of integration and continuity.

Part Three: Systems Analysis comprises four chapters that describe the concepts, methods, and applications of systems-oriented analytic approaches to the problem-solving aspects of management.

Part Four: Project Management consists of eight chapters that deal with project management in terms of its objectives, methods, techniques, and impacts. The various forms of project and matrix management are extensively discussed along with assessments of the values and limitations of the various approaches.

Part Five: Organizational Support for Systems-Oriented Management outlines the informational and cultural support that is so critical if systems-oriented management is to be successful. Strategic planning systems and project management systems will prosper only in a supportive environment. Chapter 16 deals with the information support dimension of the environment, while the last chapter deals with the cultural ambience—a less well-defined but very important aspect which can help or hinder the successful application of systems ideas to organizational management.

ACKNOWLEDGMENTS

We would like to thank Claire Zubritzky and Olivia Harris who have ably supported us through this and numerous other projects over the years. Many others have made significant contributions to the three editions; however, none have been more valuable and appreciated than the environments, cultures and motivations provided by Max Williams and Jerry Zoffes, our respective deans.

We would also like to express our thanks for the many useful comments and suggestions provided by colleagues who reviewed this text during the course of its development, especially to Professor Arthur Gagne, Northeastern University; Professor Henry Hays, North Texas State University; Professor Ted Helmer, University of Hawaii; Professor Fred Luthans, University of Nebraska; Professor James Rice, University of Wisconsin, Oshkosh; Professor Herbert Schuette, Boston University; Dr. August Smith, Texas A & M University; and Professor Michael Stahl, Clemson University.

David I. Cleland

William R. King

PART ONE

BASIC SYSTEMS CONCEPTS

CHAPTER 1

MANAGEMENT AND THE SYSTEMS CONCEPT

When we try to pick out anything by itself, we find it hitched to everything else in the universe.

John Muir

The world today is facing a set of problems and opportunities which are unprecedented. New centers of world power and influence have emerged, and the continuing change in the balance of economic, political, and military power has radically altered the rules under which nations and large organizations operate. Indeed, the nations of the world are currently more closely interrelated than were the various cultural segments of individual nations only a few decades ago.

The unique nature and magnitude of the changes that are occurring around us have been well documented and much discussed, leading to the inclusion of new phrases such as "future shock" and an "age of discontinuity"—both titles of popular books[1]—into the lexicon. In addition to documenting the radical change which is upon us and distinguishing this era of change from the more modest changes which have occurred during every period of history, authors such as Toffler and Drucker have suggested that our advanced technology can be

[1]A. Toffler, *Future Shock,* Random House, Inc., New York, 1970; and P. Drucker, *The Age of Discontinuity: Guidelines to Our Changing Society,* Harper & Row, Publishers, Incorporated, New York, 1968.

harnessed to solve "modern" problems such as energy inadequacy and racial discrimination, as well as to conquer some of the age-old enemies of humanity such as hunger and disease.

However optimistic one may be about such possibilities, to address these issues effectively one must recognize three salient characteristics of modern societal problems and opportunities that present challenges for significant human progress. The common characteristics of these challenges are *interdependency, complexity,* and *change.* Each of these challenges, wherever they are found—in transportation, education, health care, pollution, or poverty, to name a few—involves many different factors and forces, many interest groups (each with a different goal in mind), myriad organizations, uncertainty, and diverse opinions concerning the effectiveness of proposed strategies and solutions. Moreover, they involve constantly evolving issues and priorities. Thus, they are problems involving *complex systems.*

COMPLEX SYSTEMS

Most would agree with the proposition that our significant societal challenges involve interdependency, complexity, and change, but few of those who do also have a solid operational understanding of the corollary that *the solutions to these challenges will themselves involve a significant degree of interdependency, complexity, and need to adapt to change.* The words of an AT&T commercial message ("The system is the solution") are perhaps overly simplistic, but the fact is that the solutions to complex problems and issues are unlikely to be found in anything but complex systems. Thus, the complexity of societal systems dictates that the solutions to its problems will also involve complex systems.

The recognition that the critical problems of society are so complex that they defy simple solution is fairly new and novel. Within the recent past, a "war on poverty" or a return to the virtues of yesteryear could be taken seriously as an approach to resolving serious societal dilemmas. Virtually no one believes in such easy solutions today. Simplistic ideas with such differing geneses as the U.S. Peace Corps and socialistic wealth redistribution schemes have been shown—either in practice or through logical analysis—to have, at best, marginal impact in solving the human race's critical problems. Even the traditional solution of America's past—technology—is inadequate in itself to cope with this degree of complexity. Bigger, better, and faster passenger railroad trains will not improve the quality of life in a city where individuals cannot be persuaded to abandon their autos, politicians cannot agree on routes, public managers cannot agree on the appropriate technology, and rights-of-way are unable to handle trains with increased speeds.

Complex Systems and Technology

Modern societal problems are so *interdependent* that the solution to one cannot be sought without considering the others. Attacks on pollution have restricted the possible solutions available for transportation and energy problems. Both, in turn, have effects on the problems of the poor, since the availability of jobs and access to the workplace intrinsically depends on the degree to which pollution programs have driven industry from the city and the effectiveness of the transportation systems.

Thus, we are faced with complex system problems requiring complex system solutions, whose choice depends on consideration of these interrelationships and the wide array of possible consequences and impacts on diverse interest groups. Many of these solutions will have significant technological components, but these alone will not suffice.

Conversely, it is foolish even to contemplate nontechnological solutions to problems such as transportation and pollution; and it is also improbable that problems such as education and poverty will not be solved with approaches having high technological content, since just to keep track of those in school or those on welfare rolls now requires the use of some of the most advanced technology available in the form of computers.

Thus, technology will be important, if not sufficient, to solve our current problems, just as it has been in the past. However, despite our reliance on technology to a degree which few of us really perceive,[2] we have not developed the ability to *organize* the technology and to *integrate* it into systems which effectively resolve problems. In essence, we have not developed the ability to *manage technology* effectively.

Moreover, if we are ever to escape from the vicious cycle of having tomorrow's problems evolve from the technological solutions to the problems of today, we shall have to go beyond the mere management of existing technology to develop an ability to *plan* for technological change, and to forecast the effect of technological change in social and cultural, as well as economic, terms.

Complex Organizations

Another variety of complex system will also play an important role in the solution of society's complex problems. The *organization,* be it a business firm, government agency, or loosely organized pressure group, is the vehicle through which modern society operates.

Despite the familiarity of such organizations to everyone, we may not notice that organizations look quite different from the way they did only a few years ago. Members of minority groups are now more fairly represented in all aspects of organizational life. Even the U.S. Armed Forces, long a bastion of male dominance, are undergoing major changes in becoming "coed."[3]

Such changes are, and will continue to be, reflected in the *structure* of organizations as well. For instance, General Motors has adopted the "project center" concept as a way of organizing engineering resources to redesign

[2]For instance, the calculation has been made that were AT&T to return to the old "number please" noncomputer system, the required number of telephone operators to handle today's volume of calls would be fast approaching the entire population.

[3]See George Gilder, "The Case Against Women in Combat," *The New York Times Magazine,* Jan. 28, 1979.

automobile components. Project centers, composed of design teams, work on plans and engineering problems common to all divisions, such as frames, electrical systems, and steering gear. In such an organizational structure, engineers find themselves working for "two bosses." This creates a very different management situation from that which existed in the "old days" when every individual was responsible to only one superior.[4]

This variety of change in complex organizations will continue to occur. As Toffler wrote:

> The more rapidly the environment changes, the shorter the life span of organizational forms. In administrative structure just as in architectural structure, we are moving from long enduring to temporary forms, from permanence to transients. We are moving from bureaucracy to ad-hocracy.[5]

THE NEED FOR SOPHISTICATED MANAGEMENT

The title *manager* often portrays an image of an individual operating in a relatively simple environment—"worker and supervisor"—using relatively straightforward "principles" which guide that individual's behavior. While it is true that such situations still exist today, the focus of modern management is on complex organizations which deal with complex problems using sophisticated systems solutions.

The task of management in complex organizations is quite different from the simple image. Market forces are no longer the only, or perhaps even the primary, external influence with which the manager must deal. Government regulation impinges on many aspects of management that previously might have been dealt with using one's own discretion.[6]

The organization to be managed is itself changing, as are the attitudes and expectations of the people of which it is made up. The systems which the organization uses to support managers are becoming more pervasive and complex. Now most managers must routinely deal in some capacity with managerial control systems, strategic planning systems, personnel systems, etc.

The information that is available to managers as a consequence of sophisticated management systems has greatly expanded the degree to which they can obtain and use objective data, as opposed to relying on subjective judgment.[7] As a corollary, however, this information explosion has created new problems such as the need to cull out critical relevant information from the mass of less relevant data.[8]

[4]Charles G. Burck, "How GM Turned Itself Around," *Fortune,* Jan. 16, 1978.

[5]Toffler, op. cit., p. 123.

[6]For a discussion of the importance placed on government regulation by managers, see L. Fahey, and W. R. King, "Environmental Scanning in Corporate Planning," *Business Horizons,* vol. 20, no. 4, pp. 61–77, August 1977.

[7]Ibid.

[8]R. L. Ackoff, "Management MISinformation Systems," *Management Science,* vol. 14, no. 4, pp. B-147-B-156, December 1967.

The Ubiquity of Management

As management has become more complex, it has also become more pervasive. The modern theory and practice of management developed primarily in the private sector, from the vantage of the business manager, although notable contributions have been made by other sectors of society, e.g., military and ecclesiastical organizations. The business orientation was reinforced by business schools in the United States that taught management from the perspective of the business organization. In recent years, other schools in the university are teaching management from their viewpoint, e.g., courses in engineering management, public administration, public health management, hospital management, to name a few. These changing patterns for the focus of professional management education in the United States reflect a growing awareness of the need for management in virtually every aspect of modern society.

Even in business itself, the shift from a manufacturing to a service economy has created many changes in the way management must be performed. Service businesses have grown explosively in terms of revenues, employment, and complexity. The estimated proportion of the gross national product (GNP) accounted for by the service sector rose to about 67 percent in 1975. Twice as many people are employed in the service sector as in the non-service sector of the economy.[9]

This significant change has led Drucker to suggest that service institutions are:

> . . . so important that we are beginning to talk of a third sector of society—neither public (governmental) nor private in the old sense of the private business sector. The third sector is composed of institutions which are not government agencies but which are still not profit-making.[10]

Although these service organizations must certainly be managed, they present unique management challenges. The traditional business success criterion of profitability is not readily applied. Organizational results have to be measured using some criteria other than financial ones.

Thus, there is today a need for a variety of management that is quite different from that which was adequate only a short time ago. It is this sort of management to which this book is addressed.

MANAGING COMPLEX SYSTEMS

It is not our purpose here either to review the fundamentals of management or to develop a new theory of management.[11] However, in working in complex systems, we have found a number of classes of skills which appear to be

[9] *Statistical Abstract of the United States,* U.S. Bureau of the Census, 1979. Since these data ignore services provided by manufacturing firms such as warranty repairs on appliances and machines, the actual proportion is undoubtedly even higher.

[10] Peter Drucker, "Managing the Third Sector," *The Wall Street Journal,* Oct. 3, 1978.

[11] See David I. Cleland and William R. King, *Management: A Systems Approach,* McGraw-Hill Book Company, New York, 1972, for a basic systems view of management.

correlated with good management in this new variety of environment. Since these skills are not those which characterize successful traditional managers, we use them here to introduce the concept of managing complex systems.

We believe that successful managers of complex systems must possess:

1 An understanding of the technology of their "business"

2 An understanding of the "basic concepts of management"

3 An interpersonal style which facilitates their ability to get things done through others

4 An ability to conceptualize and to operate using a systems approach

Not only do we believe that these classes of skills characterize a new breed of manager, we also find that the truly successful ones possess *all* these traits woven together in a mixture which may be different by virtue of their industry, their agency, or their level in the organization, but not one of which is ever completely absent. Thus, the day of the manager who gets by on personality alone or solely on technical expertise is probably past. Today's problems and organizations require that all these facets be developed by the manager to be used in a synergistic way, whether one's "business" is public administration, university administration, or industry. This is so because the focus of management in all complex organizations has shifted from *efficiency* to *innovation*.

The Manager's Technology Base

There are many organizations which have, in the past, been effectively managed by a "pure manager"—one who knows a great deal about the universal process of management, but little about the specific context or technology of the organization. The popular literature of management is replete with stories and articles concerning professional managers who "turned around" companies by trimming away the losers, instituting financial controls, and inserting a few young MBAs into key positions. Moreover, such managers are often said to have done all of this despite the fact that they knew nothing of the technology of the particular business and had, in fact, spent all their professional lives in other industries. Thus, it is emphasized, a good manager can manage anything!

So, too, do many management textbooks tend to emphasize the universality of management. The idea of management "principles" (maxims which may be generally applied to achieve better organizational performance) is a traditional approach adopted by textbook authors attempting to distill and codify the knowledge base of management.

However much truth there may be in the concept of the generic nature of management and its universal applicability, this viewpoint is obviously more applicable to an environment where organizational improvement is more dependent on achieving *efficiency* in existing operations than it is to one where *innovation* is the key to success. For instance, management principles such as unity of command—one and only one boss for each person—have a greater degree of validity in a context where everyone understands his or her job, the environment, and the objectives being sought, than in the sort of situation where one individual's qualifications to be another's boss may not even be clear. When efficiency, neatness, and the appearance of being organized are paramount, management is indeed more universal than it is in other more uncertain contexts. But, unfortunately for those who value this superficial kind of "organization," the important problem contexts of today are filled with uncertainty and complexity.

No amount of greater efficiency in steel making and auto production will solve today's problems of pollution, and no amount of better "organization," in the simple sense, will enable the transit agencies to solve the transportation problems of the cities. These solutions are dependent on *technology* and its *effective* use—i.e., on *technology and the management of technological innovation.*

In today's rapidly changing world, managers who do not know the technological base cannot possibly foresee future developments and relationships to other technologies. If they cannot do these things, they cannot effectively participate in the integration of their technology with others to provide the system solutions which are needed. More important, without this understanding they cannot foresee the future consequences of today's decisions. Thus, they cannot participate effectively either in taking advantage of current opportunities or in anticipating the future. Without such participation, they can only nurture the existing organization activities in the fashion of caretakers.

Of course, the *technology* referred to here is more broadly defined than in the traditional engineering sense. If the organization's relevant technology is knowledge-based rather than machine-based, as in many public agencies and service industries, the requirement is equally valid, for the solution of many of our problems may well require the variety of knowledge possessed by the psychiatrist as much as that of the engineer. Indeed, the technological understanding referred to here encompasses *process* understandings as well as subject ones. For instance, an understanding of the process of innovation and development within a technological area is equally important, as is an understanding of the technology itself. Only through such process understanding can managers participate in the idea generation process and be in a position to evaluate ideas in terms of the development, production, and sales phases of their life cycle.

The Management Concepts Base

However important may be the manager's ability to understand the present and future of relevant technologies, it is as important as it was in the days of Weber's[12] bureaucracy for the manager to have an understanding of the basic concepts of management.

[12]Max Weber, *Essays in Sociology,* ed. and trans. by H. H. Gerth and C. W. Mills, Oxford University Press, Fair Lawn, N.J., 1946.

The traditional view of management recognizes that it is a process concerned with the achievement of objectives. It can be described in a number of ways, e.g., as the process of creating an environment for organized efforts to accomplish group goals, or as "the function of executive leadership anywhere."[13] Other definitions of management are fundamentally the same and most contain universal elements:

Management is a distinct process dealing with some form of *group activity.*
Objectives are involved.
A manager accomplishes organizational objectives through *working with other people* in the organization.
To do this, he must *establish effective relationships* among the human and nonhuman resources.
Decision making is pervasive in the management process.
Leadership is an integral part of the process.

Management is sometimes alternatively described as the process of administration.[14] For instance, the chief executive official of NASA is called an *administrator.* So, too, are public school officials usually called administrators. The dichotomy in a university between faculty and administrators is often used as a focal point in discussing organizational matters. Most military organizations use the term "commander" to describe the individual who gives direction to the organization. Although that title is laden with connotations of ultimate unchallengeable authority, modern military commanders find themselves in leadership positions which are not very different from those of public or corporate managers. They must work through others and motivate them to adhere to the goals which have been set. If they try simply to order things into being, they find that their *real authority* is limited and that things may not work out according to plan.

The central view of management is the same in all these contexts, whatever may be the terminology in use. Management is the process of leading organizational effort in pursuit of organizational goals.

In coordinating the achievement of organizational goals, the manager performs a number of functions. These are variously described in the literature as planning, organizing, motivating, staffing, controlling, etc.[15] Whatever may be the disagreement regarding the particular subfunctions of management, there is general agreement that the two salient functions are *deciding* and *doing.* We

[13]Ralph C. Davis, *The Fundamentals of Top Management,* Harper & Row, Publishers, Incorporated, New York, 1951, p. 6.

[14]Some have tried to distinguish between management and administration but the difference is not widely accepted. For instance, see the discussion in H. Larson, *Guide to Business History,* Harvard University Press, Cambridge, Mass., 1950, pp. 759–760.

[15]See David I. Cleland and William R. King, *Management: A Systems Approach,* McGraw-Hill Book Company, New York, 1972, pp. 117–123, for a comparison and analysis of the functional taxonomies used by various authors.

shall refer to these two functions as *planning* and *implementing*, while recognizing that any such simplistic model is an oversimplification of the complexity and diversity of the manager's function. This dichotomy is useful in conceptualizing the manager's functions and in relating the various subfunctions to one another.

Planning Virtually all management theorists agree that planning is a major element of the manager's job. However, there is no general agreement concerning precisely what constitutes planning. As the term is used here, it encompasses the activities which are variously referred to as goal setting, policy making, strategic planning, and strategic decision making. This interpretation of planning thereby involves the identification of the broad goals of the organization and the specification of strategic policies which prescribe the way in which the organization will go about achieving its goals. Thus, it encompasses such activities as:

The identification of opportunity areas for future organizational activity

The assessment of the *values* held by individuals, by the organization, and by the society

The assessment of the organization's economic and noneconomic responsibilities to society and to its clientele

The forecasting[16] of future environments and technologies which may be important to the organization

The contrivance of the strategies and the processes by which future goals can be met

Another interpretation of planning which is encompassed by the term as used here is that which incorporates the "process of preparing for the commitment of resources in the most economical fashion, and, by preparing, of allowing this commitment to be made less disruptively."[17] This interpretation of planning is usually referred to as *long-range* planning since it involves explicit consideration of the (sometimes distant) future. This is the phase of planning which is most commonly formalized in organizations, since it involves the taking of general goals and translating them into specific objectives, which objectives are consistent with the goals, together with the development of ways of achieving the objectives. The planning process therefore also encompasses the process of *strategic decision making,* which is addressed to the consideration of the alternative allocations of resources which will achieve the organization's goals and objectives to the greatest degree.

[16]The term "forecasting" as used here is not synonymous with "predicting what the future will be like." Rather, it means considering a range of possible alternative futures and, by doing so, making explicit the choices which are available. See, for example, R. Bauer, *Second-Order Consequences,* The M.I.T. Press, Cambridge, Mass., 1969, p. 29.

[17]E. Kirby Warren, *Long-Range Planning: The Executive Viewpoint,* Prentice-Hall, Inc., Englewood Cliffs, N.J., 1966, p. 21.

The roles of these various kinds of planning may be better made clear by illustrating their relationship in a specific context. A firm's top management might decide that the available opportunities and social constraints are such that it should enter a field in which social benefit, as well as profit, can be derived. For instance, the authors were recently involved in consultation with the executives of a company who were facing the strategic question of whether to enter a new product line. After extensive analysis of the risks and profitability of the product line, it was concluded that profit potential was not great, but the opportunity for making meaningful social contributions was extraordinary. The chief executive of the company ultimately decided to go ahead with the new product, noting that it would be a way to satisfy both stockholder and societal demands. Thus, the accomplishment of social good as well as profits became the goal of the company.[18]

Implementing The implementing function involves all those things that must be done to achieve objectives once the strategic choices have been made. The great importance of this function reflects the awareness of managers in even the most authoritarian environments that to order something into being does not ensure that it will actually be accomplished. Thus, the often emphasized deciding function of management is necessary, but not sufficient, to get the manager's overall job accomplished.

To accomplish objectives—even in prespecified ways—requires *organized* effort. This, in turn, requires that human beings be given specific *directions* and be *motivated* to follow them. Finally, these individuals must be *monitored* and *controlled* to ensure that they have performed their assigned functions properly.

The *organizing* subfunction of implementing thus has to do with the procurement of human and nonhuman elements, the grouping and alignment of the factors, and the establishment of authority and responsibility patterns within the overall organizational framework and within the policy and strategic limits, as prescribed by the results of the planning process. The organizing process recognizes that a complex system of informal relationships exists in any group activity. This informal organization is a network of personal and social relationships existing along with the formal organizational structure. The informal organization emphasizes people and their roles as determined by peers, in contrast to the formal organization, which emphasizes functions, positions, and specific grants of authority and responsibility.

Motivating has to do with the face-to-face leadership situation between superiors and subordinates and between peers and associates. It is in the management function of motivating that the efficacy of the organizational leader is most visible. A leader is someone who leads people along a way to the successful accomplishment, it is hoped, of personal and organizational objectives. The motivation of people depends on a myriad of environmental forces;

[18]See W. R. King and D. I. Cleland, *Strategic Planning and Policy,* Van Nostrand Reinhold, New York, 1978, for a comprehensive discussion of modern strategic planning.

perhaps most important of all is the manner in which the leader provides a cultural ambience where people can realize social, economic, and psychological satisfaction in affiliating with the organization.

Controlling is the process of making events conform to plans, i.e., coordinating the action of all parts of the organization according to the plan established for attaining the objective.

The manager performs all these functions—planning (in its various phases), organizing, motivating, and controlling—more or less continuously and regardless of organizational level, although the emphasis placed on each function is different at the different organizational levels. The operational manager who is charged with the responsibility for accomplishing a specific mission, for example, is most concerned with the control function in performing the mission itself. A staff official who is charged with the development of the overall plan of a project is more involved with the planning function than with organizing or controlling. Indeed, Drucker holds that "planning and doing are separate parts of the same job; they are not separate jobs. There is no work that can be performed effectively unless it contains elements of both . . . advocating the divorce of the two is like demanding that swallowing food and digesting it be carried on in separate bodies."[19]

Planning and Implementing It is probably not worthwhile to suggest that the planning-implementing dichotomy is of any value other than that of a general guide to use to distinguish between the deciding and doing aspects of the manager's job. In fact, it is useful to recognize that the two functions intrinsically overlap.

There is a clear planning activity involved in the implementing function, for instance. To suggest otherwise is to assume that all decisions—even the most detailed—are made before implementation is begun, and that is clearly not the case in most instances. Planning and implementing are intertwined phases of management which may be conveniently unwound to serve the purposes of better understanding and communication. In taking advantage of this convenience, one should not be lulled into thinking that the model *is* management. It is merely an abstract representation which will, hopefully, serve to cast light on the real management process.

The Manager's Personal Style Base

While both a technological and a management base are necessary for the manager, they alone will not ensure success. Whatever academics might like to believe about their ability to teach technological understanding or management skills, these "teachable" things will not suffice; there is another critical ingredient which can be readily seen through close contact with successful managers of

[19]Peter F. Drucker, *The Practice of Management,* Harper & Row, Publishers, Incorporated, New York, 1954, p. 284.

complex systems—personal style—which has been thought of as "unteachable."[20]

Managers must accomplish organizational goals by working through others. They obviously have no choice in this matter, since they can almost never achieve these goals by virtue only of their personal efforts, however salutary. Thus, to be effective, managers must develop a personal style which permits them to translate formal authority into *real authority*—the ability to exert influence on the decisions and activities of others.

Successful managers typically play a role of total involvement, high activity, and organizational, if not personal, aggressiveness. They display self-assurance in the degree to which they perceive themselves to be capable and effective in dealing with problems. These characteristics often manifest themselves in such a way that managers are perceived by others to be interested in the activities of the whole organization, not just in those for which they are responsible. Also, these individuals are perceived as "people-oriented," since they are constantly trying to accomplish successfully that elemental aspect of management—working through others.

People-oriented managers are selling the idea of cooperative effort. Thus, they each develop a working style in dealing with peers, associates, subordinates, and superiors such that they are accepted as "team leaders." People who work for them and with them recognize them as leaders—those who are accepted as informal as well as the formal leaders. They appreciate and respect the multidimensional nature of human motivation and earnestly try to provide a work environment for their people that allows the the attainment of economic and psychological satisfaction along with the achievement of organizational objectives.

It is extremely difficult to discuss personal style at anything other than the extremes—the extreme of pontification or the extreme of complex psychological and sociological jargon. However, there are some specific critical elements of style which are reasonably easy to define. One of those elements—credibility—brings to mind the negative phrase "credibility gap." The characteristic called credibility is defined as the "quality, state, or condition that produces believability or trustworthiness."[21] Depending more on the relationships among people than on contrived systems, this characteristic addresses the good faith of managers and their sense of priority in pursuing organizational purposes. Hence, it is clearly an important element of personal style.

Another such element of style is interrelated with the manager's technological base. This element is the ability of managers to fathom the professional orientations and loyalties of the technical people with whom they must effective-

[20]The American Assembly of Collegiate Schools of Business (AACSB) has initiated research aimed at determining if, and how, such "noncognitive" skills may be addressed in the curricula of business schools. Under a grant from the Westinghouse Foundation, the Graduate School of Business of the University of Pittsburgh is developing teaching materials and methods to facilitate the development of such skills.

[21]C. D. Flory and B. A. Mackenzie, *The Credibility Gap in Management,* D. Van Nostrand Company, Inc., Princeton, N.J., 1971.

ly work. It obviously means that they must be able to communicate with technical people, but it also means that they must be able to assess the natural biases and ways of thinking of these people and to "adjust" their ideas to the realities of the marketplace.

As noted earlier, there is no single personal style which will make any manager more effective. Yet, since the existence of a unique personal style is an obvious trait of virtually every successful manager, the aspirant should recognize that there are few colorless managers at the top of any kind of modern organization. This will lead to an awareness that personality is not something to hide; rather, it is something to use to advantage.

Moreover, the novice manager should recognize that some stylistic skills may be developed "on the job." This will probably lead those who wish to be general managers to develop experience in line positions where they will get the opportunity to deal with problems involving the conflicting objectives of various people and organizations, since no amount of observation of how line executives operate can replace the day-to-day hard knocks of making and implementing management decisions. People who have had experience in a staff capacity, when assigned to a line position often try to operate as advisers to their organization and to avoid being "boss." Too often, these people try to avoid the difficult and unpleasant decisions which line managers inevitably face. However, line managers cannot truly avoid the exercise of authority any more than they can avoid the responsibility for what happens to their organization.

The authors were engaged as consultants to do a "failure analysis" for a large U.S. company that experienced a precipitous financial failure in one of its operating divisions. This division was headed by a general manager who had not performed as a "general manager" in any previous capacity. He had many years of successful experience as an "assistant to" several managers. In his new capacity as general manager of the operating division, he was not able to make the necessary decisions to commit and control the organization's resources. He hedged, procrastinated, delayed, debated, and was never able to be decisive about organizational goals and objectives. One serious consequence of this avoidance behavior was that his subordinates began to make key overall organizational strategic decisions for themselves; no decision maker had the overall organizational general-manager viewpoint. The lack of general-manager experience of this individual was a major cause of the serious failure of the operating division.

There is simply no substitute for the experience of dealing with the realities of risk and uncertainty that constantly face the line manager trying to operate a "profitable" venture. An individual who has aspirations to become a general manager must be provided with an opportunity to try his or her hand at organizational responsibilities which require an analysis of uncertainty in the integration of resources toward the satisfaction of an objective. Experience as a project manager, task force leader, section leader, or department manager all provide an environment similar to that facing the general manager. Thus, these experiences are invaluable in developing the necessary skills and attitudes required for survival and growth as a line manager.

The Manager's Systems Base

Coupled with an understanding of technology, basic management concepts, and a personal style is the important aspect which alone best distinguishes modern managers from traditional ones—the ability to think and to act in a systems context. This systems view emphasizes a wholeness, or overview, of the organization or problem which is contrary to the viewpoints which may have developed through some of their formal education.

One simple but relevant view of the systems viewpoint is that it involves precisely the opposite of the traditional problem-solving technique of "cutting the problem down to size." This traditional approach tends to assume that the obvious problem is the real one when, in fact, the visible aspects of a problem may merely be symptomatic of a bigger problem. For instance, in a business firm, this means that problems which are of a marketing nature should be viewed in terms of their production, financial, and personnel implications as well. It also requires that various organizational clientele be considered in terms of their objectives, roles to be played, and possible reactions to any decision which is made.

There is also an implicit but inextricable emphasis on *analysis* in the systems base. While the analysis conducted by a manager may not be formalized operations-research modeling or "systems analysis" in the formal sense, systems-oriented managers do rely on objective *data bases* rather than relying solely on hunches or intuition.[22] However, they do not rely on analysis alone; rather, they use analysis to complement intuition and subjective judgment. Thus, in a sense, they take an overall systems approach to problem solving even in their use of the various interrelated bases which they have at their command.

Since this managerial base is so significant to the theme of this book, we shall develop the ideas more thoroughly in the next major section.

THE SYSTEMS APPROACH TO MANAGEMENT[23]

One major idea lies at the root of the modern, scientific approach to management. That idea—the systems concept—has had a substantial impact on both the planning and the implementing functions of management. This effect is best illustrated in the planning context by the increasing emphasis which is being placed on the scientific analysis of managerial decisions. Increasingly, managers are relying on decision analysts, who may call themselves "operations researchers," "systems analysts," or "management scientists," to aid in the selection of

[22]An indication of this kind of situation is described in an article in *Fortune* magazine. Uttal cites some of the difficulties suffered by RCA in decision making at the top. Robert W. Sarnoff, chief executive officer, was described as "taking great intuitive leaps from an unwarranted assumption to a foregoing conclusion." Bro Uttal, "How Ed Griffiths brought RCA into Focus," *Fortune,* Dec. 31, 1978, p. 50.

[23]Some of the material in this section is adapted with permission from a paper by one of the authors. See William R. King, "The Systems Concept in Management," *Journal of Industrial Engineering,* May 1967, pp. 320–323.

the best strategies from the myriad which are typically available. And, as they do, the basic framework by which decisions are made shifts away from the traditional pattern.

In the implementation phase of management, managers have also become increasingly independent of traditional management thinking and principles. In an effort to "get the job done," pragmatists have evolved new management approaches which are best exemplified by the current emphasis on project management[24] in executing plans and decisions.

A *system* may be defined literally as "an organized or complex whole; an assemblage or combination of things or parts forming a complex or unitary whole." The value of the systems concept to the management of an enterprise can be seen in terms of two elements of the managerial job. First, they desire to achieve *overall effectiveness* of the organization—not to have the parochial interests of one organizational element distort the overall performance. Second, they must do this in an organizational environment which invariably involves *conflicting organizational objectives.*

To demonstrate this, consider the corporate viewpoint involved in the simple decision involving which products are to be produced and in what quantities. The production department of the enterprise would undoubtedly prefer that few products be produced in rather large quantities so that the number of costly machine setups necessary to convert from production of one product to production of another would be minimized. Such a policy would lead to large inventories of a few products. The sales personnel, on the other hand, would want to have many different products in inventory so that they could promise early delivery on any product. The financial manager, recognizing that large inventories tie up money which could be invested elsewhere, would want low total inventories. The personnel manager would desire constant production levels to avoid constantly hiring new workers for short periods of peak production and laying them off in slack periods. One could go on to identify the objectives of almost every functional unit of an organization relative to this simple tactical decision problem. Obviously, all these objectives conflict to a greater or lesser degree—low inventory levels versus high inventory levels, many products versus few products, etc.

The same kind of situation can exist at every other level of the enterprise. The production department must constantly balance the speed of production with the proportion of rejects and the proportion of defective products which are not detected. The marketing function becomes involved when defective products cause complaints and lost sales. Indeed, wherever the "labor" has been divided in an organization, the management task of effectively integrating the various elements is paramount, and this can be effectively accomplished only if managers adopt the systems approach to the "system" which is their domain.

In this internal context, the systems approach involves the recognition that

[24]Sometimes referred to as "systems management" or (in the marketing context) "product management."

the organization is made up of a set of interrelated subsystems, each with its own goals. Systems-oriented managers realize that they can achieve the overall goals of the organization only by viewing the entire system, seeking to understand and measure the interrelationships, and integrating them in a fashion which enables the organization to efficiently pursue its goals.

Of course, this means that some functional unit within an organization may not achieve its parochial objectives, for what is best for the whole is not necessarily best for each component of the system. Thus, when a wide variety of products are produced in relatively small quantities, the performance of the production department may appear to be falling off; yet if this leads to greater total revenues because no sales are lost, the overall impact may be positive.

The systems viewpoint requires managers to manage their organization from the totality of the whole—from the viewpoint of how the organization fits together as an entity and how the organization fits together within a larger systems context. In this regard, Kreitzner comments:

> For many years, organizations were viewed as closed systems. More recently, an open-system perspective has gained popularity. An open-system perspective differs from the closed-system variety primarily because of its emphasis on system-environment interaction. Consequently, when one views an organization as an open system, special attention is paid to its give-and-take relationship with the surrounding environment. The "take" portion of this give-and-take relationship amounts to feedback.[25]

These simple ideas are the essence of the systems viewpoint, which has led to more effective management decisions and to organizing for the efficient execution of those decisions.

Systems and Planning

The systems concept in management decision making virtually necessitates the use of objective analysis of decision problems. The human mind can comprehend only so much, and the systems viewpoint requires consideration of the many complex interrelationships between problem elements and the objectives of numerous functional units. Moreover, even if managers were able to reduce these complexities to manageable proportions by abstracting out all but the salient aspects, they would have no guarantee that their subjective decision process was either logical or consistent.

The increasing requirements of the systems viewpoint, coupled with the previously noted changing environment, have resulted in the use of objective scientific analysis in solving decision problems. The fields of operations research, management science, and systems analysis have contributed to this end through

[25]Robert Kreitzner, "People Are Systems, Too: Filling the Feedback Vacuum," *Business Horizons,* November 1977, p. 55.

the application of scientific methods to management decision problems. Practitioners in each of these fields rely on models—formal abstractions of real-world systems—to predict the outcomes of the various available alternatives in complex decision problems.

Because these models are usually symbolic, it is possible to reduce complex relationships to a form that can be put down on paper and, using techniques of logic and mathematics, to consider interrelationships and combinations of circumstances that would otherwise be beyond the scope of any human being. Models permit experimentation of a kind which is unavailable in many environments; one may experiment on the model describing a system, rather than on the system itself.

Of course, this does not mean that decision makers cede responsibility for making decisions to some mystical scientific process or that their judgment and intuition do not play a major role in decision making. Because of the nature of the mathematics which are available, models have one of the "deficiencies" of the human brain; that is, they consider only a part of the real-world decision problem. Other parts are omitted because they are relatively unimportant or simply because they cannot be handled using existing techniques. The difference between explicit models and subjective decision making using nebulous "models" which exist in the manager's mind is one of degree. The process is similar, but explicit models formalize salient characteristics and relationships which may be blurred in the human mind. Explicit consideration is given to those aspects of the real world which should be included in the model and those which should be abstracted out. People tend to include in their mental models the first (or last) aspects which occur to them and to exclude others which stretch the bounds of their comprehension. Moreover, once the explicit model has been constructed, the objective approach has the guarantee of logic and consistency, which are not necessarily features of judgment and intuition.

The role of the manager's judgment and intuition is simply refocused by the systems approach. They are directed toward those aspects of problems which are best handled subjectively. The factors in an objectively viewed decision problem which must be handled subjectively are usually separate and distinct. This permits calm expert judgment on each aspect, rather than gross judgments encompassing wide varieties of disciplines and areas of experience.

Moreover, the manager's judgment is still of paramount importance in the process of integrating the results of objective analysis with the predicted effect of unconsidered problem elements and arriving at a decision based on the totality of available information.

In effect, the systems approach to planning may be viewed as a logically consistent method of reducing a large part of a complex problem to a simple output which can be used by decision makers in conjunction with other considerations in arriving at a best decision. It permits them to focus their attention on the aspects of the problem which are most deserving and to restrict the attention which they allocate to those things which are best handled by systems analysis. Such an integration of science and intuition permits consideration of the interrelationships of functional activities. In simple terms, it enables managers to get the "big picture" in its proper perspective, rather than requiring them to devote attention to relatively minor aspects of the total system.

Systems and Implementation

Not only has the systems concept caused great changes in the planning, or strategic decision-making, portion of the manager's function, but it has also caused revolutionary changes in the fashion in which decisions are implemented. The most striking example of this is the emergence of project-management and matrix-management concepts. Both of these related sets of ideas emphasize some form of team effort to "get the job done," however unconventionally and nontraditionally.

The *project manager* may be defined as that individual who is appointed to accomplish the task of integrating functional and extraorganizational efforts directed toward the successful performance of a specific project. The project manager (or systems manager) is confronted with a unique set of circumstances and forces with each project, and these circumstances and forces channel the manager's thought and behavior into achieving specific project goals.

The project manager's position is based on the realization that modern organizations are so complex as to preclude effective management using traditional organizational structures and relationships. Traditional philosophy is based on a vertical flow of authority and responsibility relationships and emphasizes only parts and segments of the organization. It does not place sufficient importance on the interrelationships and integration of activities involved in the total array.

Matrix management is an evolving form of project management in which the project structure is taken as a permanent aspect of the organization. In the early versions of project management, projects, and the structure within which they existed, were viewed as merely temporary.

The need for these new ways of thinking and new ways of operating organizations have evolved naturally. Top management cannot be expected to comprehend all the details and intricacies involved in the management of each activity, whether it is a weapons system under development, a product being marketed, or a client being serviced. Functional units are properly more concerned with their function than with individual products or projects. Thus, the need for a manager who can cut across traditional functional lines to bring together the resources required to achieve project goals is clear.

Just as the systems viewpoint necessitates consideration of the combined effect and interrelationships of various organizational functions in the manager's planning task, so, too, does it require integration of these functions at the implementation level. Project managers are able to operate through the various functional managers in directing the resources which are involved in effectively carrying out a project. Thus, they can focus their attention on project goals, rather than on parochial production goals or marketing goals, and they serve as instruments for implementing decisions in terms of the same structure in which they are made—the overall organizational system.

In practice, the project-management structure is superimposed on the functional organization of the company and provides a focal point for the decision-making and execution phases of a particular project. The nature of the job of managing a large project forces the integration. Management relationships have become so complex that unless someone is in control of the situation, vast amounts of resources can be exhausted before it is possible to effect a necessary retrenchment or redirection.

A SYSTEMS MODEL OF THE ORGANIZATION

When one views the organization as an open system—one made up of a collection of subsystems that interact among themselves and with their environment—it serves to facilitate systems-oriented thinking. More important, however, it also provides a basis for an operational approach to both the planning and implementing phases of organizational management. (Based on the systems model developed in this section, these operational approaches will be discussed in Chapters 2 and 3, respectively.)

Simple Systems Model

Figure 1-1 shows a generic systems model that can be applied to any system—a physical one such as a heating-cooling system, or an organizational system. Its degree of abstraction makes it seem somewhat remote from real organizations.

FIGURE 1-1
Abstract systems model of the organization.

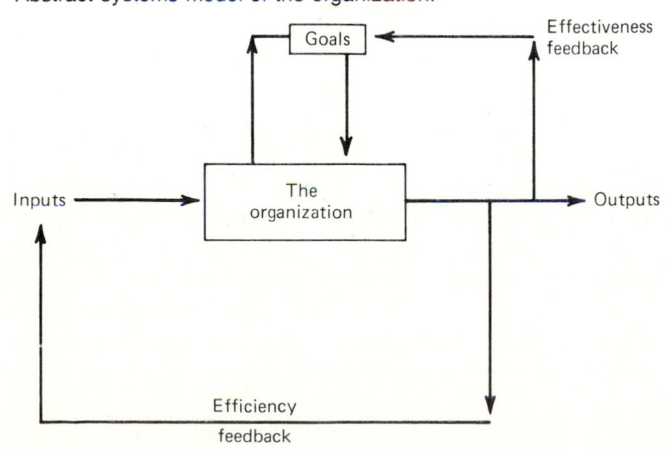

However, it will subsequently be made more detailed and less abstract, using the business organization as an example.

Figure 1-1 shows "the organization" as a block that receives inputs and transforms them into outputs. The inputs are such things as raw materials, other physical resources, human energy and knowledge, and information. The outputs are finished products and other transformations of the resources.

The figure shows a block labeled "goals." The arrows suggest that goals are both an output and an input. This means that, while goals for some organizations may be partially established externally, these organizations nonetheless have the opportunity to establish some goals. For instance, even a police department, whose goals are dictated by statute, has discretion in terms of which services it wishes to provide, how much to allocate to each service area, etc. The goals established internally feed back into the organization and serve to direct and focus its efforts, as shown in the figure.

Figure 1-1 also shows two feedback loops. The lower one, labeled "efficiency feedback," involves the comparison of *outputs to inputs.* This is a straightforward measure of the efficiency with which the organization makes the transformation. For instance, the simplest efficiency measures might relate the quantity of a product produced to the quantity of raw materials consumed. Greater efficiency would be associated with greater *yield,* i.e., more pounds produced per pound consumed.

The other feedback loop in Figure 1-1 is "effectiveness feedback." This relates *output to the goal* of the organization. Thus, while efficiency measures how well an organization is doing in transforming inputs to outputs, *effectiveness* assesses how well its output measures up to that which it wishes to achieve. In business, effectiveness measures usually involve profitability or per-share earnings assessed relative to the goal that has been previously established.

The simple systems model of the organization in Figure 1-1 is so abstract that it is not very operational. When the organization is viewed as a complex of interacting subsystems in turn interacting with an environment that is made up of interacting systems, the model can be expanded to be more useful.

Complex Systems Model

One of the most effective ways of thinking about an organization's environment is in terms of "environmental levels":[26]

Internal environment—those elements that are within the organization's official jurisdiction

Operating environment—the set of suppliers, customers, and interest groups with which the firm deals directly

General environment—the national and global context of social, political, regulatory, and economic and technological conditions

[26]Philip S. Thomas, "Environmental Analysis for Corporate Planning," *Business Horizons,* October 1974, pp. 26–38.

This is a useful way of thinking about the environment because the three levels are distinguishable in terms of the degree of control and influence which the organization can exert, and the information which is typically available to it. In general, the degree of control and the availability of information is greatest in the internal environment and least in the general environment.

Figure 1-2 shows the systems model of Figure 1-1 after it is expanded to include various illustrative subsystems from the internal, operating, and general environments. It shows the interacting and interdependent subsystems that make up the organization—here described in terms of only a few of the "departments." They are shown to overlap. For instance, the sales department and the data processing department overlap in that sales records are maintained, updated, and used through joint efforts.

The operating environment is shown to be made up of competitors, suppliers, local communities, etc., many of which are interdependent; similarly, the

FIGURE 1-2
Complex systems model of the organization.

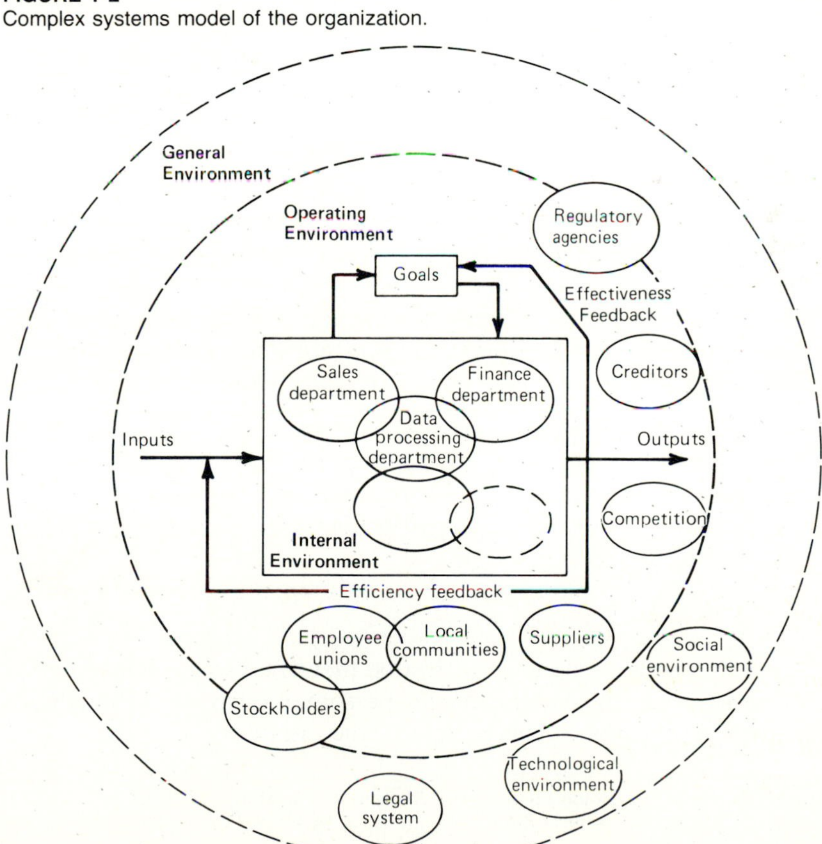

general environment reflects the broader context of government and society within which the organization operates.

The specification of these subsystems will vary for different organizations and according to particular purposes within a single organization. In one case, for instance, it may be necessary for an organization to distinguish individual regulatory agencies. For other purposes, it may be adequate to group all such agencies together so that they are thought of as a single homogeneous group.

Of course, it is not always clear into which element of the environment a particular entity falls. For example, the Federal Communications Commission (FCC) can exert a general corporate-level constraint by limiting the number of AM or FM stations owned by a single corporation, but it can also influence the operating environment at the business-unit level, as in the case of Texas Instruments' efforts to secure a waiver of certain FCC technical requirements pertaining to the linkage of home computers to TV monitors.[27] The significance of a single environmental variable in this example is dramatic—it would permit a 50 percent reduction in retail prices early in the product life cycle. Although this example of TI's "home computer problem" focuses on the regulatory issue, it should be noted that the substantive question surrounded a technical problem of potential interference with radio or TV broadcasting. Hence, there was a second level of interaction among the seemingly separate technical and regulatory subsystems.

Subsystems Models

Just as the organization and its environment can be viewed as a complex of interacting subsystems, so, too, can each subsystem be similarly decomposed. Figure 1-3 shows how *one* of the internal subsystems shown in Figure 1-2 can itself be thought of as a system—a collection of interacting subsystems. The sub-subsystems of the data-processing subsystem are depicted as overlapping circles representing the[28]

Hardware subsystem—computers and related input-output devices
Software subsystem—system and applications programs
Data base subsystem—files and their records and data elements
Procedures subsystem—instructions for the development and operation of the MIS (management information system)
Personnel subsystem—people engaged in the development or operation of the MIS

Some of these subsystems are generic, so that they are also elements of other subsystems. For instance, the data-processing personnel subsystem is simultaneously a part of the overall personnel system of the organization.

[27]"TI Gets Set to Move into Home Computers," *Business Week,* March 19, 1979, p. 37.
[28]Based on G. B. Davis, *Management Information Systems,* McGraw-Hill Book Company, New York, 1974.

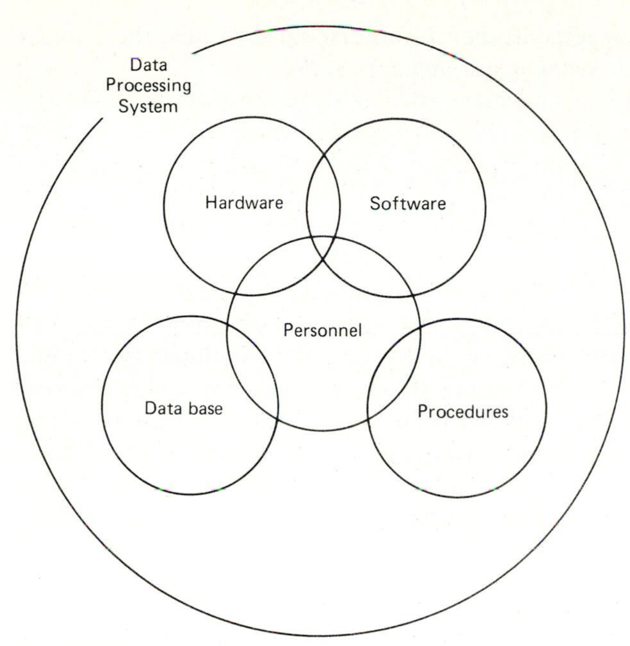

FIGURE 1-3
Data processing system subsystems.

Some of the subsystems are *management subsystems*—a key variety with which we shall deal in the remainder of the book. For example, the "procedures" subsystem dictates how the data processing will be handled and managed.

SUMMARY

The problems of our society are so complex and interdependent that they preclude simple solutions; sophisticated and complex solutions will unquestionably be required if they are to be solved reasonably, economically, and equitably.

Although these complex solutions will undoubtedly have a high technological content, technology alone will not suffice. Our ability to *manage technology*—to make *technological innovations,* to *integrate* technology into societal systems, and to *organize* these systems—is probably more critical to the advance of society than is technology itself.

The understanding and practice of effective management of complex systems is, therefore, of crucial importance to society. Such understanding may be based on the observed characteristics of successful complex-systems managers in terms of their understanding of the technology of their "business," their understanding of the basic concepts of management, their interpersonal style, and their ability to conceptualize and operate using a systems approach.

The systems approach to management focuses on a wholeness, or overall view, of the organization or problem. It is a viewpoint which may be applied in both the *planning* and *implementing* activities of management. However, it is not solely restricted to the level of a viewpoint, or "way of thinking," however important that may be. The systems approach provides a conceptual framework, a process and resulting management systems to guide and monitor both the planning and implementing phases of organizational management.

In the next two chapters, we shall demonstrate this applicability. Chapter 2 treats the planning phase of management by first developing a conceptual framework for applying systems thinking to planning. Then it describes an operational process for doing so. Finally, it outlines a *strategic planning system* which is the management system that permits systems thinking to be applied on a continuing organizational basis.

Chapter 3 then deals with the implementation phase of management in the same way. The output of the conceptual framework and the process in that context form a *project management system.*

DISCUSSION QUESTIONS

1 Systems ideas have been related to the two generic phases of management. What are these phases? How have other management authors described the phases of management?

2 What are some of the common characteristics found in modern organizations? These characteristics present problems that involve complex systems. Explain what is meant by this statement.

3 Explain what is meant by the use of the "project center" at the General Motors Corporation.

4 Discuss "open" versus "closed" systems and the concepts of "efficiency feedback" and "effectiveness feedback." Give examples from systems with which you are familiar.

5 As management has become more complex, it has also become more pervasive, extending to all sectors of the society. Is it possible to identify a modern organization where the process of management is *not* carried out to some degree?

6 There are certain characteristics that the successful manager must possess. What are these characteristics? What additional characteristics can you add to the list?

7 Rather than giving a definition of management, we have suggested that management contains certain universal elements. Identify and discuss these elements.

8 What are the several inescapable activities that the process of planning contains? Which of these activities is the most important?

9 "It is in the management function of motivating that the efficacy of the organizational leader is most visible." Explain this statement.

10 The key distinguishing characteristic that distinguishes the modern manager from traditional ones is the ability to think and act in the "systems" context. What is meant by this statement? Demonstrate that you understand this idea by selecting a situation in which a modern manager might become involved where successful management requires a systems viewpoint.

11 How should a system be defined? What is meant by the systems approach to management?
12 Is a modern business organization a closed or open system? Defend your choice.
13 What is meant by matrix management? Give an example.
14 What is meant by a systems model of an organization? Take a contemporary organization and apply the systems model suggested by the authors to that organization.

RECOMMENDED READINGS

Ackoff, R. L.: "Management MISinformation Systems," *Management Science,* vol. 14, no. 4, December 1967.

———: "Toward a System of Systems Concepts," *Management Science,* vol. 17, no. 11, July 1971.

Barnard, Chester I.: *The Functions of the Executive,* Harvard University Press, Cambridge, Mass., 1938.

Bauer, R.: *Second-Order Consequences,* The M.I.T. Press, Cambridge, Mass., 1969.

Burck, Charles G.: "How GM Turned Itself Around," *Fortune,* January 16, 1978.

Cleland, David I., and William R. King: *Management: A Systems Approach,* McGraw-Hill Book Company, New York, 1972.

——— (eds.): *Project Management Handbook,* Van Nostrand Reinhold, 1983.

Davis, G. B.: *Management Information Systems,* McGraw-Hill Book Company, New York, 1974.

Davis, Ralph C.: *The Fundamentals of Top Management,* Harper & Row, Publishers, Incorporated, New York, 1951.

Donnelly, James H., Jr., James L. Gibson, and John M. Ivancevich: *Fundamentals of Management,* 3d ed., Business Publications, Inc., Plano, Tex., 1981.

Drucker, Peter F.: *The Practice of Management,* Harper & Row, Publishers, Incorporated, New York, 1954.

———: *The Age of Discontinuity: Guidelines to Our Changing Society,* Harper & Row, Publishers, Incorporated, New York, 1968.

———: "Managing the Third Sector," *The Wall Street Journal,* October 3, 1978.

Fahey, L., and W. R. King: "Environmental Scanning in Corporate Planning," *Business Horizons,* vol. 20, no. 4, August 1977.

Fayol, Henri: *General and Industrial Management,* Pitman, London, 1949, p. 43.

Flory, C. D., and B. A. Mackenzie: *The Credibility Gap in Management,* D. Van Nostrand Company, Inc., Princeton, N.J., 1971.

Gilder, George: "The Case Against Women in Combat," *The New York Times Magazine,* January 28, 1979.

Grant, John, and William R. King: *The Logic of Strategic Planning,* Boston, Little Brown, 1982.

Gullett, C. R.: "The Systems Concept Revisited," *SAM Advanced Management Journal,* April 1971.

Kelly, Joe, and Kamran Khozan: "Participative Management: Can It Work?" *Business Horizons,* vol. 23, no. 4, August 1980.

Kerzner, Harold: "Project Management in the Year 2000," *Journal of Systems Management,* vol. 33, October 1981.

King, William R.: *Quantitative Analysis for Marketing Management,* McGraw-Hill Book Company, New York, 1967.

———: "The Systems Concept in Management," *Journal of Industrial Engineering,* May 1967.

———: "Using Strategic Issue Analysis," *Long Range Planning,* vol. 15, no. 4, August 1982.

King, W. R., and D. I. Cleland: *Strategic Planning and Policy,* Van Nostrand Reinhold, New York, 1978.

King, W. R., and C. Holloway, "Evaluating Alternative Approaches to Strategic Planning," *Long Range Planning,* vol. 12, no. 4, August 1979.

Koontz, Harold: "The Management Theory Jungle Revisited," *Academy of Management Review,* vol. 5, no. 2, 1980.

Koontz, Harold, Cyril O'Donnell, and Heinz Weihrich: *Management,* 7th ed., McGraw-Hill Book Company, New York, 1980.

Kreitzner, Robert: "People Are Systems, Too: Filling the Feedback Vacuum," *Business Horizons,* vol. 20, no. 6, December 1977.

Larson, H.: *Guide to Business History,* Harvard University Press, Cambridge, Mass., 1950.

Mee, John F.: "The Manager of the Future," *Business Horizons,* vol. XVI, no. 3, June 1973.

Miles, Rufus E., Jr.: "Miles' Six Other Maxims of Management," *Organizational Dynamics,* Summer 1979.

Nolan, R. L., and J. C. Wetherbe: "Toward a Comprehensive Framework for MIS Research," *MIS Quarterly,* June 1980.

Sales, Leonard R., and Margaret K. Chandler: *Managing Large Systems,* Harper & Row, Publishers, Incorporated, New York, 1971.

Schoderbek, Charles G., Peter P. Schoderbek, and Asterios G. Kefalas: *Management Systems Conceptual Consideration,* rev. ed., Business Publications, Inc., Dallas, Tex., 1980.

Smith, August W.: "A Systems View of Project Management, *Industrial Management,* March–April 1976.

Thomas, Philip S.: "Environmental Analysis for Corporate Planning," *Business Horizons,* vol. XVII, no. 5, October 1974.

Toffler, Alvin: *Future Shock,* Random House, Inc., New York, 1970.

Tyson, W. J., and R. L. D. Cochran, "Corporate Planning and Project Evaluation in Urban Transport," *Long Range Planning,* vol. 14, October 1981.

Uttal, Bro: "How Ed Griffiths Brought RCA into Focus," *Fortune,* December 13, 1978.

Van Gigch, John P.: *Applied General Systems Theory,* Harper & Row, Publishers, Incorporated, New York, 1974.

Warren, E. Kirby: *Long-Range Planning: The Executive Viewpoint,* Prentice-Hall, Inc., Englewood Cliffs, N.J., 1966.

Webber, Ross A.: *To Be a Manager: Essentials of Management,* Richard D. Irwin, Inc., Homewood, Ill., 1981.

Weber, Max: *Essays in Sociology,* ed. and trans. by H. H. Gerth and C. W. Mills, Oxford University Press, Fair Lawn, N.J., 1946.

CASE 1-1: Defining the Organization as a System

Choose a real organization such as a business firm, university, college, or retail store. Have each member of your team develop a complex systems model of the same organization with each working independently. Then get together and compare the models in terms of:

a how and why they are different

b the various levels of detail that are involved in each, and whether there is a single "correct" level of detail

c the ways in which each might be useful for various purposes

Consider that your purpose is to use such a model to identify the "claims" that various clientele groups such as stockholders and employees make on the organization. Choose a model that is best for this purpose and try to identify some of the claims made by these groups. Be as specific as you can.

PART TWO

SYSTEMS APPROACHES IN PLANNING AND IMPLEMENTATION

CHAPTER 2

STRATEGIC SYSTEMS PLANNING

We shall have to evolve
Problem solvers galore
Since each problem they solve
Creates ten problems more!

Source Unknown

The systems model of the organization presented in Figures 1-1 through 1-3 suggests the way in which systems thinking may be applied to both the planning and the implementation aspects of management. In this chapter, we deal with *strategic systems planning* (SSP)—the approach that may be used in the planning phase of management. (Chapter 3 will deal with the application of systems ideas to the implementation phase.)

PLANNING AND MANAGEMENT[1]

The planning function of management may be viewed in a variety of different ways. Some prefer to nebulously equate planning decision making and generally "looking ahead," since every decision involves consequences which will occur in

[1]Some portions of the discussion to follow are adapted from W. R. King and D. I. Cleland, "A New Method for Strategic Systems Planning,"*Business Horizons,* vol. 18, no. 4, pp. 55-64, August 19, 1975.

the future. Here, we choose to interpret planning as encompassing a number of activities, including the strategic one of decision making.

Taken in its broadest sense, planning is an activity consciously programmed into an organization's continuing activities and having as its focus the *objective consideration of the future.* Whether the "organization" be a company, a public agency, a nation, or humankind, there is both a need and a place for such activity; and there are evolving methodologies for effectively and objectively dealing with future-oriented activities in a fashion which heretofore was considered by many to be within the province of the occult.

This is not to say that the crystal balls or astrology charts have become a part of the organizational planner's milieu. Rather, there has come to be a recognition that, *since present actions necessarily reflect implicit anticipations and assumptions about the future, these anticipations and assumptions should be made explicit* and should be subjected to the same sort of analysis as is commonly carried out for less nebulous, and consequently often less important, immediate issues.

This modern and expansive view of planning is intended to remove it from the constraints of a pure "problem solving" reactive mentality to the larger domain of proactive thinking and acting. In accomplishing this, one is moving against a well-ingrained philosophy of our society, since reactive problem solving—"define the problem and solve it"—is a cornerstone of much of our scientific and technological thinking.

Of course, there is no converse intention to decry the utility of classical scientific problem solving as a part of the overall concept of planning. Indeed, there is a need for more—not less—formal analytical activity in the management of organizations. The key to success in the modern view of planning put forth here is to *integrate futuristic thinking with careful analysis.* Indeed, this simple statement well reflects the objectives of our entire treatment of planning.

THE SYSTEMS APPROACH TO STRATEGIC SYSTEMS PLANNING (SSP)

The approach to planning to be discussed here will be a systems view. The reasons for taking this systems approach are manifold. First, a comprehensive view of planning requires that the organization extend its attention beyond the boundaries of those things which are immediately controllable to encompass those environmental aspects which may only be influenced by organizational actions and, indeed, to those elements of the environment which must simply be accepted as "facts of life." To do this in a more than superficial way requires the conscious consideration of environmental influences which are foreign to the day-to-day activities of most managers. Hence, a systems model—one defining the organization as a subsystem of a larger system—is a necessity if one is to perform comprehensive planning. Such a model may well be qualitative and descriptive. However, it is essential to developing the understandings which are necessary to effective planning.

Second, since a comprehensive view of planning which usefully encompasses such divergent activities as "futurism" and organizational financial planning must relate these various activities in some logical way, we adopt the systems framework as an effective device for considering the myriad interrelationships and feedback loops which are inherent in an overall view.

Third, any view of planning which extends deeply into the environment of the organization almost inevitably results in the consideration of second-order social consequences. Such consequences are within the natural domain of the systems approach.

Finally, since our view of planning presumes an explicit organizational process, systems must be set up for performing and supporting the process. If these are truly systems, and not merely collections of related procedures, there is the implicit assumption that the systems approach to problem solving has been applied to their design.

A Conceptual Strategic Systems Planning Model

Jantsch[2] has proposed a model which attempts to describe the salient "dimensions of integration in a systems approach." As applied to strategic systems planning, these "dimensions" serve to define the various specific ways in which planning must be viewed in a systems context which involves continuing interactions and feedback.

The dimensions which Jantsch defines as being those which operationally define the systems approach to planning are called "horizontal," "vertical," "time and causality," and "action."

The "Horizontal" Dimensions of SSP

The *horizontal* dimension exemplifies the need in planning for continuous interaction and feedback from general to particular and back again—i.e., between aspects of systems which require description in terms of many interrelated measures and those which are best treated in terms of one, or a few, such measures. For instance, systems planners are constantly aware that they must look at the "big picture" by considering the social, cultural, economic, static, dynamic, psychological, technological, and other impacts of contemplated actions on contemplated systems designs. Yet, in doing so, they create a monster described in complex multidimensional terms. To deal with this monster, they then may proceed on the time-honored problem-solving approach of "cutting the problem down to size" by considering only a smaller set of measures, combining various measures into a lesser number of indexes, or restricting themselves to the most important consequence. In doing so, they are reducing the dimensionality of the system to make it manageable. Yet, once conclusions have been reached on the basis of this restricted version of the system, systems

[2]Erich Jantsch, "Forecasting and Systems Approach: A Frame of Reference," *Management Science,* vol. 19, no. 12, August 1973.

planners must return to the more complex description of the system to investigate the impact of their abstraction and to assure themselves that their search for a manageable model has not blinded them to reality.

Consider a specific illustration of this phenomenon. Information-systems designers have long focused on the question of assessing the value of information. Such a measure would permit the rational and economic design of information systems, since it would provide a basis for a rigorous cost-benefit analysis of the implications of various elements of information into a system. People in various information-related fields have developed a variety of disparate measures related to the worth of information—some qualitative and some quantitative.[3] However, none of these measures would appear to be sufficiently comprehensive to constitute "the" measure of information value. Each has its advantages and disadvantages and each emphasizes a different aspect of the question. A multidimensional information value concept is probably best, but it is unmanageable. Thus, systems planners are forced into a continuous interactive process between particular and general, with no universally accepted starting point.

They may, for instance, begin with a specific measure and perform analyses using it before referring to a broader concept of value; they may utilize some construct (index) to aggregate several value-related dimensions.[4] Alternatively, they may begin with multidimensional conceptualization and then successively simplify it to permit analysis. Whatever they do, they are involved in a continuous process of feedforward and feedback loops among the various levels, from particular to general, of this horizontal dimension of systems planning.

The "Vertical" Dimension of SSP

The second dimension requiring integration and feedback in systems planning involves a hierarchy of planning levels. Various authors have proposed taxonomies to describe this hierarchy. Jantsch uses an adaptation of a structure proposed by Ozbehkan[5] which sees three critical levels:

Policy planning
Strategic planning
Operational planning

which respectively deal with the "ought to," "can," and "will" aspects of strategic decision making.

At the *policy planning* level, the salient measures are the *objectives* of the

[3]See William R. King and Barry J. Epstein, "Assessing the Value of Information," *Management Datamatics,* vol. 5, no. 4, pp. 171–180, September 1976.

[4]William R. King and Barry J. Epstein, "Assessing Information System Value: An Experimental Study," *Decision Science,* January 1983.

[5]H. Ozbehkan, "Toward a General Theory of Planning," in E. Jantsch (ed.), *Perspectives of Planning,* Organization for Economic Co-operation and Development, Paris, 1969, p. 153.

organization, expressed as enduring concepts of dynamic behavior (e.g., profit maximization) rather than as specific quantitative goals to be achieved.[6]

At the *strategic planning* level, *goals*—in terms of specific outputs or functional outcomes—are the salient measures. This is the level of strategic choice which takes the objectives as given and seeks to make optimum choices among strategic alternatives.

At the *operational planning* level, activity is aimed at fixed attainable *targets*. As such, it is input-oriented in that it focuses on the allocations and arrangements of inputs to achieve fixed targets.

Just as with the horizontal dimension, the key to this dimension of systems planning is the continuous interaction and feedback among the levels which make up the dimension in question. Objectives guide the development of goals which, in turn, determine targets. Yet the input arrangements necessary to the achievement of specific targets reverberate upward to necessitate the rethinking of goals and, sometimes, objectives. Moreover, general objectives, once finally defined, must be translated into a set of specific targets. Or, alternatively, one may think in terms of the achievement of several targets which must logically aggregate to constitute movement toward a goal, several of which must, in turn, be logically equivalent to the achievement of an overall objective.

The "Time and Causality" Dimension of SSP

The planning dimension called "time and causality" involves continued feedback and interaction between *exploratory forecasting* and *normative forecasting*. This distinction is between the process of plotting alternative paths into the future and the structuring of courses of action on the normative basis of *desired* future states.

In the case of exploratory forecasting, various sets of assumptions are made concerning future developments. For instance Kahn and Wiener[7] have structured sets of assumptions, definitions, conjectures, and analyses into scenarios[8]—to develop *alternative* futures which are likely to come about under the stated conditions. Thus, the scenario is an explication of possibilities in the manner of exploratory forecasting rather than a specification of a desired state or a "point estimate" of what the future will hold.

Of course, such alternative futures become inputs into the strategic decision-making process where specific choices must be made, but it is important that

[6]We shall note a conflict in the semantics as applied in various segments of planning literature. For instance, the meaning of "policy" as used here is not the same as the meaning applied elsewhere in the literature of planning. Moreover, the distinction between objectives and goals used here is exactly the opposite of the terminology used in other related fields of literature. The problem is more apparent than real, however, and we shall attempt to explain the salient distinctions in each case rather than to impose a single unified terminology.

[7]H. Kahn and A. J. Wiener, *The Year 2000,* The Macmillan Company, New York, 1967.

[8]Daniel Bell, "Twelve Modes of Prediction—A Preliminary Sorting of Approaches in the Social Sciences," *Daedalus*, Summer 1964, p. 865.

they be made within the context of a range of objectivity-defined alternatives rather than in a vacuum. Without such guiding information, strategic decisions tend to be based on one of two extreme modes of forecasting—simple extrapolations of the present, or unrealistic idealizations. The blending together of the exploratory and normative modes of forecasting in a systems framework can prevent these extremes from being unthinkingly chosen, thus leading to more effective strategic planning.

The "Action" Dimension of SSP

The *action* dimension of strategic systems planning refers to the interaction of planning and action—or planning and "implementation," in the terminology of this book. The entire focus of planning is a learning feedback process where plans depend on the consequences of action just as intrinsically as actions reflect the results of planning.

The simple diagram of Figure 2-1 describes this feedback process in terms of mutually supportive strategic and operational decision making which results in actions. These actions influence "the world" and produce consequences which are fed back to provide the capability for revising strategic decisions. As shown by the inner feedback loop, the influence which this basic information has on operational decisions also provides part of the feedback data which are provided to support potential changes in strategic decisions.

This continuous interaction of planning and implementation in an overall strategic systems planning activity is one of the central themes of this book; i.e., one of the foundations of the interdisciplinary character of the book is that systems ideas apply equally well to planning and implementation, thus making them mutually supportive.

THE BASIS FOR THE SYSTEMS APPROACH TO SSP

The systems approach to strategic planning, as demonstrated by the conceptualization of the previous section, is a complex one. Quite literally, "everything depends on everything else" in a complicated interwoven fashion which is sometimes difficult to understand and certainly difficult to consider comprehensively. This is the case because of two major factors: the broad objectives which are set forth for strategic systems planning and the inherent complexity in the systems with which such planning deals.

The Objectives of Strategic Systems Planning

In the first regard the basic objective set for strategic systems planning is one which goes well beyond those which are a part of the systems approach to problem solving. Thus, just as the systems approach itself leads the analyst

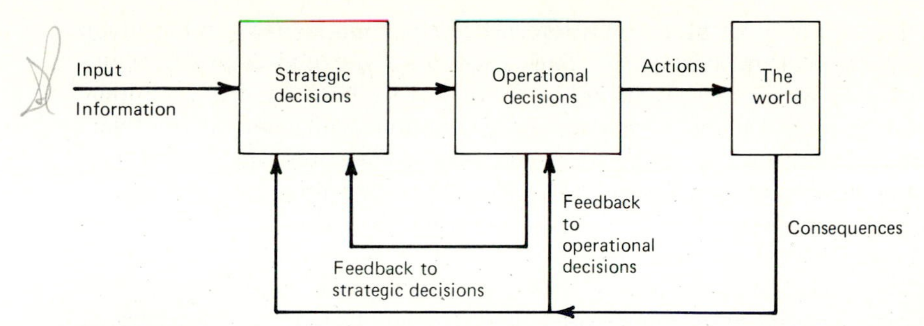

FIGURE 2-1
The action dimension of SSP.

beyond *symptoms* to underlying *problems,* strategic systems planning leads beyond problems to *opportunities.* These relationships are described in Figure 2-2 which suggests that problems create symptoms while simultaneously suggesting opportunities to be pursued.

The domain of traditional problem solving is shown in that figure to be largely that of dealing with symptoms, since the parochial nature of the analytic process and of the organizations which deal with problems are such that only narrow viewpoints are taken of most problem contexts. Of course, many may view this drawing of a correlation between problem solving and symptoms as an exaggeration, but within the context of society, there are, in reality, few problems which get dealt with at anything other than a symptomatic level—even when the processes required to do so are purely at the level of simple logic rather than of insight or clairvoyance. For instance, consider the American city which spent years in debate over an airport extension of the existing public transit system and, on finally building the system, found itself faced with a proposal for a

FIGURE 2-2
Problem solving, systems problem solving, and SSP.

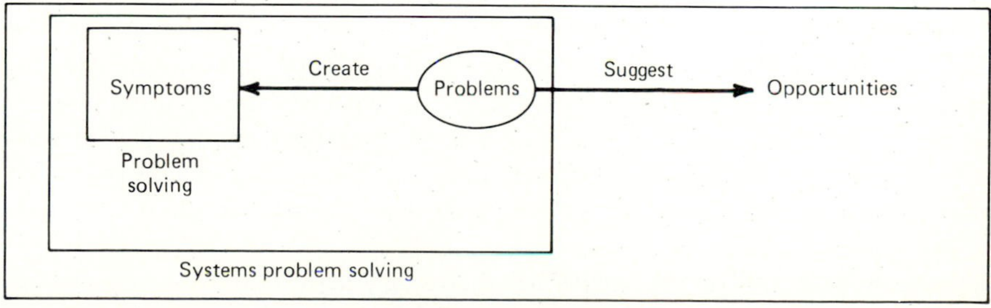

relocation of the airport based on a forecast of early inadequacy and the high cost of expansion at the existing site. Such a sequence of events clearly indicates the restrictions which are placed on problem solving both by our traditional analytical methods and by organizational constraints which do not motivate people to look beyond the bounds of "the" problem "as presented" or beyond the scope of their own organization and personal responsibilities.

Figure 2-2 also shows "systems problem solving" in terms of a focus on the broader domain of symptoms and problems. This has been the concern of systems-oriented managers and authors in recent years, and it is a clear improvement in problem solving over nonsystematic approaches. However, as the figure shows, the domain of problems can be further extended to encompass "opportunities" in the strategic systems planning process. We shall subsequently go further into the operational details of the strategic systems planning process which helps to define this transition from a focus on problems to a focus on opportunities. For now, we shall be satisfied with rather naive business illustrations in which "opportunities" are defined in terms of potential new products. For instance, a major fast-food chain is faced with a peak-load problem of tremendous magnitude. Its facilities are stretched to the breaking point at the lunch and dinner hours, and there is plenty of slack time at other hours. This problem of unused resources is viewed as an opportunity to introduce new items on its menu which will have special appeal at nonpeak hours—breakfast items, snack items, and nighttime specialties. Another company uses the major "problems" of society and business—consumerism, pollution control, etc.—as a basis for a formal study of market opportunities which may be related to those phenomena. From this study it concluded that it has the expertise to make a profit in several lines of business which will aid society and simultaneously reduce the image of business as a creator, rather than a solver, of societal problems.

In the former case, a single creative mind *could have* turned the problem into an opportunity. In the latter case, substantial study, data gathering, and analysis, beyond the capacity of any individual, were required. In both cases, the likelihood of achieving the transition from problem to opportunity was greatly enhanced by an institutionalized process of strategic systems planning.

The Inherent Complexity of Social Systems

The second major factor which dictates at least some minimum level of complexity in strategic systems planning has to do with the inherent complexity in the variety of social systems with which most organized planning processes deal. To better understand this, let us explore the basis for this complexity in terms of a number of characteristics of opportunity-oriented planning decisions.

Interactions and Interdependencies A dictionary-style definition of a system in terms of "an assemblage of interacting and interdependent parts forming a unified whole" should probably have the words "interacting" and "interdependent" in italics because, in practice, the utility of the systems viewpoint depends on these characteristics. This is so because there is no particular value to the taking of a systems view of a system of noninteracting independent parts. Such parts might just as well be treated separately since they have no effect on one another.

Thus, the interactions and interdependencies among the elements of a system are major complicating factors which necessitate an overall systems viewpoint. Such interactions and interdependencies are common in all social systems. For instance, the family is made up of a number of interacting and interdependent humans. If one were to conduct a study by trying to treat the family elements in isolation without consideration of their interactions and interdependencies, one would not achieve much understanding of the way the family operates or achieve much ability to *predict*—the ultimate objective of the systems analyst.

For instance, the teenager's objective of "having fun" is affected by the parents' desire for the education of their children and their objection to certain modes of achieving the fun goal. So, too, are parents affected by the teenager's desire to have such things as privacy. Thus, the family members interact in various ways, ranging from their personal goals, to the actions which they may take as a consequence of those goals, to the transferring of the effects of external forces applied to one family member on each of the others. What family, for instance, has not had an unpleasant circumstance at school or office for one of its members affect the lives of all its members?

But not only are the family members interactive, they are also interdependent. Quite literally, they depend on one another for support in many ways as diverse as economic and psychological, and the dependence is not one-way. The child may well depend on the father for funds, but few fathers would not admit to depending on their children in diverse, and perhaps more valuable, emotional areas.

Of course, every organization and social system displays these characteristics. The corporation, for instance, is viewed by some of its employees in terms of a particular department or function, but what marketing department could exist without a production department to create what it sells, and what production unit could exist for long without marketers to provide it with a reason for existence?

Second-Order Consequences Another important dimension of complex problems is that referred to as *second-order consequences*. These consequences are those that appear subsequent to, or as a result of, the immediate and obvious consequences of an action. One can, for instance, view the process of technological innovation as one in which many new technologies have produced immediate and obvious consequences, which were beneficial to society but which also produced subsequent consequences that were not so beneficial. The automobile produced inexpensive transportation for the masses, but it also changed society in various ways that many believe to be negative, through such things as eliminating neighborhood togetherness, providing the opportunity for greater sexual freedom among the young, and preempting the growth of mass transit. In addition to these consequences, the automobile contributed to major societal problems such as pollution, whose solution is at least partially responsible for yet another problem—an "energy crisis." This *chain of effects* in going from a problem to immediate consequences then to second-order consequences and newly created problems is one of the pervasive characteristics of modern social systems. Quite literally, in such systems everything depends on everything else and often in ways so complex and roundabout that it is difficult to understand the interrelationships.

For instance, consider the chain of effects and problems created by missionaries who replaced stone axes with steel axes as the primary cutting tools of the Australian Yir Yorunt aborigines. This account, by R. A. Bauer,[9] is a paraphrase of the original version by Sharp.[10]

> But stone axes played important functions in Yir Yorunt life beyond that of cutting wood. The men owned the stone axes, which were symbols of masculinity and of respect for elders.
>
> The missionaries had distributed the steel axes to men, women, and children without discrimination. But the older men, having less trust of missionaries, were not as likely to accept the steel axes. Soon elders of the tribe, once highly respected, were forced to borrow steel axes from women and younger men. The previous status relationships were thoroughly upset.

Time Skip Another way of viewing an aspect of second-order consequences is described by Alvin Toffler in *Future Shock* as the "time skip" phenomenon. He argues that phenomena which are rather limited in their impact when they occur may, by virtue of the simple passage of time and "natural" processes, have effects which are amplified far beyond the original impact.[11]

> The Peloponnesian War deeply altered the future course of Greek history. By changing the movement of men, the geographical distribution of genes, values, and ideas, it affected later events in Rome, and through Rome, all Europe. Today's Europeans are to some small degree different people because that conflict occurred.
>
> In turn, in the tightly wired world of today, these Europeans influence Mexicans and Japanese alike. Whatever trace of impact the Peloponnesian War left on the genetic structure, the ideas, and the values of today's Europeans is now exported by them to all parts of the world. Thus today's Mexicans and Japanese feel the distant, twice-removed impact of that war even though their ancestors, alive during its

[9]R. A. Bauer, *Second-Order Consequences,* The M.I.T. Press, Cambridge, Mass., 1969, p. 15

[10]L. Sharp, "Steel Axes for Stone Age Australians," in E. H. Spicer (ed.), *Human Problems in Technological Change,* Russell Sage Foundation, New York, 1952, pp. 69–92.

[11]A. Toffler, *Future Shock,* Random House, Inc., New York, 1971, pp. 17–18.

occurrence, did not. In this way, the events of the past, skipping as it were over generations and centuries, rise up to haunt and change us today.

Thus, the cumulative effect of things which affected a small number of people in the past can today affect everyone, just as a recently discovered arithmetic error by a debtor who failed to pay a small sum to one's ancestor can, through the magic of compound interest, make the "estate" worth millions today!

Of course, the planner must consider time skip not only in historical perspective, but also in a proactive way. What consequences, so minor as to warrant being ignored today, may grow through natural processes into major problems in the next decade? Are there population segments which, if unsatisfied today, will create more costly demands in the future? Such are the variety of time-skip considerations which must be the province of the strategic systems planner.

AN OPERATIONAL PROCESS FOR STRATEGIC SYSTEMS PLANNING

To operationalize the sometimes nebulous "systems" approach to management requires that systems concepts be applied in a systematic fashion. This may at first seem to be a play on words, but it is, in fact, true that, while some "big thinkers" can unsystematically apply systems ideas with good results, the real impact of systems ideas on organizations is through the systematic application of these ideas in the form of a specific methodology or process.

In this section, we present a process for systematically applying systems thinking to the planning phase of organizational management in terms of a five-stage process of:

1 Defining the organization in systems terms
2 Defining goals for each organizational element and claimant
3 Deciding
4 Creating management systems to instutionalize the decision-making process
5 Integrating management systems

Defining the Organization in Systems Terms

To define the organization in systems terms means that one explicitly goes through the process of "blowing up" the problem as described previously. The most important element in this process has to do with explicitly describing every relevant *clientele group* and *organizational claimant* in terms which are meaningful and hopefully, measurable.

Clientele groups—sometimes referred to as "stakeholders" to distinguish them from the legal owners of corporations—are those who have a stake in the activities and future of an organization. Thus, workers, stockholders, suppliers, and customers are invariably relevant clientele groups of a business enterprise, and the impact of the organizational decision on *all* of them must be considered in a rational systems approach to management.

Defining Goals for Organizational Claimants

Each claimant (individual, department, or group) has goals which are related to the nature of the claim on the organization and which, in turn, affect the outcome and determine the effectiveness of organizational decisions. An explicit and objective consideration of these claimants and the nature of their claims permits the systems viewpoint to be adopted and provides the opportunity for the measurement which is so critical to planning and decision making.

Table 2-1 shows how such myriad claims may be made explicit for a business firm. To use the model requires that:

1 Organizational claimants be identified
2 The nature of each claim be specified
3 Assessable measures be defined for each claim
4 Predictions of the future pattern of these measures be made in the context of the claimant's objectives
5 The impact of these predictions on organizational decisions be assessed

Deciding

The "deciding" phase of systematically applying systems concepts is the one which is often overemphasized in systems-oriented discussions. While it is of great significance, it is merely one part of such an overall systems approach.

In systematically making decisions using these systems concepts, one must relate the clientele goals to the alternative courses of action which are available to the organization. Every proposed action will affect the various clienteles in different, and often mutually contradictory, ways. The essence of decision making in such a situation involves the resolution of these conflicting impacts.

To even attempt to do this rationally requires that some model of the decision problem be developed. That model may very well be a "mental model" rather than an explicit one, but all of us use such models to abstract out the many confusing aspects of a situation and reduce it to its salient elements. When this is accomplished, mortals can comprehend complex problems, because they are dealing with *representations* of reality rather than with reality itself.

The area of thought which has come to be identified with the conduct of this process in a structured and systematic way is called *systems analysis.* After considering the broader role of systems in implementation in Chapter 3, we shall discuss the specifics of the systems-analysis process in subsequent chapters.

TABLE 2-1
ORGANIZATIONAL CLAIMANTS AND THEIR CLAIMS

Stockholders	Participate in distribution of profits, additional stock offerings, assets on liquidation; vote of stock, inspection of company books, transfer of stock, election of board of directors, and such additional rights as established in the contract with the corporation.
Creditors	Participate in legal proportion of interest payments due and return of principal from the investment. Security of pledged assets; relative priority in event of liquidation. Participate in certain management and owner prerogatives if certain conditions exist within the company (such as default of interest payments).
Employees	Economic, social, and psychological satisfaction in the place of employment. Freedom from arbitrary and capricious behavior on the part of company officials. Share in fringe benefits, freedom to join union and participate in collective bargaining, individual freedom in offering up their services through an employment contract. Adequate working conditions.
Customers	Service provided the product; technical data to use the product; suitable warranties; spare parts to support the product during customer use; R&D leading to product improvement; facilitation of consumer credit.
Supplier	Continuing source of business; timely consummation of trade credit obligations; professional relationship in contracting for, purchasing, and receiving goods and services.
Governments	Taxes (income, property, etc.), fair competition, and adherence to the letter and intent of public policy dealing with the requirements of "fair and free" competition. Legal obligation for businesspeople (and business organizations) to obey antitrust laws.
Union	Recognition as the negotiating agent for the employees. Opportunity to perpetuate the union as a participant in the business organization.
Competitors	Norms established by society and the industry for competitive conduct. Business statesmanship on the part of contemporaries.
Local communities	Place of productive and healthful employment in the local community. Participation of the company officials in community affairs, regular employment, fair play, local purchase of reasonable portion of the products of the local community, interest in and support of local government, support of cultural and charity projects.
The general public	Participation in and contribution to the governmental process of society as a whole; creative communications between governmental and business units designed for reciprocal understanding; bearing fair proportion of the burden of government and society. Fair price for products and advancement of the state-of-the-art in the technology which the product line offers.

Creating Management Systems

Once a choice has been made through the use of systematic analysis, the manager is faced with a further responsibility. In order to progress, there is a need to *institutionalize the problem-solving process*. In this way, problems can be dealt with in what is, to some degree, a routine fashion.

Of course, many problems will never be dealt with routinely, but there is an element to every problem-solving situation which can be systematized in such a way as to make the next occurrence of a similar problem more routine.

This systematization may well be in terms of the model which is used. It may even involve only the simple explication of claimants and the nature of their claims as shown in Table 2-1; it may involve the creation of new *data bases* which provide important information concerning a class of problems. At the furthest extreme is the sort of problem which can be literally routinized because a definitive model has been developed. For instance, stock level inventory models, which consider costs of ordering, holding inventory, and "stock outs," have been routinized to the degree that the resulting decision rules are applied "automatically" by computer systems. This is even carried to the extent of having the computer monitor stock levels and type out purchase orders when replenishment is needed.

However, this high degree of routinization is clearly only appropriate for fairly simple, well-understood problems which have been adequately modeled. Most significant management problems are not of this variety, yet almost all have the characteristic of being susceptible to a degree of systematization once they have been solved.

Integrating Management Systems

For managers to develop a system for problem solving without giving due consideration to other management systems would be tantamount to ignoring systems concepts as applied to their own domain. Many organizations have made this mistake at great cost, and so managers who wish to be effective should be aware of the potential dangers.

For instance, consider the bank which has institutionalized a wide variety of transactions through the development of computer systems for checking, credit cards, loans, etc., only to find that these distinct systems do not provide the capability for providing salient management information. The marketing vice president, in fulfilling the role of developing new customers, would like to conduct a campaign to sell new services to people who are currently customers for some subset of services. The vice president reasons that a checking customer has a higher potential for sale of a credit card than does someone with no existing ties with the bank. However, the various systems have no low-cost capability for providing a list of the services used by each customer. This is so because the various systems were developed independently, without concern for such interactions. Moreover, since individuals may be listed in various systems in terms of various forms of their names—e.g.,"last, first, middle" versus "last, first, middle initial"—there is not even any reliable way of performing manual analysis to provide reasonably accurate information.

The cause of this incongruous situation is the ignoring of systems integrations during the design of the various subsystems, each of which may itself be productive and effective. The difficulty is in the system interactions, so that the manager who wishes to avoid such pitfalls should be cognizant of these interrelationships and their potential for both great harm and great benefit.

A STRATEGIC PLANNING SYSTEM[12]

The most effective use of strategic systems planning comes when it is applied in a systemic fashion, i.e., when systems are developed for the application of systems ideas through strategic systems planning. Such systems planning has five primary elements.

1 A system of plans
2 A planning process
3 A decision subsystem
4 A management information subsystem
5 An organizational culture and management subsystem which facilitates planning

Since the basic thesis is that all these elements are intrinsically interrelated—both in concept and in the design and implementation of operating systems—it is impossible to discuss them in a logical sequence which leaves no loose ends. For instance, one cannot discuss a system of plans without making certain assumptions about a facilitative organizational structure. However, we shall take up each of these elements in an order which is designed to minimize the feedback requirements of the presentation.

A System of Plans

The output of an effective planning process is, typically, a series of documents called *plans*. The various plans should be interrelated to form a *system of plans* which can be used to guide the organizational system.

A conceptual model for a system of plans is schematically displayed in Figure 2-3. Much of the terminology used in that figure and in the subsequent discussion is directly appropriate to the business firm. However, the functions and purposes of the various plans are appropriate to any organization. Indeed the authors have themselves applied this system of plans in a wide variety of nonbusiness contexts.

Project plans are the basic building blocks of the system of plans in Figure 2-3. Each echelon of plans receives guidance from prior plans and further specifies such direction by focusing on groups of activities having a common purpose.

This exemplifies both the horizontal and vertical dimensions of strategic

[12]Much of the material in this section is adapted from D. I. Cleland and W. R. King, "Decision and Information Systems for Strategic Planning," *Business Horizons,* April 1974.

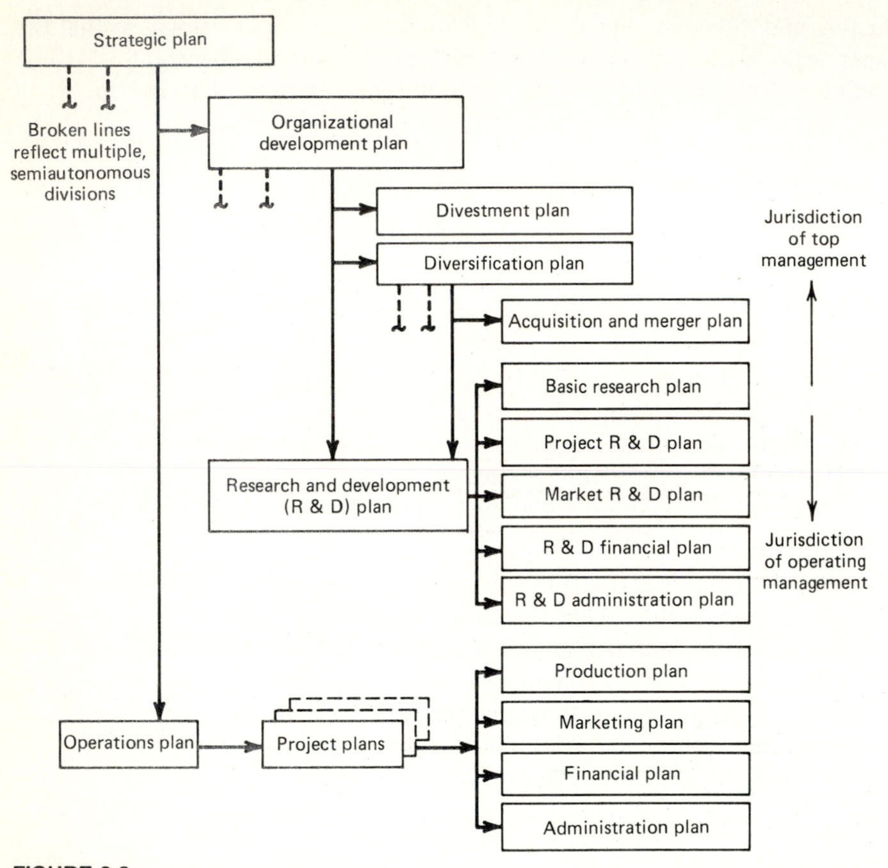

FIGURE 2-3
A system of plans. (Adapted with permission from *A Framework for Business Planning,* Report no. 162, Stanford Research Institute, California, February 1963.)

systems planning as outlined earlier, since the horizontal dimension of Figure 2-3 represents a transition from general to particular (left to right), and the vertical dimension represents a transition from the policy and strategic levels to the operational level. Of course, the feedback loops and interactions are not shown in this figure. However, they are apparent in the planning process to be described subsequently.

The various elements of Figure 2-3 are generally self-explanatory and in any case, are meant here solely to be illustrative of the important interrelationships among the various outputs of the planning process.

The system of plans given in Figure 2-4 emphasizes categories of plans, the specific content of plans, and planning horizons. Figure 2-4 has been prepared using terminology applicable to an educational system to emphasize the broad applicability of the ideas beyond the business organization.

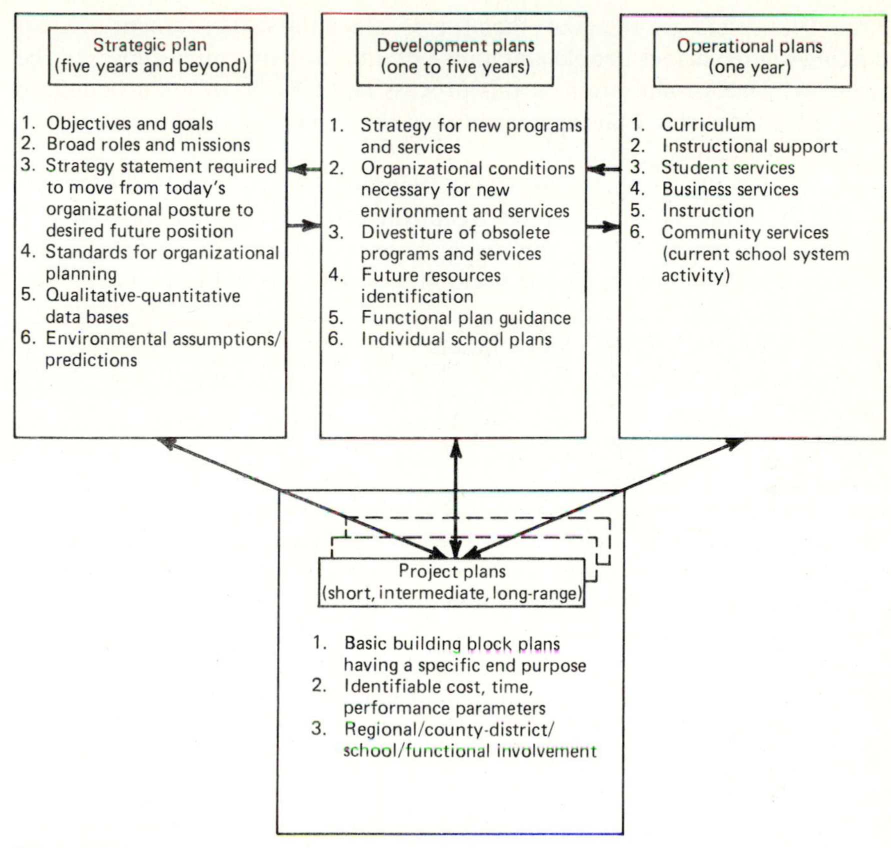

FIGURE 2-4
A system of plans for an educational system.

A Planning Process

Various descriptions of planning processes are available in the literature of planning. The one presented here is meant to illustrate some critical aspects of effective planning rather than to serve as a definitive statement of what the planning process "should" be.

The various phases of the planning process may be thought of as:

1 Establishing general goals
2 Information collecting and forecasting
3 Making assumptions
4 Establishing specific objectives
5 Developing plans

The first phase of this process—that of establishing general goals—can be more important than are the goals themselves, since the establishment of goals emphasizes where various people and units "fit into" the overall purposes of the organization. The formalization of this process can also be of great benefit in motivating individuals to pull together toward the achievement of a set of goals which they have helped establish.

The basic information necessary for a planning process involves a critical assessment of the current status of the organization together with a forward look at the environments which may reasonably be anticipated for the future. These assessments and forecasts should be done in at least four major areas:

1 Major products, services, and markets
2 External environment
3 Internal environment
4 Competitive environment

The assessment of products, services, and markets should be focused toward developing a redefinition of the "business" of the organization in terms of opportunities as well as existing status. The external environment assessment should include consideration of government actions, social mores, politics, the international situation, and other externalities which may affect the organization's future.

An assessment of the internal environment may take the form of a "strength-weakness" analysis of the organization., When this is contrasted with a similar assessment of major competitors, it provides a vehicle for the assessment of tentative areas of opportunity for the organization.

Assumptions concerning the behavior of markets, competitors, and the environment also play an important role in a planning process, since they, along with formal forecasts of the future, provide a basis for the development of the plans which are the output of a planning process.

Objectives—tentative statements of specific desired accomplishments—may be developed on the basis of all the information in the previous steps. These objectives will undoubtedly be revised as the planning process proceeds. The development of objectives at one organizational level, which can be broken down into subobjectives for lower organizational levels, provides a concrete beginning point for the development of plans.

As plans are developed, many of the previous steps need to be refined or reconsidered. This is especially true of the preliminary objectives which have been established, since they may prove to be infeasible. Also, an organization will often find that its assumptions will have to be revised, and that proposed actions will need to be considered in the light of a range of contingencies rather than a single assumption concerning the future. The "time and causality" dimension of strategic systems planning thus comes directly into play in this process, since a series of feedback loops must exist between exploratory forecasts, normative forecasts, and the retesting of proposed actions against a background of a variety of alternative futures. When such a continuous feedback process is concluded, the results should be a well-tested and well-integrated system of plans.

A Decision Subsystem

The planning process just described proposes that the last step in the planning process is the "development of plans." While this is true—plans *are* the documents which are the output of the planning process—it does not accurately reflect the significance and complexity of that phase of the overall process.

Planning, at least in this restricted sense of formally developing plans, is a decision-making function. The "plans" represent *choices made among alternatives.* In effect, one can conceive of all the possible plans which might be developed being laid out, and the planner's job being the "simple" one of choosing the one best plan from the array of possible plans.

Of course, it is really not done this way. The planner makes choices along the way which are eventually aggregated into the set of choices which is referred to as *the plan.* However much simpler this process may appear to be than that of selecting a "whole" plan from a large array of possibilities, it presents difficult problems of its own. These problems, in part, reflect the old systems cliché "everything depends on everything else," since a choice judged to be best at one level of the development of the plan may be affected by a choice made subsequently. Thus, what is deemed best at one point may be relegated to a less desirable role by choices made later.

The subsystem which permits such choices to be made in a systematic fashion is one of the major areas of concern of this book. That area—referred to as *systems analysis*—deals with such decisions and with rational ways to make them. Part Three of the book will be devoted to this topic (beginning with Chapter 4), and so we shall curtail further discussion here.

Suffice it to say at this point that complex strategic systems planning necessarily involves sophisticated decision problems which are interrelated and interdependent. Such problems typically involve few obvious measures of effectiveness which may be used to evaluate their worth. It is the task of systems analysis to deal with such complexity and interdependency so that decisions may be made rationally and objectively.

Management Information Subsystem

In discussing information collection as a part of the planning process, the focus was on general sources of critical information and information-gathering processes, such as strength-weakness analysis, which are essentially of a one-time nature. The need for a continuously operating and continuously updated information *system* to support planning is also acute.

Many planning failures are due to a lack of supportive information relevant to planning. Most of the information, or data, which is currently processed by the organization's information system is descriptive of the *past* history of the *internal* organizational subsystem. Most such information is outdated and inward-directed. To be useful for strategic planning, such information must be prospective and focused toward those environmental and competitive elements of the organization which will most critically affect its future.

This sort of management information is often referred to as "intelligence" or "environmental information." It can have many different forms; indeed, the diversity of such information is one of the primary arguments used by those who say that no information system can be developed for the support of traditional long-range planning, much less for innovative strategic systems planning.

To even attempt to address such broad diversity of intelligence without systematization would be utter folly. No individual or group could reasonably be expected to be aware of the broad range of competitive, industry, customer, legal, financial, technical, and other information which forms the crucial data base for strategic systems planning. Therefore, without a system, decisions will be made in the absence of relevant information.

Even subjective, personally derived information needs to be systematized. The "slip of the lip" will stand a much greater chance of being caught and passed on to the key decision makers if it can be evaluated, put into its proper context, and highlighted as useful intelligence.

To systematize business intelligence really means only that questions such as those that follow be answered and the answers made optional.

1 What needs to be known?
2 Where can the data be obtained?
3 Who will gather the data?
4 How will the data be gathered?
5 Who will analyze and interpret the data?
6 How will extracted information be stored most efficiently for equally efficient future retrieval?
7 How can extracted intelligence be disseminated to the proper parties at the right time for consideration?
8 How will the system be protected from "leakage" and from sabotage?

Unless the intelligence problem is focused toward a single program or objective and can therefore be approached on a somewhat ad hoc basis, reasonable answers to these questions generally dictate that a computer system be utilized for effective storage and retrieval operations. Dissemination and display over and above periodic briefings and responses to specific requests can be facilitated by assembling user-interest profiles and feeding them into the computer. Then, as additions are made to the file, match-up with these profiles can automatically trigger intelligence outputs which might have been missed by the human data interpreter. This is a common approach taken by military intelligence units which will permit a double check on the process and ensure rapid dissemination of vital information to the proper users.

The Integration of the Decision and Information Subsystems

The idea of a *decision support system* (DSS) that integrates aspects of both the decision and information subsystems is prominent in strategic systems thinking. Such a system incorporates a number of elements that facilitate the use of the computer in helping the planner or manager to address the unstructured and ill-structured problems which must be dealt with when planning for complex systems. These decision support systems are discussed in greater detail in Chapter 16.

An Organizational Culture and Management Subsystem

Strategic systems planning does not just happen; it must be motivated. An important part of motivation is the attitude that managers create and the culture that exists in the organization. Since it is people who perform planning, the strategic planning process must itself be structured and managed.

Part of the planning process includes giving attention to the organizational climate necessary for creative planning. An effective way of enhancing the climate for innovative planning is to encourage widespread participation in planning at all levels. Individuals can be encouraged to submit their own ideas for planning in terms of product modification, new products, new organizational arrangements, new strategy for the organization, and so forth. Such ideas should have enough justification and documentation to enable analysis groups to perform an initial appraisal and to see if the idea is worthy of further investigation.

Another important factor in creating a suitable climate for strategic planning is the stressing of the idea that change is normal and is to be expected as the organization faces a dynamic environment. Top executives must not only be change seekers; they must convince other people in the organization of the inevitability of change. Part of this convincing can be accomplished by drawing up organizational policies which reflect the official attitude toward change and planning, but there is more to it than just the paper work. In this respect Irwin has noted:

> Top managers must be change seekers. Their leadership role is to provide a climate for rapid improvement toward excellence. The success their business achieves in the future will be in geometric proportion to their understanding of, planning for, dedication to, personal involvement in, and self-motivation toward the implementation of purposeful change. For many companies this demands a reorientation in the thinking of senior executives. It means honest commitment to a new concept. Insincerity or lip service will soon destroy confidence.[13]

Of course, the only truly effective way of creating a proper climate for

[13]Patrick M. Irwin and Frank W. Langham, Jr., "The Change Seekers," *Harvard Business Review,* January–February 1966, p. 83.

strategic systems planning is to permeate the organization with planning, to demonstrate that it works, and to make use of it. When this pragmatic test of results has been passed, skeptics will be stilled and the organizational climate will be ripe for the institution of strategic systems planning.[14]

When strategic systems planning has been introduced, *it must be managed like any other organizational activity.* For some reason, many organizations have behaved in a fashion which suggests that they have a fear of applying the same hard-nosed management to planning and information processing that they apply to all other areas of their enterprise. This attitude will lead to poorer, rather than better, performance in both functions.

For instance, in planning, top-level analysis of all aspects of proposed plans should be conducted by objective analysts who have no vested interests in the various organizational elements who are the "proposers." This ensures that challenges to overly optimistic assumptions and forecasts will be made. Such objectivity can be acquired by having top-level planning analysts formally involved in the planning process and by having them rewarded in terms of measures related to planning rather than in terms of corporate or divisional results. Thus, whereas operating managers naturally tend to focus on achieving short-run results (often at the expense of the future), the corporate analysts' personal advancement should depend on their taking exactly the opposite view. If the future is sacrificed by a division for today's results, the corporate planning analysts should bear a portion of the blame. Similarly, if the future is well cared for, the corporate analysts should be rewarded.

Another important aspect of accountability for strategic planning lies in top-level comparison of present plans with past plans and with actual results. No one would expect results and past plans to be in perfect agreement, or even expect past plans and present plans for the same time period to agree. However, these deviations should be explained. If goals are changed, top management should be told the reason, or if forecasts have been radically changed on the basis of newly available information, this fact should be brought out into the open and its impact discussed.

The greatest benefit of a review of new plans, old plans, and actual results may well be in the implied accountability for planning. Too frequently the "ideal" strategic planning system deteriorates into a valueless exercise when managers find that their payoff is solely on the basis of short-run performance. If some accountability for strategic planning is applied to managers as well as to analysts, the likely result is better planning.

SUMMARY

The systems approach to strategic systems planning emphasizes the complexity and sophistication of the overall planning process in such terms as the utilization

[14]Chapter 17 expands on the ideas of an "organizational culture" and a "planning culture."

of an overall environmental framework. The incorporation of divergent planning activities ranges from futurism to financial analysis, the consideration of the second-order consequences of proposed actions, and the development of systems for strategic systems planning.

A conceptual strategic systems planning model based on one explicated by Jantsch[15] uses four dimensions to define ways in which planning may be viewed in a systems context. These dimensions are labeled *horizontal, vertical, time and causality,* and *action.*

The horizontal dimension exemplifies the continuing necessity of feedback between the general level and the particular level and back again in a strategic systems planning process. The vertical dimension deals with similar feedback loops within the hierarchy of strategic systems planning levels.

The time and causality dimension emphasizes similar continuing interactions between the exploratory forecasting required to develop alternative futures and the normative forecasting which is inherent in choice situations. Finally, the action dimension emphasizes the unity of the planning and implementation aspects of management which is one of the central themes of this book.

Of course, this systems approach to strategic systems planning is itself sophisticated and complex. This is so because of the broad objectives which are set forth for strategic systems planning—which focus on the identification and evaluation of opportunities rather than just problem solving—and because of the complexity of the systems with which strategic systems planning deals. These complexities—in the form of interactions, interdependencies, second-order consequences, chains of effect, and such phenomena as *time skip*—are inherent in the opportunity orientation of strategic systems planning.

Nonetheless, strategic systems planning may be effectively conducted by an organization at two primary levels. The first level—that of perception, awareness, and utilization—may lead the organization to awarenesses which permit it *to take advantage of the complexities of the systems* with which it is dealing. In a more analytical fashion, the organization may conduct cost-benefit analyses of second-order systems effects or perform such focused analyses as technology assessments—assessments of the second-order consequences associated with the introduction of new technologies.[16]

At a higher level, the organization may develop systems for strategic systems planning in terms of a system of plans, a planning process, a decision subsystem, a management information subsystem, and an organizational culture and management system which facilitates planning. These subsystems involve the systematization of the output (plans), process, decision-making and informational support activities, and the climate of strategic systems planning.

Thus, the systems approach, as discussed in Chapter 1, is shown in this chapter to provide a conceptual framework and a process for the planning phase

[15]Op. cit.

[16]Kasper, R. G. (ed.), *Technology Assessment: Understanding the Social Consequences of Technological Applications,* Praeger, New York, 1972.

of management. As well, a strategic planning system can be developed to guide and monitor the ongoing process of planning in the organization.

In Chapter 3, this same approach is taken to the implementation phase of organizational management.

DISCUSSION QUESTIONS

1 Taken in its broadest sense, what is meant by the planning function of management?
2 We have taken the systems view in discussing the planning function of management. What is the justification for taking this viewpoint?
3 Explain what is meant by the "horizontal" and "vertical" dimensions of strategic systems planning.
4 What does the planning dimension called "time and causality" involve?
5 A model of the action dimension of strategic systems planning has been presented. Identify and explain this model.
6 What is meant by second-order consequences? Give some examples. What is meant by time skip? Explain.
7 What is the process for systematically applying systems thinking to the planning phase of an organization? Explain this process.
8 Select a contemporary organization. Identify the claimants and the nature of their claims for such an organization.
9 Identify and explain the primary elements of strategic systems planning. Select a contemporary organization and apply these elements to that organization.
10 What are the appropriate phases of the planning process? Could these phases be applied to planning for an individual's future?
11 Many planning failures can be attributed to the lack of supportive information relevant to planning. Explain what is meant by this statement. What kinds of information must the strategic planner be concerned about?
12 Define what is meant by an organizational culture. How might the nature of the organizational culture motivate the process of strategic planning in that organization?
13 Senior managers are change seekers—constantly involved in abandoning that which has been successful in the past. What is meant by this statement?
14 Select a contemporary organization and try to determine the extent to which it has been able to cope with changes in its marketplace through the process of strategic planning. Identify a contemporary organization that is in difficulties because of lack of strategic planning.
15 "Success ultimately breeds failure" in modern organizations. What is meant by this statement? Can any example be developed from contemporary organization history to support this notion?

RECOMMENDED READINGS

Abel, Derek E.: *Defining the Business: The Starting Point of Strategic Planning,* Prentice-Hall, Inc., Englewood Cliffs, N.J., 1980.

Ackoff, Russell L.: *Redesigning the Future: A Systems Approach to Societal Problems,* John Wiley & Sons, Inc., New York, 1974.

Alter, S.: *Decision Support Systems: Current Practice and Continuing Challenges,* Addison Wesley, 1980.

Ansoff, H. I.: "The Concept of Strategic Management," *The Journal of Business Policy,* no. 2, 1972.

Ansoff, H. Igor, Roger P. Declerck, and Robert L. Hays (eds.): *From Strategic Planning to Strategic Management,* John Wiley & Sons, Inc., New York, 1976.

Argenti, J.: *Corporate Collapse,* McGraw-Hill Book Company, New York, 1976.

———: "Corporate Planning and Corporate Collapse," *Long Range Planning,* December 1976, pp. 12–17.

Bauer, R. A.: *Second-Order Consequences,* The M.I.T. Press, Cambridge, Mass., 1969.

Bell, Daniel: "Twelve Modes of Prediction—A Preliminary Sorting of Approaches in the Social Sciences," *Daedalus,* Summer 1964.

Chambers, J. D., et al.: "Catalytic Agent for Effective Planning," *Harvard Business Review,* January–February 1971.

Charan, Ramm, and R. Edward Freeman: "Planning for the Business Environment of the 1980s," *Journal of Business Strategy,* vol. 1, no. 2, Fall 1980, pp. 9-19.

Cleland, D. I., and W. R. King: "Decision and Information Systems for Strategic Planning," *Business Horizons,* April 1974.

———: *Strategic Planning and Policy,* Van Nostrand Reinhold, 1978.

Cohn, A. M.: "Planning the Business Data Environment," *Journal of Systems Management,* vol. 32, September 1981.

Ellis, L. W.: "Effective Use of Temporary Groups for New Product Development," *Research Management,* January 1979.

Emshoff, J. R., and R. E. Freeman: "Stakeholder Management: A Case Study of the U. S. Brewers Association and the Container Issue," in Schultz, R. (ed.), *Applications of Management Science,* vol. I, JAI Press, 1981.

Fahey, L., V. Narayanan, and W. R. King: "Environmental Scanning and Forecasting in Strategic Planning: The State of the Art," *Long Range Planning,* vol. 14, no. 1, February 1981.

Gerrifield, D. B.: "Selecting Projects for Commercial Success," *Research Management,* vol. 24, November 1981.

Gilmore, Frank F.: "Formulating Strategies in Smaller Companies," *Harvard Business Review,* vol. 49, no. 3, May–June 1971.

Grant, John, and William R. King: *The Logic of Strategic Planning,* Little Brown, 1982.

Hoole, R. W.: "Systems Planning for Rapidly Growing Companies," *Journal of Systems Management,* vol. 32, August 1981.

Irwin, Patrick M., and Frank W. Langham, Jr.: "The Change Seekers," *Harvard Business Review,* January–February 1966.

Jantsch, Erich: "Forecasting and Systems Approach: A Frame of Reference," *Management Science,* vol. 19, no. 12, August 1973.

Kahn, H., and A. J. Wiener: *The Year 2000,* The Macmillan Company, New York, 1967.

King, William R.: "Strategic Planning for Management Information Systems," *MIS Quarterly,* vol. 2, no. 1, March 1978.

———: "Environmental Analysis and Forecasting: The Importance of Strategic Issues," *Journal of Business Strategy,* vol. 1, no. 3, Winter, 1981.

King, William R., and Barry J. Epstein: "Assessing the Value of Information," *Management Datamatics,* vol. 18, no. 4, August 19, 1975.

———: "Assessing Information System Value: An Experimental Study," *Decision Science,* January 1983.

King, W. R., and D. I. Cleland: "A New Method for Strategic Systems Planning," *Business Horizons,* vol. 18, no. 4, August 19, 1975.

Levitt, Theodore: "Marketing Myopia," *Harvard Business Review,* vol. 38, no. 4, July–August 1960.
Linneman, Robert E.: *Shirt-Sleeve Approach to Long-Range Planning for the Smaller, Growing Corporation,* Prentice-Hall, Inc., Englewood Cliffs, N.J. 1980.
Lorange, Peter: *Corporate Planning: An Executive Viewpoint,* Prentice-Hall, Inc., Englewood Cliffs, N.J. 1980.
Miles, R. E., and C. C. Snow: *Organizational Strategy, Structure and Process,* McGraw-Hill Book Company, New York, 1978.
Murray, Edwin A., Jr.: "Strategic Choice as a Negotiated Outcome," *Management Science,* vol. 24, no. 9, May 1978.
Nenbauer, F. F., and N. B. Solomon: "A Managerial Approach to Environmental Assessment," *Long-Range Planning,* vol. 10, April 1977.
Ozbehkan, H.: "Toward a General Theory of Planning," in E. Jantsch (ed.), *Perspectives of Planning,* Organization for Economic Cooperation and Development, Paris, 1969.
Polak, F. L.: "Crossing the Frontiers of the Unknown," in A. Toffler, *The Futurists,* Random House, Inc., New York, 1972.
Prahalad, C. K.: "Strategic Choices in Diversified MNCs," *Harvard Business Review,* vol. 54, no. 4, July–August 1976.
Radford, K. J.: *Strategic Planning, An Analytical Approach,* Reston Publishing Company, Inc., Reston, Va., 1980.
Rondinelli, D. A.: "Public Planning and Political Strategy," *Long-Range Planning,* vol. 9, no. 2, April 1976.
Shank, J. K., et al.: "Balancing Creativity and Practicality in Formal Planning," *Harvard Business Review,* January 1973.
Terry, P. T.: "Mechanisms for Environment Scanning," *Long-Range Planning,* vol. 10, June 1977.
Toffler, A.: *Future Shock,* Random House, Inc., New York, 1971, pp. 17–18.
"The New Planning," *Business Week,* December 1978.

CASE 2-1: Organizing for Strategic Planning in the Matrix Management Context

The International Organization in the XYZ Corporation has existed for almost two years. At a management council meeting soon after the new international organization was announced, the international president described how matrix management would be expected to operate in this environment. The four "cornerstones" of this matrix management system would be: shared decision making, shared accountabililty, shared financial results, and the fundamental concept that the "country" would be a primary building block of the international organization.

Success in the international marketplace starts with successful strategic planning. The "cultural ambience" for strategic planning in this company has improved considerably over the past several years. Strategic planning in the international marketplace poses particular challenges to XYZ matrix managers and professionals. The "in-country" manager will be responsible for the strategic planning encompassing a particular country.

There is a growing appreciation of the need to know how best to organize "in-country" strategic planning, particularly in the sharing context of matrix management.

Small Work Group Assignment: As a geographic manager or product manager, develop an approach of how a geographic manager should best organize resources to accomplish strategic planning. Items you might wish to consider in developing this organizational approach include:

- A consideration of the strategic planning process
- Use of a task force to accomplish strategic planning
- Delineation of authority and responsibility patterns in organizing for strategic planning
- Work breakdown structure for strategic planning

What problems would you expect in doing strategic planning in the matrix environment? In particular, what do you think will be some of the major problems in doing planning for a country? What recommendations do you have to help solve (or alleviate) these problems? Prepare a summary briefing of your small work group ideas and select a representative to present your conclusions.

CHAPTER 3

IMPLEMENTATION USING THE SYSTEMS APPROACH

Nothing chastens a planner more than the knowledge that he will have to carry out the plan.[1]

This chapter deals with the application of the systems approach to the implementation phase of management. As with its application to the planning phase, as discussed in Chapter 2, there are three parts: a conceptual basis, a process, and a "system" that enables implementation to be accomplished on an ongoing basis. In the case of the planning phase, the "system" is a *strategic planning system.* In this implementation phase, the system is a *project management system.*

THE CONCEPTUAL BASIS FOR SYSTEMS-ORIENTED IMPLEMENTATION[2]

The literature of planning is replete with descriptions of plans that have been painstakingly developed and then "placed in a file" while critical program and project decisions in the organization are made without reference to the plan. In such cases, the strategic plan is not being effectively implemented, since programs and projects are the key activities through which a plan must be carried out.

[1]James M. Gavin, *On to Berlin,* The Viking Press, Inc., New York, 1978.

[2]Some material in this section is adapted from W. R. King and D. I. Cleland, *Strategic Planning and Policy,* Van Nostrand Reinhold Co., New York, 1978, chap. 6.

A recent audit of the existing and planned programs in the central research laboratory of a major diversified firm found:[3]

a Programs and projects that could not be associated with any business or corporate objective or strategy
b Programs and projects that apparently fell outside the stated mission of the corporation or the charter of the laboratory
c Projects whose funding levels could not reasonably be justified in terms of the expected benefits to be produced

Such observations have so frequently emanated from less formal analyses in other companies as to suggest the existence of a faulty linkage between corporate plans and strategy and the programs and projects through which they should be implemented.

The conceptual framework through which such implementation failures may be resolved involves a systems-oriented model of the strategic choice elements that the organization possesses, i.e., the strategic choices that it makes. These choice elements form a system in that they are interrelated and interdependent. Only when those interrelationships are understood and appropriately defined can the organization's strategy be effectively implemented.

The Choice Elements of Strategy

Because of the semantics jungle which exists in the area of strategy, it is necessary to define rather precisely the terms to be used. The *choice elements of organizational strategy*—those choices that must be explicitly or implicitly made in the organizational strategic planning process—are its:

Mission—the "business" that the organization is in
Objectives—desired future positions or roles for the organization
Strategy—the *general direction* in which the objectives are to be pursued
Goals—specific targets to be sought at specified points in time
Programs/projects—resource-consuming sets of activities through which strategies are implemented and goals are pursued
Resource allocations—allocations of funds, personnel, etc., to various units, objectives, strategies, programs, and projects

These informal definitions are meant to provide a common framework for communication rather than to define the "correct" terminology. Various organizations may use different terminology, but none can escape the need to make choices of each variety.

Most organizations conduct planning processes which are aimed at explicitly choosing all or some of these strategic choice elements. However, many firms fail to deal with all of the choice elements in the detail and specificity that each deserves.

[3]See W. R. King, "Implementing Strategic Plans Through Strategic Program Evaluation," *OMEGA: The International Journal of Management Science*, vol. 8, no. 2, pp. 173–181, 1980.

Often, for instance, missions are dealt with implicitly, as in the case of the firm that responds to the mission concept by stating its mission to be: "We make widgets." Such a product-oriented view of the organization's business ignores new market opportunities and perhaps, the firm's generic strengths. It is these opportunities and strengths which form the most likely areas for future success. Thus, it is these opportunities and strengths, rather than the current product line, which should define the mission.

Strategies are almost always explicitly chosen by firms, but often strategies are thought of in output, rather than input, terms. In such instances, strategies may be described in terms of expected sales and profits rather than in terms of strategic directions such as product redesign, new products, or new markets.

Thus, the elements of strategic choice are inescapable in the sense that the avoidance of an explicit choice about any of the elements means that it is chosen implicitly. However, many firms make poor or inappropriate choices, both explicitly and implicitly, because they do not have a clear awareness of the relationships among the strategic choice elements and their innate interdependence.

Relationships among the Strategic Choice Elements

One of the most important conditions for the effective implementation of plans has to do with the relationships among the strategic choice elements. If these relationships are well defined and carefully analyzed and conceived, the plan is likely to be implemented. If they are not, the plan is likely to be a voluminous document which requires substantial time and energy to prepare, but which is filed on the shelf until the next planning cycle commences. Indeed, many plans are so treated precisely because they do not carefully spell out the relationships among various strategic choice elements. They therefore do not provide the appropriate information necessary to guide the many decisions which must be made to implement the plan and to develop and manage the projects and programs which are the operational essence of the plan.

Figure 3-1 shows the elements of strategic choice in the form of a triangle which illustrates that the mission and objectives are the highest-level elements. They are supported by the other elements—the strategies, goals, and programs/projects. The strategic resource allocations underlie each of these elements.

Figure 3-2 shows an illustration of these concepts in terms of a business firm. The mission chosen is that of "supplying system components to a world-wide non-residential air-conditioning market." Note that while this mission statement superficially appears to be product-oriented, it identifies the nature of the product (system components), and the market (worldwide nonresidential air conditioning) quite specifically. By exclusion, it guides managers in avoiding proposals for overall systems and strategies that would be directed toward residential markets. However, it does identify the entire world as the company's territory and (in an elaboration not shown

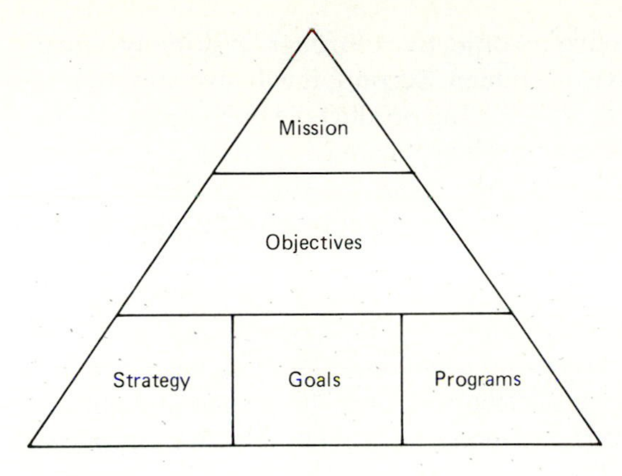

FIGURE 3-1
General relationship among strategic choice elements.

here) defines air conditioning to include "air heating, cooling, cleaning, humidity control and movement."

Supporting the base of the triangle are strategies, goals, and programs. The firm's strategies are stated in terms of a three-phase approach. First, the

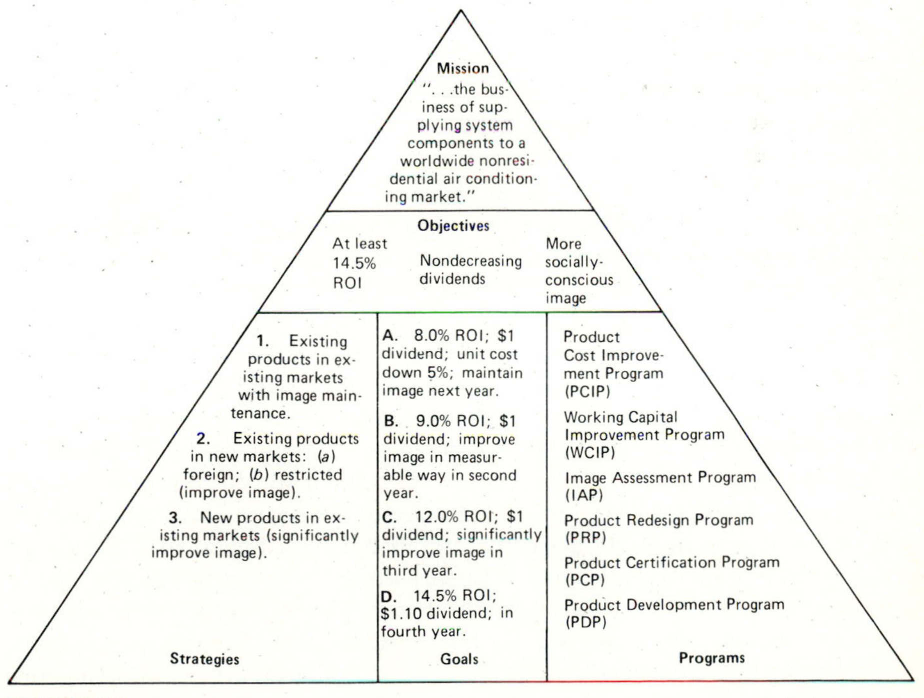

FIGURE 3-2
Illustrative strategic choice elements.

company will concentrate on achieving its objectives through existing products and markets while maintaining its existing image. Second, it will give attention to new markets—foreign and restricted—for existing products, while improving the company's image. "Restricted" markets may be thought of as those that require product-safety certification before the product can be sold in that market. Finally, it will focus on new products in existing markets while *significantly* improving its image.

Clearly, this is a staged strategy: one that focuses attention first on one thing and then on another. This staging does not imply that the first strategy element is carried through completely before the second is begun; it merely means that the first element is given primary and earliest attention, then the second and third in turn. In effect, the first element of the strategy has its implementation *begun* first. This will be made more clear in terms of goals and programs.

At the right-hand base of the triangle, a number of the firm's programs are identified. Each of these programs is made up of a variety of projects or activities. Each program serves as a focus for various activities having a common goal. For instance, in the case of the Product Cost Improvement Program, the associated projects and activities might be as follows:

Quality-control project
Production planning improvement project
Production-control system development project
Plant layout redesign project
Employee relations project

All of these projects and activities are focused toward the overriding goal of product cost improvement.

In the case of the Working Capital Improvement Program, the various projects and activities might include a "terms and conditions" study aimed at revising the terms and conditions under which goods are sold, and an "inventory reduction project." Each of the other programs would have a similar collection of projects and activities focused on some single well-defined goal.

The goals are listed in the middle-lower portion of the triangle in Figure 3-2. Each goal is stated in specific and timely terms related to the staged strategy and the various programs. These goals reflect the desire to attain 8.0 percent return on investment (ROI)—a step along the way to the 14.5 percent objective—next year, along with a $1 dividend (the current level) and a unit cost improvement of 5 percent, while maintaining image. For subsequent years, the goals reflect a climb to 14.5 percent ROI, a steady and then increasing dividend, and an increasing and measurable image consistent with the staged strategy that places image improvements later in the staged sequence. This is also consistent with the program structure, which includes an Image Assessment Program, a program designed to develop methods and measures for quantitatively assessing the company's image.

Figure 3-3 shows the same elements as does Figure 3-2, with each being indicated by number, letter, or acronym. For instance, the block labeled "1" in Figure 3-3 represents the first stage of the strategy in Figure 3-2, the letter "A" represents next year's goals, and so on.

The arrows in Figure 3-3 represent *some* illustrative relationships among the various objectives, strategy elements, goals, and programs. For instance, the arrows (a), (b), and (c) reflect direct relationships between specific timely goals and broad timeless objectives:

a A, next year's goals, primarily relate to the objective of nondecreasing dividends.
b B, the second year's goals, relate to the "more socially conscious image" objective.
c D, the quantitative ROI figure, is incorporated as a goal in the fourth year.

Of course, each year's goals relate implicitly or explicitly to all objectives. However, these relationships are the most direct and obvious.

Similarly, arrow (d) in Figure 3-3 relates the first year's goals to the first element of the overall strategy in that these goals for next year are to be attained

FIGURE 3-3
Relationships among illustrative strategic choice elements.

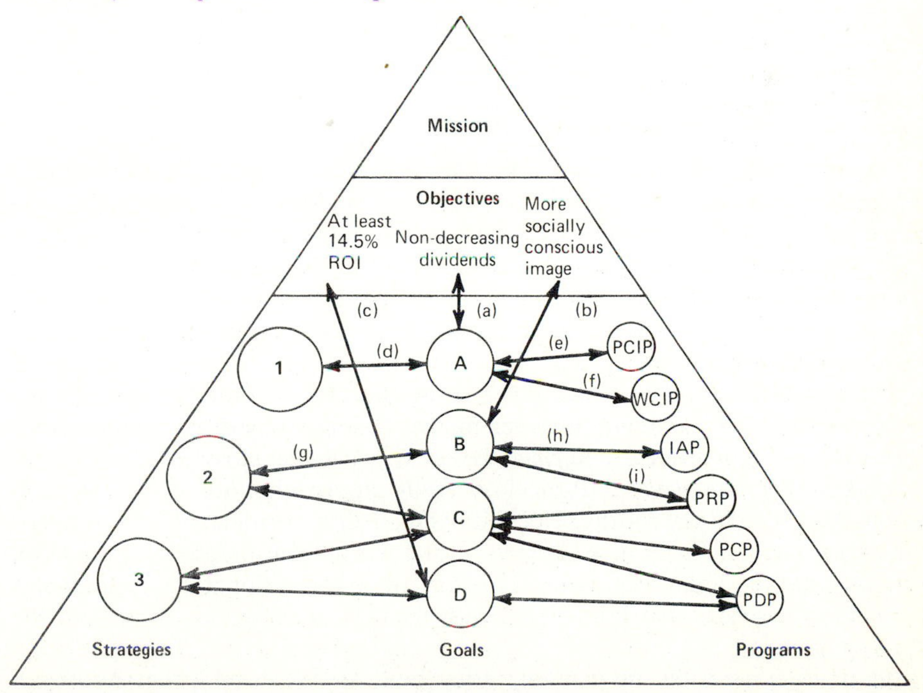

primarily through the strategy element involving "existing products in existing markets." However, arrows (e) and (f) also show that the Product Cost Improvement Program (PCIP) and the Working Capital Improvement Program (WCIP) are also expected to contribute to the achievement of the first year's goals.

The second year's goals will begin to reflect the impact of the second strategy element (existing products in new markets), as indicated by arrow (g) in Figure 3-3. The effect of the Product Redesign Program (PRP) is also expected to contribute to the achievement of these goals (arrow [i]), as is the Image Assessment Program (IAP), which is expected to provide an ability to measure image by that time, as indicated by arrow (h). The other arrows in Figure 3-3 depict other rather direct relationships whose interpretation is left to the reader.

From this figure, relationships among the various strategic decision elements can be seen:

1 Goals are specific steps along the way to the accomplishment of broad objectives.
2 Goals are established to reflect the expected outputs from strategies.
3 Goals are directly achieved through programs.
4 Strategies are implemented by programs.

Thus, the picture shown in Figure 3-3 is that of an interrelated set of strategic factors that demonstrate *what* the company wishes to accomplish in the long run, *how* it will do this in a sequential and sensible way, and *what performance levels* it plans to achieve at various points along the way.

THE STRATEGIC IMPLEMENTATION PROCESS

An understanding of the systems interrelationships among the various strategic choice elements is itself important. However, the enhancement of implementation comes through a process of strategic program evaluation.

Program/project evaluation, when performed in the overall systems context of the strategic choice elements, can ensure effective implementation of planning decisions. This requires a formal process because the interrelationships are so complex and voluminous.

The strategic program evaluation process thereby becomes the integrating factor for the many strategic choice elements. It does so by *directly utilizing the results of the higher-level strategic choices to evaluate alternative programs, projects, and funding levels.* "Project selection" approaches are well known and widely used in industry for the selection of engineering projects, R&D projects, and new product development projects. However, if program/project evaluation is to be the key link in unifying the array of organizational strategic choice elements, the evaluation framework must itself be an integral element of the strategic plan.

Figure 3-4 indicates in general terms how an idealized program/project evaluation process can serve as an integrating factor for the firm's array of

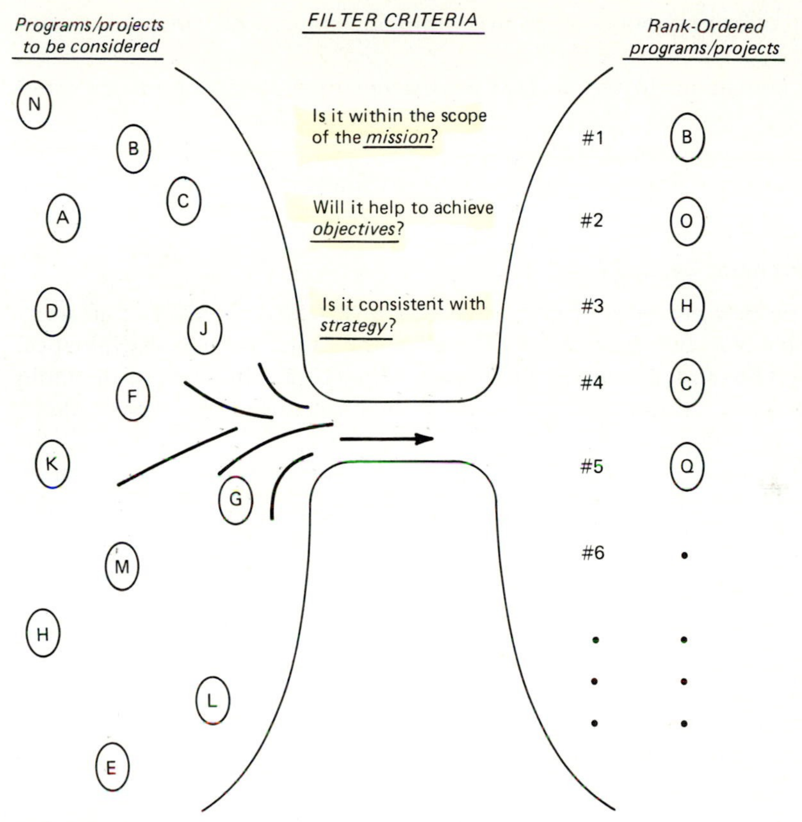

FIGURE 3-4
A "filter" process for intergrating strategy and programs.

strategic choices. It shows a wide variety of potential projects and programs being "filtered" through the application of strategic criteria that are based on the higher-level choices that have previously been made—the organization's mission, objectives, and strategy. The output of this filtering process is a set of rank-ordered program and project opportunities that can serve as a basis for the allocation of resources.

Other important criteria must come into play in implementing this evaluation process. These criteria are those that are *implicit* in a good specification of the organization's mission, objectives, and strategy. However, they must be *specifically addressed* if the program and projects are to truly reflect corporate strategy. These criteria are:

1 Does the opportunity take advantage of a *strength* that the company possesses?

2 Correspondingly, does it avoid a dependence on something that is a *weakness* of the firm?

3 Does it offer the opportunity to attain a *comparative advantage* over competitors?
4 Does it contribute to the *internal consistency* of the existing projects and programs?
5 Is the level of *risk* acceptable?
6 Is it consistent with established *policy guidelines*?

A Strategic Program Evaluation Illustration

A strategic program/project evaluation framework based on these criteria is shown as Table 3-1. In the leftmost column of the table is a set of evaluation criteria that relates to the example in Figures 3-2 and 3-3. The body of the table shows how a proposed new program to begin manufacturing of system components in Europe might be evaluated.

The "criteria weights" in the second column of the table reflect their relative importance and serve to permit the evaluation of complex project characteristics within a simple framework. A base weight of 20 is used here for the major criteria related to mission, objectives, strategy, and goals. Weights of 10 are applied to the other criteria.

Within each major category, the 20 "points" are judgmentally distributed to reflect the relative importance of subelements or some other characteristic of the criterion. For instance, the three stages of strategy and the four subgoals are weighted to ensure that earlier stages and goals are treated to be more important than later ones. This implicitly reflects the *time value of money* without requiring a more complex "present value" discounting calculation.

The first criterion in Table 3-1 is the "fit' with mission." The proposal is evaluated to be consistent with both the "product" and "market" elements of the mission and is thereby rated to be "very good," as shown by the 1.0 probability entries at the upper left.

In terms of "consistency with objectives," the proposal is rated to have a 0.2 (20 percent) chance of being "very good" in contributing to the ROI element of the objectives (see Figure 3-2), a 60 percent chance of being "good," and a 20 percent chance of being only "fair," as indicated by the probabilities entered into the third row of the table. The proposed project is rated more poorly with respect to the "Dividends" and "Image" elements.

The proposal is also evaluated in terms of its expected contribution to each of the three stages of the strategy as outlined in Figure 3-2. In this case, the proposed project is believed to be one which would principally contribute to stage 2 of the strategy. (Note that only certain assessments may be made in this case, since the stages are mutually exclusive and exhaustive.)

The proposal is similarly evaluated with respect to the other criteria.

The overall evaluation is obtained as a weighted score that represents the sum of products of the likelihoods (probabilities) and the 8, 6, 4, 2, 0 arbitrary level weights that are displayed at the top of the table. For instance, the "consistency with objectives—ROI" expected level weight is calculated as

TABLE 3-1
A STRATEGIC PROGRAM/PROJECT EVALUATION.

Program/project evaluation criteria		Criteria weights	Very good (8)	Good (6)	Fair (4)	Poor (2)	Very poor (0)	Expected level score	Weighted score
"Fit" with mission	Product	10	1.0					8.0	80
	Market	10	1.0					8.0	80
Consistency with objectives	ROI	10	0.2	0.6	0.2			6.0	60
	Dividends	5		0.2	0.6	0.2		4.0	20
	Image	5			0.8	0.2		3.6	18
Consistency with strategy	Stage 1	10					1.0	0	0
	Stage 2	7	1.0					8.0	56
	Stage 3	3					1.0	0	0
Contribution to goals	Goal A	8					1.0	0	0
	Goal B	6	0.8	0.2				7.6	45.6
	Goal C	4		0.8	0.2			5.6	22.4
	Goal D	2					1.0	0	0
Corporate *strength* base		10				0.8	0.2	1.6	16
Corporate *weakness* avoidance		10				0.2	0.8	0.4	4
Comparative advantage level		10	0.7	0.3				7.4	74
Internal consistency level		10	1.0					8.0	80
Risk level acceptability		10				0.7	0.3	1.4	14
Policy guideline consistency		10			1.0			4.0	40

Total score 610

$$0.2(8) + 0.6(6) + 0.2(4) = 6.0$$

This is then multiplied by the criterion weight of 10 to obtain a weighted score of 60. The weighted scores are then summed to obtain an overall evaluation of 610.

Of course, this number in isolation is meaningless. However, when various programs and projects are evaluated in terms of the same criteria, their overall scores provide a reasonable basis for developing the ranking shown on the right side of Figure 3-4. Such a ranking can be the basis for resource allocation, since the top-ranked program is presumed to be the most worthy, the second-ranked is the next most worthy, etc.

It can readily be seen that such a strategic program evaluation process will enhance the implementation of the choices made in the planning phase of management.

The critical element of the evaluation approach is its use of criteria that ensure that programs will be integrated with the mission, objectives, strategy, and goals of the organization, as well as of criteria that reflect critical bases of strategy, such as business strengths, weaknesses, comparative advantages, internal consistency, opportunities, and policies.

PROJECT MANAGEMENT SYSTEMS

Once an appropriate array of projects has been selected and resources allocated, the planning phase of management is largely completed. Of course, as noted in the previous sections of this chapter, this must be done in a manner which ensures consistency with organizational missions and objectives and which enhances the likelihood of successful implementation. The *project management system* is the management subsystem designed to ensure successful implementation on a day-to-day ongoing basis.

A *project* may be more precisely defined than it previously has been: A complex effort to achieve a specific objective within a schedule and budget target, which typically cuts across organizational lines, is unique, and is usually not repetitive within the organization.

Project management is called by many different names. Sometimes it is called *program management; systems management; satellite management; task force management; team management; ad hoc management; "micro-company" management,* among other names.

All of these forms, together with the more permanent form usually termed "matrix management," emphasize some form of team effort to integrate and synergize the application of resources to organizational problems and opportunities. Matrix management had its genesis in several ways: product management, international management, and project/program management.

Project management emerged in an unobtrusive manner, starting in the early 1960s. The term "project (or program) management" was used to describe a type of structure which already existed in various forms. No one can claim to have invented project management; its beginnings are often cited as the ballistic missile program of the space program of the United States. The origins of project management can be found in the management of large scale ad hoc endeavors such as the Manhattan Project, large construction projects, or the use of naval task forces.

In the early 1960s we began to recognize project management for what it is. As early as 1961 Fisch, writing in the *Harvard Business Review,* spoke of the obsolescence of the line-staff concept and the growing trend of the use of a "functional teamwork" approach to organization.[4] In 1961 IBM established system managers with overall responsibility for various computer models across functional divisional lines. In the early 1960s and 1970s a wide variety of organizations experimented with the use of alternative project-management organization forms. Project management has reached a degree of maturity and has been the precursor of today's matrix management approach.

Since that time, the terms "matrix management" and "matrix organization" have come to be used to describe both project-driven two-dimensional organizations and organizations that have "permanent" matrix forms. Although the project-driven matrix is the prime focus in this book, we shall from time to time use illustrations of both "temporary" and "permanent" matrix forms.

In the mid-1970s Chase Manhattan's corporate bank reshaped itself from a geographic form of organization into one that assembled officers into teams, each team organized to focus onto a single industry (such as drugs or electronics). Chase at present is steadily moving toward the matrix form of organization. According to *Fortune:*

> A short time ago, an organization chart of the bank's international department would have shown it almost entirely divided along geographical lines, with only merchant banking roped off and functioning more or less worldwide. Today the bank has several other cross-border operations, to which it may add still more, that give its structure an unmistakable matrix look: international institutions (which primarily means correspondent banks), export and trade financing, and private banking (for well-heeled individuals).[5]

Such structures are perhaps most visible in multinational corporations. Product managers are located in the profit center of the organization; geographic managers are responsible for a geographical portion of the market. The use of "in-country" managers is a common mechanism to provide geographic focus; in-country manufacturing divisions and sales operations report in some way to *both* product managers and geographic managers. In this matrix, key decisions are shared in the three basic organization elements.

In such a company, there are usually two coordinated avenues of planning: by product, and by geography. Since decisions are shared, accountability for results is also shared in terms of product and geographic profitability through profit-center mechanisms on a product and geographic basis. International companies (particularly in Europe and Japan) emphasize group performance. This is

[4]Gerald G. Fisch, "Line-Staff is Obsolete," *Harvard Business Review,* September 1961.

[5]"It's a Stronger Bank that David Rockefeller is Passing to his Successor," *Fortune,* Jan. 14, 1980, p. 44.

reflected in their compensation policies. Financial visibility by product and geography is the norm in the multinational company.

In all these cases, the organizational interfaces develop a cultural ambience where key managers share key decisions; authority and responsibility are complementary. Successful managers in the matrix organization have learned to work key decisions through the complementary interface managers. This sharing of decisions, results, and accountability is illustrated in Figure 3-5.

A basic organizational factor that is significantly affected by the matrix organization is the concept of the profit center and the delegation of authority to one manager who is held responsible for producing profitable results. Everything counts at the profit-center level, everything is measured there, and people are rewarded accordingly. For those managers who have operated successfully for years in a decentralized profit-center mode, the sharing of decisions and results with some other manager outside the parent hierarchy can be a cultural shock. Certain key decisions are traditionally considered the profit-center manager's "territory," such as:

Product pricing
Product sourcing
Human resources
Facility management
Cash management

In the matrix organization, the profit-center manager must share these key decisions with others. For example, in product pricing in the international market the profit-center manager will find it necessary to work with an in-country manager to establish price. Product sourcing decisions may be made by senior marketing executives at corporate headquarters rather than by the profit-center manager. In practice, decision authority is complementary.

Clearly, all of this will not just happen in an organization by having someone order it into being. Such an approach requires day-to-day monitoring, new information and reporting methods, etc. In Part Four of this book, we shall

FIGURE 3-5
Sharing in the matrix organization.

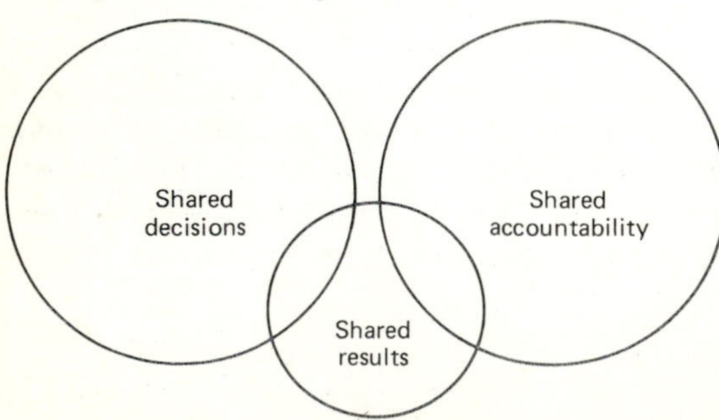

develop the ideas of how this may be done. Here we need only provide the broad outline of the system that can be instituted to do this.

The *project management system* operates to ensure that projects which have been evaluated, selected, and funded are appropriately executed on a day-to-day basis. It begins when a focal point is established for pulling together the work of several different organizational elements, and an individual is designated the *project manager.* When this focal point is established, and an individual is designated to provide leadership, a new integrating role has been created. Project management involves the use of a team for the accomplishment of some specific purpose. The project is not a permanent entity but, rather, an activity whose purpose is to work itself into ultimate dissolution after the objectives of the project have been accomplished. These objectives are threefold:

1 To accomplish the project on schedule
2 To accomplish the project within budget
3 To finish the project so that its technical performance objectives are achieved

Project management is based upon several key concepts: *first,* the designation of an organizational position to bring into focus the expenditure of organizational resources to support a desired end purpose; *second,* project planning and control techniques to be accomplished through the use of a team; *third,* the assignment of the work to be accomplished on the project to various people representing the disciplines necessary to support a project. Project management requires that the organization's managers abandon some of their traditional views on how organizations are structured and managed.

Subsystems of a Project Management System[6]

The project management system may be defined in terms of its subsystems:

The Facilitative Organizational Subsystem is the organizational arrangement that is used to superimpose the project teams on the functional structure. The resulting "matrix" organization portrays the formal authority and responsibility patterns and the personal reporting relationships aimed at providing an organizational focal point for starting and completing specific projects. Two complementary organizational units tend to emerge in such an organizational context: *the project team* and *the functional units.*

The Project Control Subsystem provides for the selection of performance standards for the project schedule, budget, and technical performance. This subsystem deals with information feedback to compare actual progress with planned progress, and with the initiation of corrective action as required. The rationale for a control subsystem arises out of the need for monitoring the various organizational units that are performing work on the project in order to deliver results on time and within budget.

The Project Management Information Subsystem contains the *intelligence* essential

[6]Portions of this section are adapted from David I. Cleland, "Defining a Project Management System," *Project Management Quarterly,* vol. 8, no. 4, pp. 37–40, December 1977.

to the effective control of the projects. This subsystem may be informal in nature, consisting of periodic meetings with the project participants who report information on the status of their project work, or a formal information retrieval system that provides frequent "printouts" of what is going on. This subsystem provides the intelligence to enable the project team members to make and implement decisions in the management of the project.

Techniques and Methodology is not really a subsystem in the sense that the term is used here. This subsystem is merely a set of techniques and methodology, such as: PERT, CPM, PERT-Cost, and related scheduling techniques as well as other management science techniques which can be used to evaluate the risk and uncertainty factors in making project decisions. (These are discussed in Chapter 15.)

The Cultural[7] *Ambience Subsystem* is the subsystem in which project management is practiced in the organization. Much of the nature of the cultural ambience can be described in how the people—the social groups—feel about the way in which project management is being carried out in the organization. The emotional patterns of the social groups, their perceptions, attitudes, prejudices, assumptions, experiences, and values, all go to develop the organization's cultural ambience. This ambience influences how people act and react, how they think and feel, and what they say in the organization, all of which ultimately determines what is taken for socially acceptable behavior in the organization.

The Planning Subsystem recognizes that project control starts with project planning, since the project plan provides the standards against which control procedures and mechanisms are developed. Project planning starts with the development of a work breakdown structure which shows how the total project is broken down into its component parts. Project schedules and budgets are developed, technical performance goals are selected, and organizational authority and responsibility are established for members of the project team. Project planning also involves identifying the material resources needed to support the project during its life cycle. "What are we aiming for and why?" is the key question which project planning answers.

The Human Subsystem involves just about everything associated with the human element. An understanding of the human subsystem requires some knowledge of sociology, psychology, anthropology, communications, semantics, decision theory, philosophy, leadership, and so on. Motivation is an important consideration in the management of the project team. Project management means working with people to accomplish project objectives and goals. Project managers must find ways of putting themselves into the human subsystem of the project so that the members of the project team trust and are loyal in supporting project purposes. The artful management style that project managers develop and encourage within the peer group in the project may very well determine the success or failure of the project. Leadership is the most important role carried out by a project manager. Project-management leadership is the leading of people along a way to the delivery of the project results on time and within budget.

Figure 3-6 shows a project management system (PMS) diagrammatically, with

[7]The term *culture* was introduced in a strategic planning context in David I. Cleland and William R. King, "Developing a Planning Culture for More Effective Strategic Planning," *Long-Range Planning*, vol. 7, no. 3, September 1974.

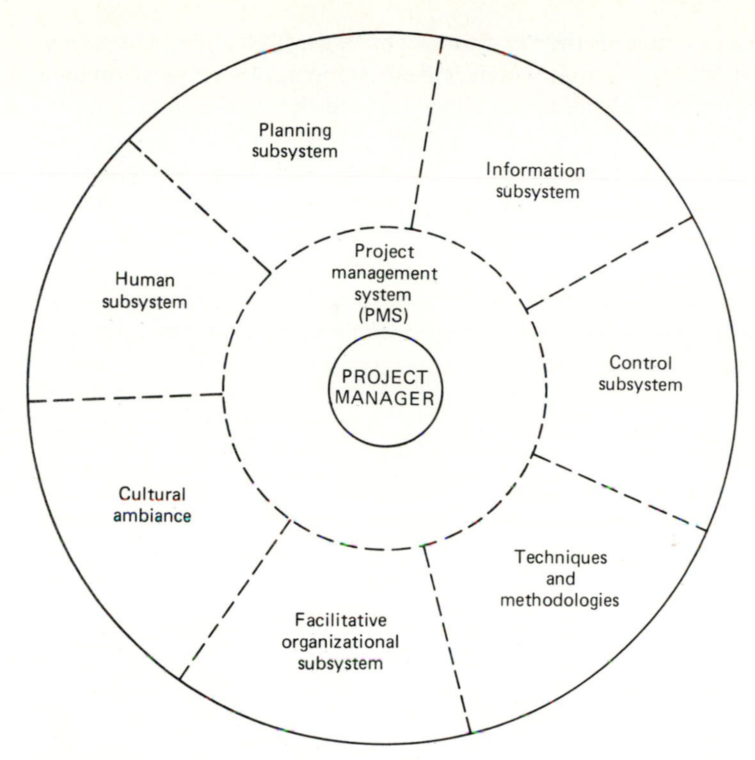

FIGURE 3-6
A project management system.

the project manager as the focal point of the interactive subsystems. The PMS clearly reflects the systems approach to management in that it undertakes the integration of a system of multiple independent subsystems. These subsystems work toward the development of a synergy to produce an objective on time and within budgeted cost.

SUMMARY

This chapter develops the conceptual basis, process, and system for the implementation phase of management using the systems approach.

The conceptual basis is an understanding and clear depiction of the interrelationships among the strategic choice elements of the organization—its mission, objectives, strategy, goals, and programs/projects.

The conceptual basis may be operationalized through a process of strategic program evaluation that ensures that the projects and programs that are selected, or given high priority, are those that best implement the strategic choices that have previously been made in the planning phase.

The implementation phase is then institutionalized through a *project management system* that ensures that that which is desired, in terms of performance, cost, and time, will be directly addressed and pursued by the organization on a day-to-day basis.

Part Three of the book, beginning with Chapter 4, deals with the "technology," called *Systems Analysis,* of the planning phase of management. In effect, it deals with the technical basis for doing the things that have previously been outlined in Chapter 2.

Part Four of the book performs the same function with respect to the framework for the implementation phase of management that has been presented here in Chapter 3.

DISCUSSION QUESTIONS

1 What are the key activities through which a strategic plan is carried out? We refer to a "faulty linkage" between corporate plans (strategy) and strategic programs. What is meant by this?

2 Identify and discuss the choice elements of organizational strategy. Why must these choice elements be identified in the corporate strategic planning process?

3 Strategies are almost always explicitly chosen by firms, but often strategies are thought of in output, rather than input, terms. What is meant by this statement?

4 Develop a model showing the relationship of strategic choice elements in the strategic planning process. Explain what is meant by this model.

5 What is meant by the implementation phase of strategic planning? What is the importance of this phase relative to survival and growth in the marketplace?

6 Explain in general terms how an idealized program/project evaluation process can serve as an integrating factor for the organization's array of strategic choices. What are the strategic criteria through which the potential programs and projects can be "filtered"?

7 How might a strategic program evaluation framework be developed? What are some of the criteria to consider in developing such a framework?

8 Identify and explain the subsystems involved in a project management system. Why is it important to take a systems view in describing project management in a contemporary organization?

9 What are some of the different names by which project management has been called? Are there really any differences in these titles?

10 What are some of the key characteristics of matrix management?

11 Why did project management emerge in management theory? What are some of its distinguishing characteristics?

12 Matrix management is sometimes used in multinational corporations. Why are such corporations using matrix management? What are some of the advantages and disadvantages of matrix management in the multinational corporation?

13 Identify the key elements in the sharing context of the matrix organization. How might this sharing context affect the way executives manage the organization? Does this sharing context have any implications for the professional in the matrix organization?

14 The development of matrix management is an outgrowth of the size, complexity, and rate of change in modern organizations. Defend or refute this statement.

15 Project management is based upon several key concepts. What are these concepts? Select a contemporary organization and explain briefly how these concepts might apply.

RECOMMENDED READINGS

Alavi, M., and J. C. Henderson: "Evolutionary Strategy for Implementing a Decision Support System," *Management Science,* vol. 27, November 1981.

Anshen, Melvin: "The Management of Ideas," *Harvard Business Review,* July–August 1969.

Cleland, David I.: "Defining a Project Management System," *Project Management Quarterly,* vol. 8, no. 4, December 1977.

———, and William R. King: "Developing a Planning Culture for More Effective Strategic Planning," *Long-Range Planning,* September 1974.

Drake, Rodman L., and Lee M. Caudill: "Management of the Large Multinational: Trends and Future Challenges," *Business Horizons,* vol. 24, no. 3, May–June 1981.

Fisch, Gerald G.: "Line-Staff Is Obsolete," *Harvard Business Review,* September 1961.

Harell, R. D.: "Five Principles for Successful Computer System Implementation," *Governmental Finance,* vol. 10, September 1981.

Janger, Allen R.: *Matrix Organization of Complex Businesses,* The Conference Board, Inc., New York, 1979.

King, William R.: "Implementing Strategic Plans Through Strategic Program Evaluation," *OMEGA: The International Journal of Management Science,* vol. 8, no. 2, 1980.

———, and David I. Cleland: *Strategic Planning and Policy,* Van Nostrand Reinhold Co., New York, 1978, chap. 6.

——— (eds.): *Project Management Handbook,* Van Nostrand Reinhold, 1983.

Meads, Donald E.: "The Task Force at Work—the New Ad-Hocracy," *Columbia Journal of World Business,* November–December 1970.

Menzies, Hugh D.: "Westinghouse Takes Aim at the World," *Fortune,* January 14, 1980.

Schultz, R. L., and D. P. Slevin (eds.): *Implementing ORLMS,* Elsevier, New York, 1975.

Smith, Robert F.: *The Variations of Matrix Organization,* Special Study No. 73, The Presidents Association, The Conference Board, Inc., New York.

Williams, Earle C.: "Matrix Management Offers Advantages for Professional Services Firms," *Professional Engineer,* February 1978.

Wolff, Michael F.: "The Joy (and Woe) of Matrix," *Research Management,* March 1980.

———: "Knowing When The Horse Is Dead," *Research Management,* November 1981.

"Format Fears at Philips," *Management Today,* August 1978.

Zmud, R. W., and J. F. Cox: "The Implementation Process: A Change Approach," *MIS Quarterly,* vol. 3, June 1981.

CASE 3-1: Project Management System Start-Up

Situation: You manage one of several engineering sections of an engineering design department.

As the result of the "need for enhanced effectiveness and productivity," project management is being introduced in your section for the first time. At a recent all-employee meeting, the engineering department manager has outlined the need for project management and explained its value. At that meeting, it was stated that the manager of the projects section, which has been partially in place for some time now, "will increase in size and take on full responsibilities beginning immediately."

Several of the project engineers who report to you have asked to meet with you to discuss project management. These engineers have each had total responsibility for "their" own projects, and they are asking specifically, "What should we do differently now that project management is taking over?" Specifically, they would like to know the difference between "project management" and "project engineering."

As a concerned manager, you have a strong interest in ensuring that the transition to project management works effectively. You therefore want to have such a meeting and use it not only to answer the specific question, but also to encourage your project engineers to (1) support the project management concept, and (2) take whatever actions they can to help make it work.

Your Small Work Group Assignment: As a group, outline the agenda for the meeting which you will conduct with your people in such a way as to contribute to a speedy and successful start-up.

Specify all questions, issues, and probable concerns which your people might have, and outline how you would plan to handle each during the meeting. If there are any questions, issues, or concerns which will require additional action on your part, specify and outline your plan for resolving each.

CASE 3-2: Operationally Defining a Project

Situation: An industrial organization is involved in establishing a *project management system.* Important to such an undertaking is the operational definition of a "project" within that organization.

Your Task: Develop a standard operational definition of a "project" for use in such an organization's policy and procedure documentation. In such a definition, you might want to consider such things about a project as: duration, dollar amount, frequency of project team meetings, number of "cost centers" involved, record-keeping requirements, priority, and complexity of work breakdown structure.

CASE 3-3: Evaluating Strategic Programs

Situation: The criteria, criteria weights, levels (good, fair, etc.), and level weights (8, 6, 4, 2, 0) in Table 3-1 on p. 69 may be considered to be the *overall framework* which an organization has decided to use to evaluate all strategic programs. The probability entries in that table represent judgments as to how a *particular* proposed program will meet each of the criteria.

Your Task: As an exercise, make some arbitrary (but logical and consistent) judgments concerning another program and use the criteria, criteria weights, levels, and level weights from Table 3-1 to evaluate this other program. Compare this program's overall score with that given for the program that is evaluated in that table. Interpret this comparison in overall terms and in terms of the organization's strategic choice elements.

Have team members formulate their personal criteria, criteria weights, levels, and level weights to evaluate alternative "programs" in terms of each individual's personal "strategic choice elements." For instance, you might state "financial independence" as one of your personal objectives and "having fun while preparing for the future" as one of your strategies. Evaluate several alternative programs, such as

- **a** Go to graduate school full-time
- **b** Go to graduate school part-time and have a part-time job
- **c** Get a full-time job and go to evening part-time graduate study
- **d** Get a full-time job
- **e** Take a year off

in terms of your criteria, using a process such as that of Table 3-1. Compare the strategic program evaluations done independently by each team member. Discuss them in order to arrive at an understanding of how you differ. For instance: Do you have different goals? or Are your priorities for the goals different? or Do you evaluate various programs differently by how you will perform each goal?

PART THREE

SYSTEMS ANALYSIS

CHAPTER 4

SYSTEMS ANALYSIS

Ultimately all policies are made . . . on the basis of judgments. There is no other way, and there never will be. The question is whether those judgments have to be made in the fog of inadequate and inaccurate data, unclear and undefined issues, and a welter of conflicting personal opinions, or whether they can be made on the basis of adequate, reliable information, relevant experience, and clearly drawn issues. In the end, analysis is but an aid to judgment. . . . Judgment is supreme.[1]

Since all planning is intrinsically decision making, the decision subsystem of the system for strategic systems planning is of great importance. In this chapter and the following two chapters, we shall discuss the makeup of this subsystem in terms of the methodology called *systems analysis*.

In discussing systems analysis, one must immediately be aware that a semantic "jungle" exists; i.e., different people use different terms to express the same thing. For example, many managers and analysts use the terms "systems analysis," "policy analysis," "cost-benefit analysis," among others, almost synonymously.

There has been a general trend to broaden the focus and definition of systems analysis since it was introduced in the 1950s. At that time, the definition given in *Webster's New Collegiate Dictionary* was applicable:

[1]Alain C. Enthoven, as quoted in *Business Week,* Nov. 13, 1965, p. 189.

> The act, process, or profession of studying an activity (as a procedure, a business, or a physiological function) typically by mathematical means in order to define its goals or purposes and to discover operations and procedures for accomplishing them most efficiently

The emphasis in this definition on mathematical means and efficiency was largely descriptive of the methodologies that were introduced into U.S. Department of Defense decision making by RAND Corporation analysts in the early 1960s.[2]

However, since then the field has developed to encompass non-mathematical means of analysis and a greater concern with *effectiveness* rather than mere efficiency. As the scope of analysis has changed, there have been attempts to distinguish the "new" domains from the old by the introduction of new names for the activity.[3] For instance, the term "policy analysis" is typically used when the domain of concern is largely social and political rather than economic.[4] In such a case, it is difficult to "capture" all relevant aspects of a system in terms that may be dealt with mathematically. Hence, in policy analysis, non-mathematical analysis is more emphasized than it is in the "traditional" systems analysis of the 1950s and 1960s.

Whichever term is used to describe the activity, the two salient concepts that are inherent in the term "systems analysis" are necessary ingredients of it. The two concepts are the *systems approach* and *analysis.* The former concept has been discussed in detail previously. The idea of "analysis," particularly as it applies to organizational decision problems and their solution, requires further exploration.

DECISION PROBLEMS

Any decision problem, whether within the confines of an organization in the design process for a complex system, or in the daily life of an individual, involves several important elements. First, someone or some group must be faced with the problem—the *decision maker.* The term "decision maker" as used here should not be interpreted to mean a forceful dynamic activist, as opposed to one who procrastinates. The scientific meaning of the term implies nothing about the personal qualities of the person who may fill the role. In the formal sense, *a decision maker is an entity, either an individual or group, who is dissatisfied with some existing state or with the prospect of a future state and who possesses the desire and authority to initiate actions designed to alter this state.*

For example, the marketing vice president who is dissatisfied with a downward sales trend is potentially a decision maker in this sense. This person must,

[2]See, for example, E. S. Quade (ed.), *Analysis for Military Decisions,* Rand McNally & Company, Chicago, 1964.

[3]See, for example, K. A. Archibald, "The Pitfalls of Language, or Analysis Through the Looking Glass," Chap. 11 in G. Majone and E. S. Quade (eds.), *Pitfalls of Analysis,* John Wiley & Sons, Inc., New York, 1980.

[4]See, for instance, Y. Dror, *Design for Policy Sciences,* Elsevier/North Holland, New York, 1971.

of course, actually possess both the desire and the authority to alter promotional expenditures or take other actions designed to increase sales if he or she is to function as a decision maker.

The decision maker's desires to achieve some state of affairs—objectives—are the reasons for the existence of a problem. These objectives have (or should have) some relationship to the overall strategic goals of the organization. The strategic goals provide the policy framework and criteria from which ancillary objectives are formulated.

Often, objectives are expressed as a wish for either the attainment of a new state, such as higher profits, or the retention of an existing one, such as "our image as the industry's leader." Usually, however, the objectives of the decision maker are expressed as some combination of achievable goals and retentive constraints on the pursuit of those goals, e.g., to increase ROI while simultaneously maintaining an image level as in the example in Chapter 3.

To pursue objectives meaningfully, the decision maker must have available *alternative actions* which can promote the state of affairs desired. These available alternatives, together with a *state of doubt* as to which one is best, constitute the heart of any decision problem.

Perhaps the most common error made by potential decision makers (those dissatisfied individuals with authority to act) who base their thinking solely on subjective experience, judgment, and intuition is that they fail to recognize the existence of alternatives. Theirs is the failure to perceive and consider alternatives which have never been used before, and which are therefore likely to be beyond the scope of experience and sound subjective evaluation. In effect, potential decision makers who are dissatisfied but who recognize no alternatives to their present methods abdicate responsibility, since the existence of alternatives implies the potential existence of a problem. The failure to recognize the opportunities, and hence the problem, has the same result as would a conscious choice of the status quo from among a range of possible alternatives.

The significant aspect of the decision maker's failure to consider alternatives is that it is perfectly natural. The judgment of all of us is conditioned by our range of past experience, and most of us do not conduct our daily lives in a fashion which is conducive to the generation of new alternatives. Thus, if decision makers take the same approach to organizational problems that they do to personal ones, they are unlikely to be fully aware of the range of opportunities which may be available.[5] What they need in the organizational context is a set of concepts and procedures which will aid in defining and developing opportunities. The ideas and techniques of systems analysis provide such a basis.

We shall investigate these basic concepts and techniques later. At this point, we shall simply note that the decision maker must constantly be on the alert for

[5]We shall not investigate the implications of this statement concerning the way in which we should conduct our personal lives.

decision problems which may appear from nowhere with the development of new techniques or policies. It should always be borne in mind that one never really shirks a decision, since the state which may be retained by the apparent avoidance of alternatives is itself an outcome of the unrecognized problem. As such, it is always a prime candidate for subsequent comparison with outcomes which might have been realized from unrecognized opportunities.

When the specification of problems, objectives, and alternatives is thought about in this general way, the essential unity of problem solving of various different levels becomes clear. In Chapter 2 we stressed the *differences* between traditional problem solving and the creative process of strategic systems planning. Here we emphasize that traditional problem solving, if approached properly, should entail the same inventive aspects as does the superficially more creative process involved in strategic systems planning.

We should also emphasize that this view of problem solving applies equally well to the tactical problems which the managers face in their implementation function and to the more strategic problems which are dealt with in the planning function. Moreover, many of the concepts and techniques of systems analysis are applicable to those tactical problems, to be discussed in Part Four.

SOLUTIONS TO DECISION PROBLEMS

To formally solve a decision problem, it is necessary that decision makers choose the *best* of the available alternatives. In very simple decision problems, this is equivalent to saying that they should choose an alternative which leads them to a state which is at least as good as all other states. In more complex problems, the idea of a best alternative is somewhat more subtle. We may define a *problem solution* as the best of a set of alternative actions.

This apparently simple and straightforward statement has tremendous practical ramifications. What is meant by "best," for instance? How is the best alternative to be found? Is the alternative we consider best necessarily the same one our superior would choose? We shall investigate such questions as these in the remainder of this chapter and in subsequent chapters.

The concept of searching for and selecting the best alternative, however defined, is itself subject to controversy at the practical level. Professor Herbert Simon of the Carnegie-Mellon University has proposed in his *principle of bounded rationality* that people seldom attempt to find the best alternative in a decision situation.[6] Rather, they select a number of "good enough" outcomes and an alternative which is likely to achieve one of them. In searching for a new product, for instance, marketing executives make no attempt to enumerate *all* possible products so that they can select the best one; rather, they select one which is likely to satisfy their desires.

[6]For instance, see J. M. Roach, "Simon Says . . . Decision Making is a 'Satisficing' Experience," *Management Review*, vol. 68, no. 1, January 1979, pp. 8–17.

This descriptive concept (how people *do* act) has normative implications (how people *should* act), for one might argue the irrationality of complete rationality; i.e., the completely rational person should evaluate *all* alternatives and choose the best one, and yet it would be irrational to do so if this would involve the expenditure of vast amounts of time and money.

Of course, the idea of selecting a best alternative is questionable in another respect. Not only might it be irrational to try to investigate every alternative, but in most complex problems, it is also impossible to do so. Our understanding of the underlying structure of most complex systems is incomplete, and we are often unable to understand the interrelationships of all the factors bearing on the decision problem in question. To expect optimization in such a state of knowledge would be utter folly. In fact, as we shall illustrate, one view that may be taken of systems analysis is that its prime role is in helping the decision makers to better understand the structure of the problem.

THE SYSTEMS-ANALYSIS PROCESS

Systems analysis is a scientific process, or methodology, which can best be described in terms of its salient problem-related elements. The process involves:

1 Systematic examination and comparison of those alternative actions which are related to the accomplishment of desired objectives

2 Comparison of alternatives on the basis of the costs and the benefits associated with each alternative

3 Explicit consideration of risk

Examination and Comparison of Alternatives

The process of examining and comparing the alternatives which are deemed relevant to the accomplishment of objectives is not as simple as it might at first appear. At one level, there is the problem of examining a prescribed set of actions. For example, when a corporation attempts to compare several potential new product ideas in order to select the best one, or when the Air Force compares several proposed designs for a counterinsurgency aircraft, the range of alternatives to be considered is rather well defined.

On the other hand, it may be necessary or desirable to invent new alternatives. If the obvious alternatives are all inadequate to achieve a given qualitative objective, it will be necessary to develop new ones.[7] Often the obvious

[7]The great decision makers of history have often been so regarded because of their ability to see beyond the range of alternatives presented to them. Solomon, for example, pretended to choose an alternative which he had invented (to have the child's body severed). In reality, he chose an alternative involving the gathering of information (in the form of the reactions of the women claiming to be the baby's mother). So, too, did Napoleon invent an alternative by accepting the crown from the Pope's hand and placing it on his head. He thus avoided the two poor alternatives presented to him: to be crowned by the Pope or to refuse to be so crowned.

alternatives will achieve an objective but will involve a prohibitive cost, so that new alternatives are necessary. In either case, the new alternatives to be considered may simply be different combinations of the same controllable aspects. For instance, if the objective is to give the United States some defined capability to move troops and supplies, and if the basic alternatives considered are land, sea, and air transport, it may not be possible to achieve a given capability with any one mode. Fast air transport might not be able to move sufficient quantities, land transport is obviously restricted when bodies of water are encountered, and sea transport is restricted to water, but some combination of the three might well serve to achieve a capability for land and sea movement in the desired quantity. Of course, this is obvious, and no rational person would conceive that only one of the media of transport—land, sea, or air—should be considered exclusively. Yet this is exactly the general sort of thing which has sometimes been done in the past in planning for the achievement of objectives along parochial lines.

An Illustration from the Public Sector[8] New York City's fire-fighting system was subjected to the scrutiny of systems analysis. The system was plagued with increasing demands (calls) for service, false alarms, obsolete communications systems, and union demands for the hiring of more fire fighters to achieve more tolerable work loads.

Clearly, the fire department cannot always be at the scene of a fire an instant after it is reported. Some reasonable response time is necessary. As the work load (number of calls per hour) increases, the response time increases as well, because units tend to be already occupied elsewhere. At some demand level, such a system breaks down to the point that calls go "unanswered" (the fire has done maximum damage by the time that the fire-fighting unit arrives). As this "breakdown" point appears to be approaching, the natural demand is for more resources in the form of fire trucks, fire fighters, and stations to avert the impending disaster.

However, in a complex system, there are generally alternative ways of achieving objectives such as "reduced response time." For instance:

a Changes in communications systems
b Changes in decision-making rules concerning the order in which calls are to be answered
c Changes in the "force level" (number of units) initially dispatched to each call

The impact of any specific alternative, such as that of increased resources,

[8]This situation is discussed in "An Institute for Policy Modeling: The New York City-Rand Institute," Chap. 7 in M. Greenberger, M. A. Crenson, and B. L. Crissey, *Models in the Policy Process,* Russell Sage Foundation, New York, 1976.

may seem to be clear, but in a complex system there are so many interactions that one's intuition is often unreliable. For instance, a "response time model" demonstrated that the "obvious" solution—increased resources—would not reduce the work load in the fashion anticipated by unions who were demanding that fire fighters' work loads be controlled. What it would do is enable the department to better achieve its goal of sending more than one fire truck to every call. (The model showed that, because of the demands being served, this objective was not currently being met well, so that increased resources would enable it to be better met, but would not reduce the fire fighters' work load.)

Thus, the *degree of impact* of "obviously good" alternatives may not be easy to determine in a complex system. Clearly, "more fire fighters" seems to be an obviously good idea in a situation in which work loads are becoming unbearable. Yet the real questions to be asked are:

How much impact would this have?
How much would it cost?
Are there alternative ways of achieving the same impact at the same cost?

An Illustration from Industry The search for a new product begins with the generation of ideas. The ideas may reflect changes in existing products which are redesigned to make them appeal to new or expanded markets, or they may involve entirely new classes of products. Usually, there is some attempt to relate the product idea to some perceived need and/or some resource of the organization in question. Therefore, one might examine each product idea in terms of questions such as:

1 Can the present production facilities be utilized?
2 Can present raw materials be utilized?
3 Can it be distributed through the present organization?
4 Does it make use of the know-how of our present organization?

However, there are no simple good or bad answers to such questions. In general, positive responses might appear to be favorable for most companies. Products which fit into the existing raw-material, production, and distribution scheme would appear to be preferable to those which do not. Hence, an oil company might consider products which use oil as a raw material, require refining, and can be sold through service stations to be preferable to products not meeting these criteria. In fact, however, the list of potential products meeting all these criteria is rather limited. As a result, many major oil companies have broadened this limited view of raw materials, production facilities, and the distribution organization to include their clerical and managerial resources and correspondence with credit-card customers. One result has been the marketing of travel insurance policies. This new product utilizes many of the human resources already available to the companies as "raw material" and "production capacity." In addition, the sale of these policies through existing postal correspondence with credit-card customers represents a broad view of distribution facilities. The possibility of direct insurance marketing through the standard outlets—the service stations which sell the company's brand of gasoline and oil products—has also not been omitted from consideration.

This example illustrates how a broader view of the function of an organization can lead to feasible new product ideas. The oil company thinks of itself as being in the travel business rather than the oil business. The result is travel insurance and arrangements with food processors to distribute hot meals to weary travelers at their service stops. The insurance company, on the other hand, thinks of itself as being in the finance business rather than the insurance business and begins to invest its capital in new ventures and to develop new insurance policies to meet the changing needs of an affluent society.

In any case, the answers to questions concerning the feasibility of new product ideas, such as those posed, should be interpreted in two ways. First, are the definitions of the present organization broad enough? Second, do positive responses to the questions really reflect preferences which are consistent with the organization's objectives? If diversification is a primary organizational goal, positive answers to questions concerning the compatibility of a new product with the existing system might reflect negatively on the product idea. Rather than the more restricted view of always finding alternatives which fit in well with the existing system, the primary overall consideration must always be: *What are the objectives of the organization, and of the system being developed, and how compatible is the alternative with them?*

Interdisciplinary Teams Both of these illustrations lead one to the conclusion that the generation of alternatives other than those which are obvious or those which are obvious combinations of basic alternatives is not a simple task which can be handled by a single individual. To generate feasible transportation alternatives requires a variety of technical skills, as does the generation of new product alternatives. Usually, no single individual possesses all the requisite skills.

The concept of an *interdisciplinary team* has been found to be beneficial in this and other phases of systems analysis. An interdisciplinary team is a working group made up of people with varying backgrounds and skills each of whom brings his or her own point of view and experience to bear on the problem, often with results which are significantly superior to those which a single individual might be expected to produce.

In addition, the nature of most significant decision problems implies alternatives which have psychological, sociological, and physical aspects. Hence, what better way to study complex systems made up of interrelated parts than with teams made up of individuals who can bring their knowledge of the related disciplines to bear?

However, there are problems that have come to be identified with decision making in groups. Some have referred to these as the problems of "groupthink."[9] Therefore, when one wishes to generate alternatives, there is a clear need for a group process that will facilitate, rather than hinder, creativity.

Brainstorming One device which has been successful in various forms is the technique of *brainstorming*. It stimulates a form of free thinking—something daring, perhaps way out. In a brainstorming session, a group comes together and is encouraged to discuss alternative ways of dealing with a situation, solving a problem, or taking advantage of an opportunity. "Blue-sky" thinking out loud is encouraged. No criticism of an idea put forth by a participant is allowed. Each idea is identified and it is hoped that each idea may lead to another and eventually to an idea that has promise of dealing with the situation under discussion. The interaction of individual members of the group is believed to stimulate the emergence of ideas. A synergistic effect often results from the interaction of the brainstorming session. For a successful brainstorming session, a well-trained, highly skilled leader is crucial. This leader establishes a modus operandi for the group, keeps the participants from breaking operating procedures, and encourages reticent participants.

The following operating rules are believed to improve the probability of success of brainstorming sessions:

Concentrate efforts in well-defined problems/opportunities.

Encourage the emergence of all ideas that are in some way related to the situation.

Do not criticize any idea, even those which might appear outrageous.

Keep the dialogue on the topic material.

Wait until the ideas have been collected before beginning their evaluation.

Often, during a later analysis most of the recorded ideas put forth will be discarded; a few ideas which merit further evaluation should have emerged. Some who have participated in apparently fruitless sessions have felt that, at the very least, the sessions served to stimulate discussion and thinking about the situation under consideration.

Brainstorming sessions are normally conducted in a sequence. They typically begin with the statement of a situation, or the presentation of a specific topic. Typically, five to eight people participate in these sessions. They can be from different organizational elements and with different backgrounds, and they may or may not know each other.

The first session is dedicated to the suggestion of ideas about the proposed topic under the direction and encouragement of the leader. It usually takes about two to three hours to complete the initial session, ending with a list of

[9] C. W. von Bergen, Jr., and R. J. Kirk, "Groupthink: When Too Many Heads Spoil the Decision," *Management Review*, March 1978, pp. 44–49.

ideas. The next sessions are usually shorter than the first, and consensus is sought to evaluate the ideas. After the most appropriate ideas have been chosen, their implications are studied and acceptable ways to implement them are discussed. The result of a well-conducted set of brainstorming sessions is either a set of solutions to specific problems, a set of alternatives to achieve a particular goal, or an assessment of future conditions.

Brainstorming sessions can be used to evaluate new product proposals, develop strategies, solve production problems, identify market opportunities, and for a host of other uses which can help to improve organizational effectiveness.

Brainstorming can be used to stimulate unrestrained thinking in any group situation if the ambience is similar to that described above as the prerequisite for a successful session. A manager who has the trust and confidence of workers and colleagues can encourage them to participate in a free-wheeling session at a regular staff meeting. For example, a president of a small steel company holds weekly luncheon meetings with his manufacturing manager and hourly rated production workers. At these meetings any topic relevant to the manufacturing activity can be thrown out for open-minded discussion. Any participant in these discussions can suggest a topic. Over several years, these luncheons have uncovered some critical manufacturing problems and have led to improvement in the fabrication of steel products.

Thus, in the idea of brainstorming as well as that of interdisciplinary teams, we see the team ideas of project and matrix management playing a role in the planning and problem-solving phases of management. We shall elaborate on this in a later chapter in dealing with the myriad applications to which these "team concepts" have been put.

The Basis for Comparison

In order to determine which of several alternative ways of accomplishing some objective is the best, they must be evaluated and compared on the basis of costs and benefits.

Costs are the resources—dollars, people, machines, etc.—which, when allocated to one alternative, cannot be used for other purposes. *Benefits* are those worthwhile elements which are derived as a result of some action. Thus, if an advertising manager decides to spend $100,000 on TV "spots" (the cost) and this results in $300,000 in additional sales revenue (the benefit), both aspects describe the overall *worth* of the alternative. And although there is an apparent single measure which one might apply to evaluate this alternative—benefit minus cost—it is clear that another alternative including the expenditure of $110,000 and resulting in an increased revenue of $310,000 might not be equally good, even though both have a benefit-cost difference of $200,000.

We shall go more deeply into a discussion of the appropriate combination of costs and benefits to use in evaluating alternatives later. Here, the important point is that both aspects must be considered.

This simple principle is certainly not always followed in practice. In any situation in which "requirements" and costs are separately considered, for example, the requirements that are generated normally exceed the funds that are available. This is true at the national level when "defense requirements" are assessed independent of costs, and it is true in a business firm when the "information requirements" of computer system users are requested without any cost or quantity limitation.

In such instances, when "needs" and budgetary funds are unequal, a conflict inevitably occurs, and the normal process is to reduce the "requirements" to the level of available funds. Given the complexity of the systems with which we deal and the myriad interrelationships that exist among the various elements of each subsystem, this can lead to results which appear to be silly, but which can nonetheless occur in complex systems and organizations.

For instance, one manager who had stated that a five-digit Standard Industrial Classification (SIC)[10] code designator was required for each customer was forced to "get along" with a three-digit SIC number rather than a five-digit one, after some arbitrary approaches were applied to resolving the "requirements versus dollars" dilemma. Unfortunately, the three-digit code was entirely useless to him, but he found that he could do nothing to influence the arbitrary decision which had been made to ensure "fair" treatment to all of those managers who had to have their "information requirements" cut back.

To avoid this sort of ridiculous outcome, one must be sure that in *assessing the overall worth of an alternative, both cost and benefit are simultaneously considered.* Alain C. Enthoven some years ago discussed a "homey example" of the consequences of failing to do so in the building of a new home.[11]

> Suppose that I want to buy a house and, instead of using the cost effectiveness approach, I do it in the more traditional way. First, I determine my housing requirements without any consideration of costs. I count up the rooms I require: I need a bedroom for myself, one for each of my children, and one for my parents or other guests who come to visit us occasionally. I need a study because I occasionally bring some of my work home with me and need a quiet place to work. My wife needs a sewing room. I need a pool in the basement because my doctor has told me that I must swim every day if I don't want to have another operation on my back. Now you might laugh when I say that I have to have a pool in my basement, but I can validate that requirement. I can argue for it very convincingly. I can produce a Doctor's certificate, and you can't prove to me that I don't need that pool. Moreover, I work at the Pentagon and I work long hours. Therefore, I need to live within five minutes drive of the Pentagon. When I put this all together, I find that I have established a requirement for a house that costs a hundred thousand dollars. Having done that, I review my financial situation and find that I am only able to spend about $30,000. So what do I

[10]The SIC code system describes firms in terms of broad industrial categories and successively finer subcategories, through the use of a numerical designator. A code number with a larger number of digits more finely defines a firm's particular industrial subcategory.

[11]Alain C. Enthoven, address before the Aviation and Space Writers Convention, Miami, Fla., May 25, 1964.

> do? If I am operating under the old concept, I take the $100,000 design and I slice off 70 percent of it and what's left is my house.
>
> Now, clearly that's not a very sensible way to design a house. I might find that I left off the bathroom, or included the bathroom, but left off the plumbing that is required to make it work. Yet that's a pretty fair description of the way that the Department of Defense did its business. We found in 1961 that we had Army Divisions without adequate airlift or other means of mobility and with far from adequate supplies of equipment. We had tactical air wings without supplies of nonnuclear ordnance, and numerous other similar problems. In effect, we had bought a lot of houses without the bathrooms or the plumbing.

Thus, as Enthoven describes, if one does not take explicit account of costs and benefits, but rather thinks solely in terms of requirements, one is likely to be left holding the bag. If it takes a $100,000 house to meet my requirements and I can only afford a $30,000 house, I clearly should not just "chop off" some features of the $100,000 house until it costs $30,000. Rather, I should begin to plan with the $30,000 in mind. Enthoven described this process as follows:[12]

> The rational economic way of buying a house, or of buying a defense program, is to consider alternative balanced programs each of which yields the most effectiveness possible within a budget that corresponds approximately to the availability of our resources. If I think that I have about $30,000 to spend on a house, I should consider several alternative houses each optimized for my purposes within financial limits such as $28,000, $30,000, $32,000, and perhaps $34,000 and then I should ask myself whether the extra advantages associated with the more expensive houses are worth the extra financial sacrifices I would have to make to pay for one. It's altogether possible that they might be. For example, a larger house might have a recreation room, and this might enable me to economize elsewhere on recreation.

Risk and Uncertainty[13]

Most organizational decision problems which are of the complexity of those faced by strategic planners in government and industry involve great uncertainy. Strategic planning and decision making necessarily involve consideration of the future course of events, and the future is inherently uncertain. Virtually all decision making—whether it is subjected to formal analysis or not—involves the uncertain future, of course. The important aspect of the systems-analysis approach is that it gives *explicit* consideration to uncertainty. Many approaches to decision making—whether subjective or objective—do not have this characteristic. The common assumption in these is one of certainty—that each action leads uniquely to a specified outcome. A simple example will serve to illustrate the implications of the certainty assumption.

[12]Ibid.

[13]The terms "risk " and "uncertainty" are used here in the manner of common usage. The technical distinction between the two is unnecessary. For those interested, see the classic work: R. D. Luce and H. Raiffa, *Games and Decisions,* John Wiley & Sons, Inc., New York, 1957, Chap. 2, or any book on "decision theory."

Suppose that two people are available to be assigned to two tasks—one to each. Table 4-1 gives the anticipated time which each person will require to accomplish each task; i.e., person A will take two hours to perform task 1 (if assigned to it), and person B will take three hours to do it (if assigned to it). The decision maker in this situation has two alternatives:

Alternative 1: (Assign A to 1 and B to 2)
Alternative 2: (Assign A to 2 and B to 1)

These two alternatives involve the selection of either the two circled entries in the table (alternative 1) or the two uncircled entries (alternative 2).

The total number of person-hours required to perform both tasks is eight (i.e., 2 + 6) in the case of alternative 1 and seven in the case of alternative 2. Hence, the second alternative appears the better on this basis. However, if we recognize that the entries in the table are *predictions* of certain future events, we may wish to reconsider this evaluation. In deciding on the superiority of the second alternative, we have assumed that we know precisely (with certainty) that each person will perform each task in a given amount of time. In fact, probably neither of the people has performed the tasks so often that we could confidently predict the time that will be required. Suppose, for example, that the second task is of such a nature that we (unknowingly) always overestimate the required time by 100 percent. If this were so, the true table corresponding to Table 4-1 would be Table 4-2 (assuming that we are perfectly accurate in estimating for the first task).

From Table 4-2, it is clear that we have no basis for a preference between the two alternatives; both are identical in terms of the information and criteria. Of course, there may well be other factors we would wish to consider, but no such information is given here.

The point, then, is that even with formal analysis, it is easy to fall prey to the certainty assumption, and such an assumption can lead to very poor results. In this example, it really does not matter which alternative we choose, since both are equal in every way. However, we must recall that decision makers do not

TABLE 4-1
ANTICIPATED TIME (HOURS) REQUIRED BY TWO PEOPLE IN TWO TASKS

		Tasks	
		1	2
People	A	②	4
	B	3	⑥

TABLE 4-2
ACTUAL TIME (HOURS) REQUIRED BY TWO PEOPLE IN TWO TASKS

		Tasks	
		1	2
People	A	2	2
	B	3	3

have knowledge of this sort; i.e., they never have available to them anything other than imperfect information of the kind contained in Table 4-1. Information such as that shown in Table 4-2 is always unknown.

The problem of the decision maker and the decision analyst is to take prior account of the inherent uncertainty in strategic decisions. Systems analysis attempts to provide a vehicle for doing this, and although it is much easier to talk about considering uncertainty than it is to actually do it, explicit consideration of uncertainty is one of the basic reasons for the success of the systems-analysis approach.

THE ROLE OF SYSTEMS ANALYSIS

From the preceding discussion, one might reasonably infer that a strong "sales pitch" is being made for systems analysis. Indeed, this is the case, but in order not to make the mistake of the supersalesperson who leaves customers dissatisfied, we should carefully consider the appropriate role to be played by systems analysis, and its limitations.

Should All Problems Be Analyzed?

First, we must consider the sort of problem to which systems analysis is applicable. Is it applicable to all decision problems?

The answer is that while it is conceptually applicable to the strategic problems encountered in the planning phase of management, to the tactical problems of the execution phase, and indeed to any problem, be it organizational or one of the day-to-day personal problems which we all face, a reasonable person would not wish to apply systems analysis to all problems.

Some problems are relatively simple in nature and need not be subjected to extensive analysis of any sort. The decision maker who is standing in the middle of a street with a car bearing down on him at 60 miles per hour is in a problem situation; he wants to avoid injury, and he has at least two alternatives—to run to the right curb or to run to the left curb. Any prolonged contemplation or analysis on his part, however, will lead to an outcome which is clearly not the one he desires. The need in this situation is for a quick and accurate decision of the variety which is the forte of some traditional managers and which is the very cornerstone of combat management in the military, where battles may be won or lost simply as a result of initiative rather than the choice of the best alternative. In such "time-sensitive" tactical military contexts, the speed with which the decision is made is critical; i.e., the eventual outcome depends more on the speed with which action is taken than on the specific alternative which is chosen. A tactical commander in such a situation must therefore necessarily rely largely on a framework of experience and intuition to solve his decision problems.

In strategic decision problems—in either the public or the business sphere—the environment is more "knowledge sensitive"; i.e., the eventual outcome is usually influenced much more by the chosen alternative than by the speed of action. This sort of decision involves great uncertainy, a myriad of alternatives, and the employment of resources in a time frame which extends into the (perhaps distant) future. If analytic resources are available, the "Damn the torpedoes; full speed ahead" philosophy, which can lead to brilliant military accomplishments, is quite inappropriate to the environment in which the alternatives are new products to be marketed, various missile systems to defend our nation, or social programs to aid the needy.

Of course, there are other sorts of decisions—between the time-sensitive and knowledge-sensitive extremes—which also do not require extensive analysis. Routine decisions made on a day-to-day basis within the framework of established policy exemplify those decisions which require neither inordinate haste nor the expenditure of significant analytic resources. In the chapters dealing with the planning function of management, we shall concentrate on those knowledge-sensitive strategic decisions which clearly warrant analysis.

Quantitative and Qualitative Analysis

Since systems analysis is an outgrowth of traditional scientific method, and since measurement in science is ideally quantitative in nature, systems analysis itself has come to be viewed as quantitative. While it is true that systems analysts often draw on mathematics as an aid in formulating and solving problems, the complementary role of quantitative and qualitative analysis should be made clear. In fact, since sound qualitative analysis was being performed before most quantitative analysts were born, there is some need to justify the use of quantitative approaches in decision making.

To justify the use of quantitative analysis as an aid to decision making which is preferable to witchcraft or coin tossing, one must argue on some basis other than the inherent "goodness" of science. Science, like everything else in this complex world, has both positive and negative aspects and therefore is not justified simply by the fact that it exists.

One can, of course, point to the successful applications of the scientific method to public and industrial decision problems and argue the likelihood of further success as greater knowledge is gained. Indeed, the gains achieved in the past as a direct result of scientific analysis are impressive, and perhaps this is sufficient evidence to warrant further attention.

To proceed a bit further along a pragmatic line, one might recognize that the same objective input information is available to the systems analyst as is available to the witch, the gambler, or the "intuitive" decision analyst. The scientific approach has the additional virtue of guaranteed logic and consistency, and the totally subjective process has no such guarantee. In addition, if one views the scientific process as serving as a complement to the subjective processes of the decision maker, nothing is lost and something may be gained by making use of it.

Another value of systems analysis is its reproducibility. Systems analysis is a logical process which is well suited to being carried out with pencil and paper (and sometimes desk calculator or electronic computer). In any event, the assumptions, logical steps, and conclusions are always clearly spelled out and recorded. Thus, the analysis may always be resurrected after the decision to which it contributed is made and the results observed, and the *analytic procedure itself can be evaluated.* If this "testing" proves the worth of the analysis, the same procedure can be applied again with greater assurance that the results will be desirable. A purely qualitative approach has no such permanent value aside from its contribution to the trial-and-error learning process of the individual.

However, the importance of judgment in decision making is not reduced by the potential significance of systems analysis. One of the primary features of the scientific approach is its degree of abstraction—the omission of certain aspects of the real-world problem which the decision maker faces. These omissions mean that only a part of the real-world problem is treated scientifically. Decision makers must then integrate the results of scientific analysis with the significant intangibles which are not part of the formal analysis in order to arrive at a best decision. In doing so, they must call upon the same levels of judgment, intuition, and experience which are used by traditional managers. The difference between the two approaches is that scientifically oriented decision makers distinguish between subjective analysis and objective analysis and apply each to the areas in which it is most useful.

Human morale is a good example of the kind of factor which is often difficult to incorporate into formal analysis. In any decision problem, the impact of an action on the morale of the people involved may well have as much to do with the organization's effectiveness as anything else, or perhaps more. Yet impact on morale is difficult both to measure and to evaluate. In such a case, the analyst might formulate and solve a problem without giving consideration to the morale level of the organization. It would then be the province of the decision maker to integrate this solution with the judgment made concerning its impact on morale and to determine an alternative that is considered to be best overall. This may or may not be the same alternative that the formal analysis produced. The value which the analysis had, however, was that it considered everything other than morale and allowed the decision maker to focus judgment on that element which required judgment. If the analysis had not been performed, the decision maker could easily have been so concerned with the obviously important factors that not much attention would have been given to the one element—morale—which could not be adequately handled on any basis other than human judgment[14]

The view which we shall take of strategic decisions and of the role which

[14]Even the state of morale in an organization is amenable to some degree of scientific analysis through a morale survey. In today's large organizations, the human relations environment is too complex for an executive to learn how employees feel simply by observing them and then making an intuitive value judgement about the state of their morale. In a morale survey, checklists, attitude surveys, or similar devices are used to gather and interpret the data and place a numerical value on the results. Appraisal then involves the exercise of judgment about what the measurements mean in light of the total organizational situation. See, for example, D. Q. Mills, "Human Resources in the 1980s," *Harvard Business Review,* vol. 57, no. 4, July–August 1979.

scientific analysis should play in them is therefore a simple one. The process of systems analysis is viewed as a logical and consistent method of reducing a large part of a complex decision problem to simple outputs which the manager can use, in conjunction with other factors, in arriving at the best decisions. It permits the manager to focus the analytic resources which are at his or her disposal on the aspects of the problem where they are most effective. The manager is therefore able to utilize efficiently both scientific and nonscientific analysis to best advantage. Such an integration can hardly be worse than, and is potentially far superior to, a purely subjective approach to decision making.

Optimization

The formal solution of a decision problem involves the determination of the best available alternative. The process of seeking the best is called *optimization*; i.e., best alternatives are optimum alternatives.

In many complex decision problems involving great uncertainty, the "state of the art" of systems analysis is such that optimization cannot be meaningfully achieved or even sought. In other words, even though we think in terms of achieving best alternatives, we often cannot do so.

Alain Enthoven eloquently states the importance of judgment in decision making in the quotation which introduces this chapter. If only because *all* judgment cannot be quantified, systems analysis necessarily cannot seek global optimization. Only a part of a problem is even quantified, and if one is led to believe unquestioningly that the solution to the abstract model which has been analyzed is also necessarily the solution to the real-world problem, one is doomed to failure.

Moreover, since the systems dealt with by systems analysis are complex, the *structure* of the system may not be well understood. Perhaps the best illustrations of this come from the areas of government fiscal policy and advertising decisions. No economist would claim a comprehensive understanding of the impact of federal government fiscal policy on the economy. At least for any economist who did, one could easily find ten others who considered the first one's conclusions completely erroneous—as is illustrated each time the prime interest rate is changed, the necessity for a tax increase or decrease is proposed, etc. In media advertising, the same lack of understanding exists. No one would claim to understand the precise relationship between the expenditure of dollars on television commercials and the benefit obtained in terms of additional sales of the advertised product.

Yet in both cases—fiscal policy and advertising—the fact that we do not completely understand the structure and operation of the system *does not* mean that these decision problems cannot be subjected to analysis. What it does mean is that to perform the analysis, it is necessary to make assumptions, to omit some aspects of the problem which we do not understand, and to make other abstractions of the real world.

In doing so, the systems analyst must realize that the problem which has been formulated and constructed on paper is not the real problem. It is a fictitious one which is (hopefully) closely related to the one existing in the real world, and (hopefully again) the solution to the fictitious problem will be helpful to the decision maker in solving the real-world problem. But the differences between the two are of obvious importance, and the decision maker or analyst who seeks to apply "paper solutions" directly to real problems is likely to be in for a rude shock.

In fact, often the abstraction process which has just been described may not even be feasible. Analysts may find that their understanding of the system's structure is so limited that any formalization they might make would bear little resemblance to the real problem. Caught in such a quandary, they may be unsure as to the best course. However, when you realize that the alternative to an attempt at explicit analysis is a completely subjective approach, your apprehensions concerning the *relative value* of formal analysis—properly applied—begin to disappear.

Human beings are not known for their ability to comprehend complex problems involving many interacting factors. Any formal analysis—or attempt at formal analysis—is usually valuable since it serves at least to make decision makers think about the right things. Although systems analysis may not, in the final analysis, be able to unerringly tell decision makers the "right" thing to do, it does require them to enumerate the alternatives, to ask themselves what it is that they are trying to achieve, etc. Moreover, decision makers are presented with a precise statement of what they should know in order to make a rational decision. Even if they do not know all that they should or have all the necessary information, a knowledge of just what they should have will usually provide them with a better basis for making a decision, such as to be wary and to choose conservatively or to err on the positive rather than the negative side of an issue. And, of course, if the necessary information is not available, the recognition of its significance can lead decision makers to obtaining it—if not for use in the problem at hand, at least for use in future, similar problems.

G. H. Fisher, of the RAND Corporation, expressed this very well when he said:[15]

> The conclusion itself may not be the most useful thing to the decision-maker. . . . most high-level decision-makers are very busy men, with the result that they do not have time to structure a particular problem, think up the relevant alternatives (especially the *subtle* ones), trace out the key interactions among the variables of the problem, and the like. This, the analyst, if he is competent, can do and should do. And it is precisely this sort of contribution which is most useful to the decision-maker. The fact that the analysis reaches a firm conclusion about a preferred alternative may in some instances be of secondary importance.

[15]G. H. Fisher, "The Analytical Bases of Systems Analysis," address before a symposium on systems analysis in decision making sponsored by the Electronics Industries Association, Washington, D.C., June 23, 1966.

As the field of systems analysis has matured, the emphasis on "optimization" has diminished. Indeed, the need to go beyond those elements of a system or problem that can be readily quantified, and therefore optimized, has always been one of the features that distinguished systems analysis from "operations research."[16]

An Implementation Criterion for Systems Analysis

One of the motivations for the movement away from optimization analysis is the issue of *implementation.* In the early days of systems analysis, many systems analysts found that their elegant systems analyses were being ignored ("not implemented") by the decision makers that they were trying to serve. Concerns with lack of implementation of systems analyses have led to the adoption of a dual set of criteria for the judging of systems analyses:

> The quality of a systems analysis study can be assessed in terms of two distinct but related sets of criteria: internal, process-oriented criteria of adequacy relating to the technical competence of the work; and external, outcome-oriented criteria of effectiveness relating to the impact of analysis on the policy process.[17]

It has been argued that both of these varieties of criteria must be applied at *every stage of the problem-solving process.*[18] This is quite unlike earlier approaches in which, *first, and primarily,* technical criteria were applied, and then, almost as an afterthought, implementation issues were addressed. The modern view of systems analysis deals with both sets of criteria simultaneously.

When this is done, the technical validity of the problem-solving approach may indeed be compromised to some degree, but there is much greater likelihood that the results of the analysis will have real-world impact. Since systems analysis is inherently an "applied" activity, the importance of having its results actually used cannot be overemphasized.[19]

This "implementation criterion" complements the traditional technical criteria that have been applied to systems analysis. Moreover, it also provides a linkage between the planning and implementation stages of management. Using this criterion, the phases are no longer separate and distinct. Rather, those things done in the planning phase are evaluated, in part, in terms of the impact that they will have in the subsequent implementation phase. In this fashion, a true systems view of the overall management process becomes viable.

[16]K. A. Archibald, op. cit.

[17]G. Majone and E. S. Quade, (eds.), *Pitfalls of Analysis,* John Wiley & Sons, Inc., New York, 1980, Chap. 1.

[18]W. R. King, "Methodological Optimality in Operations Research," *OMEGA: The International Journal of Management Science,* vol. 4, no. 1, pp. 9–12, 1976.

[19]See R. L. Schultz and D. P. Slevin (eds.), *Implementing Operations Research/Management Science,* Elsevier, New York, 1975, and R. Doktor, R. L. Schultz, and D. P. Slevin (eds.), *The Implementation of Management Science,* TIMS Studies in the Management Sciences, vol. 13, 1979.

The Role of Judgment in Systems Analysis

We have already discussed the importance of human judgment in decision making—even in those decision problems which are analytically solved. The role of judgment is not confined to the complementary function which has been described, however. The conceptual framework for analysis, which is known as *decision theory* and is discussed in Chapter 5, directly incorporates human judgment into the formal analysis of the decision problem.

Human judgment is incorporated into formal analysis in two ways—in the context of *likelihood judgments* and in the context of *value judgments*. Thus, in most decision problems involving the uncertain future, it is necessary to assess the likelihood of future events—"It will rain tomorrow"; "Our product's sales will be over a million"; etc. Such assessments are often best made on the basis of experienced judgment. So, too, it is necessary to evaluate outcomes as a part of the formal analysis of decision problems. If, for example, it is predicted that one level of marketing effort will lead to a profit of $100,000 coupled with a market share of 25 percent, while another level will lead to a profit of $110,000 and 23 percent of the market share, the question of the relative worth of the two outcomes can be answered only through human judgment. In one case involving, say, a newly introduced product where great concern is given to penetrating the market, the larger market share might predominate over a few thousand dollars of immediate profit, while in the case of a mature product, immediate profit might be of primary concern. Such a problem of relative worth can be resolved only through the judgment of the decision maker.

One of the values of systems analysis is that these two varieties of judgment are treated separately and distinctly from each other, whereas in the mind of a decision maker they may often become confused. In the next chapter, we shall discuss the conceptual framework for systems analysis and demonstrate this point further.

SUMMARY

An understanding of systems analysis and the role which it can play in strategic decision making revolves around the concepts of a decision problem, the important elements of a decision problem, and the idea of a problem solution.

Systems analysis is a methodology for analyzing and solving decision problems through a systematic examination and comparison of alternatives on the basis of the resource cost and the benefit associated with each. As a part of the examination, explicit consideration is given to the uncertainties involved in decisions which will be implemented and have impact in the future.

While systems analysts often make use of logic and mathematics in solving problems, there is no necessary connection between systems analysts and sophisticated mathematics. Complex problems are often analyzed and solved without resort to anything more complex than high school math.

The role of human judgment in systems analysis is often misunderstood.

Systems analyses are complementary to the experienced judgment and intuition of the decision maker. Moreover, in some ways, systems analysts can make use of judgments more effectively than can the individual decision maker. In the next chapter, we shall discuss the conceptual basis of systems analysis and develop the complementary role of analysis and judgment further.

DISCUSSION QUESTIONS

1. Policies and strategies in modern organizations are ultimately made on the basis of judgment. If this is so, then an executive who always exercises good judgment will be successful in the management of the organization. Defend or refute this statement.
2. Trace the evolution of the meaning of *systems analysis* in the last two decades in this country. Can the notion of systems analysis be applied to any system, however complex or simple?
3. There are two key concepts inherent in the term *systems analysis*. What are these concepts? Which is the more important?
4. What is a decision problem? What are the several important elements in a decision problem? What is a *decision maker* in the decision problem? What is that individual's role?
5. Describe the kinds of costs and benefits that might be associated with the following decisions: (a) selecting a spouse, (b) selecting a new home, and (c) selecting a public program to aid business firms in recessionary times.
6. Systems analysis is a scientific process, or methodology, which can best be described in terms of its salient problem-related elements. What does the process involve?
7. The process of examining and comparing the alternatives which are deemed relevant to the accomplishment of organizational objectives is a relatively simple and straightforward process. Defend or refute this statement.
8. What are the advantage and disadvantage of using interdisciplinary teams for the solution of strategic problems in an organization?
9. What is brainstorming? What are the requirements for its successful use in an organization? Under what conditions might it best be used?
10. Relate the "Illustration from Industry" given on pp. 89–90 to the evaluation process described using Table 3-1 on p. 69.
11. Alternatives must be compared in order to select the one most suited to complement the objectives and goals of the organization. What is the basis for such comparison?
12. Since systems analysis is a rigorous methodology to use in problem decisions, the approach should be used for all problems. Defend or refute this statement.
13. Quantitative and qualitative assessments can both be used in the context of systems analysis. How might decision problems that have a high content of qualitative factors be analyzed using a systems-analysis approach?
14. Many elegant systems-analysis solutions are either ignored or fail in the implementation stage. Why does this happen? What might the systems-analysis expert do to better ensure that the solution will be accepted and effectively implemented in the organization?
15. Intuition plays a small role in the use of systems analysis in the solution of complex problems. Defend or refute this statement.

RECOMMENDED READINGS

Ackoff, R. L.: *The Art of Problem Solving,* John Wiley & Sons, Inc., New York 1978.

———, and E. Vergara: "Creativity in Problem Solving and Planning: A Review," *European Journal of Operational Research,* vol. 7, 1981.

Archibald, K. A.: "The Pitfalls of Language, or Analysis Through the Looking Glass," Chap. 11 in Majone, G. and E. S. Quade (eds.), *Pitfalls of Analysis,* John Wiley & Sons, Inc., New York 1980.

Beltrami, E. J.: *models for Public Systems Analysis,* Academic Press, New York, 1977.

Burton, R., D. Dellinger, and W. R. King: "Legislative Implementation of Public Policy Analysis," *Implementation of Management Science,* TIMS Studies in the Management Sciences, North Holland, New York, 1979.

———: "Alternative Strategies for Legislative Analysis of Public Policy," *Urban Systems,* vol. 3, 1978.

Churchman, C. W., R. L. Ackoff, and E. L. Arnoff: *Introduction to Operations Research,* John Wiley & Sons, Inc., New York 1957.

Dannenbring, David G., and Martin K. Starr: *Management Science: An Introduction,* McGraw-Hill Book Company, New York, 1981.

Doktor, R., R. L. Schultz and D. P. Slevin (eds.), *The Implementation of Management Science,* TIMS Studies in the Management Sciences, vol. 13, 1979.

Dror, Y.: *Design for Policy Sciences,* Elsevier, New York, 1971.

Ernst, M. L.: "Operations Research and the Large Strategic Problems," *Operations Research,* vol. 9, no. 4, July–August 1961.

Greenberger, M., M. A. Crenson, and B. L. Crissey: *Models in the Policy Process,* Chap. 7, Russell Sage Foundation, 1976.

Grant, John H., and William R. King: *The Logic of Strategic Planning,* Little Brown, Boston, 1982.

Hill, W. A., and D. Egan (eds.): *Readings in Organizational Theory: A Behaviorial Approach,* Allyn and Bacon, Inc., Boston, 1966.

King, W. R.: "Methodological Optimality in Operations Research," *OMEGA: The International Journal of Management Science,* vol. 4, no. 1, 1976.

Lave, C. A., and J. G. March: *An Introduction to Models in the Social Sciences,* Harper & Row, New York, 1975.

Luce, R. D., and H. Raiffa: *Games and Decisions,* John Wiley & Sons, Inc., New York, 1957.

Majone, G., and E. S. Quade *Pitfalls of Analysis,* John Wiley & Sons, Inc., New York 1980.

McGann, J. A.: "Where Are the Alternatives?" *Jornal of Systems Management,* vol. 32, May 1981.

Quade, E. S. (ed.): *Analysis for Military Decisions,* Rand McNally & Company, Chicago, 1964.

———: *Analysis for Public Decisions,* 2d ed., Elsevier/North Holland. New York, 1982.

Radford, K.J.: *Complex Decision Problems: An Integrated Strategy for Resolution,* Reston Publishing Company, Inc., Reston, Va., 1977.

Rivett, P.: *Model Building for Decision Analysis,* John Wiley & Sons, Inc., New York, 1980.

Roach, J. M.: "Simon Says . . . Decision Making is a 'Satisficing' Experience," *Management Review,* January 1979.

Schultz, R. L., and D. P. Slevin (eds.): *Implementing Operations Research a Management-Science,* Elsevier/North Holland, New York, 1975.
Summers, J., and D. E. Whik: "Creativity Techniques: Toward Improvement of the Decision Process," *Academy of Management Review,* April 1976.
Turner, B.A.: "The Organizational and Inter-Organizational Development of Disasters," *Administrative Science Quarterly,* vol. 21, September 1976.
Von Bergen, C. W., Jr., and R. J. Kirk: "Groupthink: When too Many Heads Spoil the Decision," *Management Review,* March 1978.
White, D. J.: *Decision Methodology,* John Wiley & Sons, Inc., New York, 1975.
Wilmutte, R.M., and G. G. Crotts: "Bridging the Gap Between Analysis and Decision Makers," *Management Review,* January 1979.

CASE 4-1: Selecting an Information System

Top managers perceive that a business firm needs a new information system to support the three levels of management—operational control, management control, and strategic planning. Operational-control information is that which is directly created by operations such as manufacturing, order processing, and billing. Management control information is that which facilitates the allocation of fixed resources to various tasks, activities, or suborganizations. Strategic-planning information is that which is used to develop *new* activities, resources, products, etc.

Your project team is asked to investigate the following alternatives:

1 A highly computerized operational-control system, with the higher levels being served by manual information processing

2 A system that would heavily computerize both operational-control and management-control information, but which would leave strategic-planning information to be handled manually

3 A system that computerizes all three levels

Your team should specify the general nature of the costs, benefits, advantages, disadvantages, etc., of the three varieties of systems, as well as recommend *how* the choice should be made. (Since you are not operating in a real firm, it is probably impractical to actually make the selection.)

CHAPTER **5**

A CONCEPTUAL FRAMEWORK FOR SYSTEMS ANALYSIS

Jack and Jill went up the hill
To fetch a pail of water
Jack fell down and broke his crown
And Jill came tumbling after.

Jack could have avoided that awful lump
By seeking alternative choices
Like installing some new pipe and a great big pump
And handing Jill the invoices.[1]

The salient aspects of a decision problem, which were discussed in the last chapter, must be considered in the light of a theoretical framework for analysis. This framework provides the basis for thinking about a decision problem in a logical and consistent fashion. It shows the way in which the various elements of a problem relate to one another, and it does so in a manner which permits us to take each individual piece and put it into its proper perspective.

The conceptual framework for analysis which we shall develop here is simple and straightforward. As we shall demonstrate, in addition to its theoretical value it also has a great deal of intuitive appeal. In fact, we shall argue that it is simply a formalization of the way in which we all know we should make decisions or,

[1]Stacer Holcomb, OSD (SA), as quoted in *The C/E Newsletter,* publication of the Cost-Effectiveness section of the Operations Research Society of America, vol. 2, no. 1, January 1967.

perhaps, the way in which all rational people believe they *do* make decisions. However, most of us, on contemplating the conceptual framework, will realize that it is virtually impossible for people to do all that they know they should do (or think that they actually do) in solving a problem. The scope and complexity of real-world problems are simply too great for most human beings to comprehend without the explicit guidance of a framework for decision-problem analysis.

The conceptual framework may be viewed as a way of reminding us what is important and what is not or, better, as a way of reminding us to ask questions about the relative importance of various facets of the problem. Its basic objective is to have the decision maker and analyst *ask the right questions,* for it is axiomatic that a precise answer to the wrong question is undesirable.

The biggest difficulty with a discussion of the conceptual framework for analysis is that it is exactly that—a *conceptual* framework. In some respects, it is unrealistic in that it cannot be rigorously and explicitly applied in detail to all real-world problems, or, at least, it appears at first glance that it is not directly applicable. You will recognize immediately that many of your real-world problems could not be directly attacked in the fashion to be described. As a result, you may rush to the conclusion that you are again witnessing the gap between the ivory tower of the printed word and the complexities of the real world. To prevent hasty conclusions which may lead to the neglect of useful analytic methods, we hasten to point out that the framework for analysis in this chapter need not always be directly applied in all its detail for it to be a useful analytic tool. The introduction of the framework is accomplished here in terms of overly simple situations which do not reflect the complexity of real-world problems. This is done so that the essential simplicity of the ideas can be demonstrated. After the basic framework has been discussed, we shall look into its values and limitations and attempt to demonstrate the role it can play in managerial planning.

DECISION-PROBLEM FORMULATION

The first step in solving a decision problem is in formulating it. "Formulation" of a problem means identifying it precisely and defining its boundaries and critical components. This may seem to be a straightforward thing to do, since the very fact that one is doing it implies that someone is dissatisfied with an existing state of affairs or the prospect of some future state. Hence, "the problem" is generally thought to be "obvious."

However, "problems," as initially perceived by someone, may be only the symptoms of the underlying "real" problem. For instance, in the New York City Fire Department example used in Chapter 4, the systems analysts were initially presented with a "problem" of a need for improved communications to permit improved response time, and hence an improved work-load situation. When they discovered that improved communications would be unlikely to affect the work load significantly, they "redefined" the problem in terms of the size of the work load itself. Thus, the work-load "symptoms" which were initially thought to reflect a "communications problem" were subsequently assessed to be, in fact, the "real problem." (This complex relationship among symptoms, actual problems, and problem formulations has been treated in Chapter 2.)

However, problem-formulation difficulties even go beyond this level of complexity because of the nature of the problems that systems analysts are increasingly being called on to deal with. Their problems have been termed "wicked" or "messy" problems because they are ill-defined even though they are presumed to exist. Such problems may really be a cluster of interrelated problems (a "problematique"), or they may involve so many diverse and conflicting elements—political, social, economic, technological, etc.—as to be difficult to define. Alternatively, there may be so many conflicting existing constraints—laws, past practice, organizational policies, technical limitations, etc.—as to make a solution to the problem appear to be infeasible even before "the problem" has been well formulated.

Decision-problem formulation is therefore the art or craft, as opposed to scientific, element of systems analysis. It involves the establishment of the boundaries of the problem or system to be studied; the definition of the criteria that will be used to judge whether the problem has been "solved," or whether the situation will be improved by a proposed "solution"; the making of basic assumptions about the environment of the problem or system in question; and the identification of the "stakeholder," or "claimant," groups to be considered.

The nature of decision-problem formulation has been made clear by Checkland:[2]

> Problems . . . in the world are unbounded, in the sense that any factors assumed to be part of the problem will be inseparably linked to many other factors. Thus an apparently technical problem of transportation becomes a land-use problem, which is seen to be part of a wider environment-conservation problem, which is itself a political problem. Do these then become part of the original problem? Can any boundary be drawn? How can the analyst justify the limits that practicality forces him to impose?

There are clearly no precise rules to direct the systems analyst in this regard. In discussing "pitfalls in problem formulation and modeling," Quade has provided some useful guidance, however. These are briefly discussed below.[3]

Recognize the Importance of Formulation

One great pitfall is to be in such a hurry to get started on the "real work" of analysis that one pays insufficient attention to formulation and ends up working on a problem

[2]P. B. Checkland, "Formulation of Problems for Systems Analysis," paper submitted for *Handbook of Applied Systems Analysis,* as quoted by E. S. Quade, "Pitfalls in Formulation and Modeling," Chap. 3 in G. Majone and E. S. Quade (eds.), *Pitfalls of Analysis,* John Wiley & Sons, Inc., New York, 1980.

[3]Adapted, with wide interpretative latitude taken by the authors, from E. S. Quade, "Pitfalls in Formulation and Modeling," Ibid. All unreferenced quotes in this section are from that source.

that has little relation to the real issue, or on which no real progress can be made with the time and resources available, or in which no one has any interest.

Many analysts, who are trained in *analysis* rather than formulation, feel that no progress is being made until calculations are being made and the "real" job of analysis is under way. In paying relatively little attention to formulation, they risk the very real danger of working vigorously on the "wrong problem."

Question the Decision Maker's Statements of Goals and Constraints The statement of objectives that is made by a decision maker may well presume the wrong problem, or it may confuse means and ends, such as would be the case if a problem were stated in terms of "the need for a new facility." The systems analyst must therefore question the statements to ensure the real goal is understood.

The same is true of constraints. Often, constraints are artificial—such as those created by fiat without a full understanding of their implications. When the implications are made clear—such as by pointing out that the last 10 percent in desired performance may increase the cost by 50 percent—artificial constraints are often relaxed.

Moreover, all of this assumes that the decision maker is clearly identified; indeed, it assumes that there is a decision maker! In many complex organizations and systems, decision making is so diffuse that it is difficult to identify *the* decision maker. In such a case, an individual may choose to behave like *the* decision maker in discussions with the systems analyst. If the analyst assumes that the goals and constraints that are so identified are real, he or she may later discover that they reflect only the viewpoint of one of many interested parties.

Consider Alternative Measures of Effectiveness The analyst who assumes that "profit" is the "correct" goal for a business enterprise may be surprised to find that non-profit-justified actions are often taken. This is because even such a "simple" enterprise as a business has a complex array of goals, all or any of which may be operational in any given decision context.

In public institutions, this multiple-goal situation is even more vivid and may be exemplified by the conflicting goals of the various claimant groups, all of which "should" be, ideally, satisfied. Of course, this ideal is impractical, so the analyst must work with the decision maker to develop *appropriate measures of the relevant goal subset.*

Often, proxy measures will be necessary, such as is the case of assessing the systems performance of a health-care program by using mortality rates or hospitalization rates of the population being served. Clearly such measures omit certain aspects of the objective of "improved health," since they do not consider such elements as persons who may be ill at home.

Bound the Problem in a Realistic Fashion The analyst who is faced with a complex system, a multitude of conflicting objectives, and all of the other attributes of "wicked" problems may set out to solve the "whole" problem by drawing system boundaries that are so all-encompassing as to defy analysis and solution.

The formulated problem should be more abstract, less complex, and therefore more tractable, than the unformulated one.

> An analyst can set out to tackle far more than he can handle with the time and resources available. He can trap himself if he does not realize that he is not going to give a complete solution to the decision maker's problem. At best, he will produce evidence that other analysts will find hard to refute.

THE DECISION-MAKING PROCESS

To gain further insight into decision problems, we shall briefly discuss the process one must go through—either explicitly or implicitly—in solving a decision problem. This process serves as a focal point for the subsequent discussion of the conceptual framework for problem analysis.

Figure 5-1 outlines a series of activities, beginning with the collection and analysis of input data, which describes the general decision-making process. After the basic data have been collected and assembled, the decision maker predicts the outcomes which may result from each of the various alternatives available. Having done this, the decision maker evaluates the outcomes in terms of their worth and compares the alternatives on the basis of the outcomes to which each may lead. The best alternative is selected and an action taken which impacts on the "world"—the uncontrollable factors which affect the actual outcome of the decision situation. When the results of the action are known, the decision maker measures the results and compares them with the predicted outcome. The process can then begin anew with new predictions which incorporate the lessons learned in comparing the previous predictions with the actual outcome.

We shall discuss each of the activities which are implicit in the decision-making process in subsequent sections. One of the objectives of the conceptual

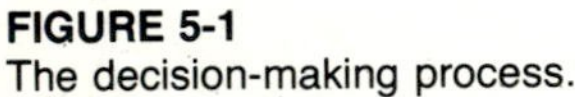

FIGURE 5-1
The decision-making process.

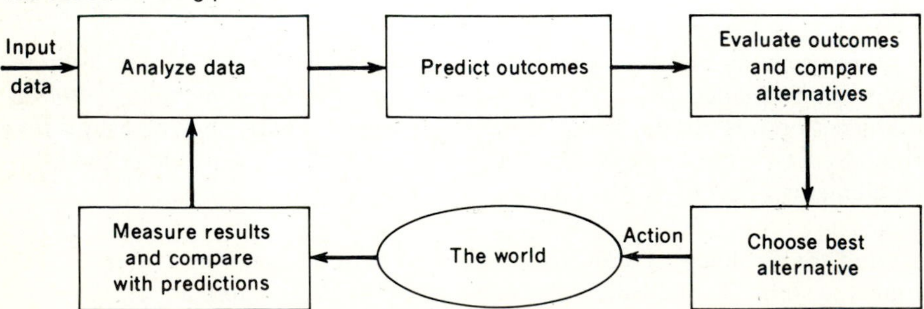

framework will be the explicit analysis and consideration of those aspects of the decision-making process which are too frequently left implicit in subjective problem solving.

DECISION-PROBLEM ANALYSIS

The alternative which the decision maker comes to select in a decision situation—whether it involves weapon selection, the pricing of a product, or the development of a social program for the indigent—interacts with the "general state of things." This interaction, together with the actions which may be taken by other rational beings, determines the outcome of the decision situation—the state of affairs which is attained as a result of the chosen alternative, outside influences, and competitive actions. In the pricing of a product, for example, the chosen price interacts with the economic status of consumers and the prices of competitors to result in an outcome which might be described in terms of the number of units sold, the share of market attained, revenue, or any number of other possible *outcome descriptors*.

Outcomes of decision situations represent the states by which success or failure of the chosen action is gauged. In truth, the aggregate of the outcomes of the many decision situations which face a business enterprise or other organization during a single year determines the success or failure of operations for that year and, in the long run, the success or failure of the organization itself.

The range of outcomes which may result from any particular decision situation is often quite wide. When the Department of Defense sets about to choose a strategic retaliatory force for the nation, it is not inconceivable that the range of outcomes might run from a mix so obviously powerful that it would deter aggression for decades, to one which would be recognized as so unreliable that the nation would be forced to rush in and invest billions to augment it with reliable weapons. Social programs for preschool children illustrate the wide range of possible outcomes even more clearly—at one extreme are those boondoggles which have no noticeable impact, and at the other are projects which significantly affect thousands of children.

Because of this wide range of possible outcomes, there is great pressure on the decision maker to choose the best, or at least a good, alternative, whether it is a new product, a weapons system, or a social program. To make such a choice intelligently, the manager must be able to sift the important factors which bear on the problem from the maze of complex trivia with which all significant decision situations abound. This is precisely the role in which the conceptual framework can be most valuable.

The outcome of a decision situation is determined by the interaction of the aspects which are controllable by the decision maker and the factors which are not subject to control—the uncontrollables.

When an airline is selecting a new model of airplane that will be its primary passenger carrier for many years, it must select from alternative designs available from various manufacturers. It should also specifically consider the alternative of "do not replace existing equipment." Among the many uncontrollables are the demand level for air travel for many years in the future, the price of fuel, and the world situation. All of these factors will have an influence on the outcome, or consequence, in a decision problem situation.

Both types of factors—controllable and uncontrollable—interact to produce the outcome, or consequence, in a decision problem. In order to aid decision makers in their choice, the totality of the *possible* outcomes may be described in the form of an outcome array.

The Outcome Array

Decision theorists use the terms "strategy" and "state of nature" to describe, respectively, alternatives available to the decision maker and the environmental or competitive factors which bear on the outcome of a decision problem.

In the general context, the selected strategy and the existing state of nature interact to produce a single outcome of the decision situation. These interactions may be conceptualized in a two-dimensional outcome array in which each element represents the outcome associated with a *particular combination* of a specific strategy and a specific state of nature. The entire array displays all outcomes which might result from the various combinations of the two interacting elements.

Table 5-1 shows the general form of an outcome array involving three alternative strategies, four possible states of nature, and hence twelve (3 × 4) possible outcomes. For convenience, the strategies are labeled S_1, S_2, and S_3, and the states of nature are denoted as N_1, N_2, N_3, and N_4, and the outcomes are symbolized by the letter "O" with two subscripts—the first indicating the corresponding strategy and the second indicating the corresponding state of nature.

We should be careful to recognize in a situation such as that described in Table 5-1 that *only one of the states of nature will occur;* i.e., either N_1, N_2, N_3, or N_4 will occur in this case. All four must be considered because we do not know in advance which one will occur. Hence, we must plan for all possible *contingencies*

TABLE 5-1
OUTCOME ARRAY

	N_1	N_2	N_3	N_4
S_1	O_{11}	O_{12}	O_{13}	O_{14}
S_2	O_{21}	O_{22}	O_{23}	O_{24}
S_3	O_{31}	O_{32}	O_{33}	O_{34}

(states of nature). We do this on the basis of the set of outcomes which are predicted for each strategy–state of nature combination.

Of course, the symbolism in Table 5-1 is for ease of discussion only. In practice, the outcome array would consist of descriptions of each of the possible outcomes.

Suppose, for example, that there are two manufacturers of appropriate airplanes, and that each has two versions that might be reasonable for an airline to consider purchasing. There are four alternatives:

Mfg. A—Model 1
Mfg. A—Model 2
Mfg. B—Model 1
Mfg. B—Model 2

There are therefore a total of five alternatives to be considered, since one should always evaluate proposed options versus the status quo, or "do nothing," alternative.

The relevant states of nature in this example involve demand levels for airplane travel and fuel price levels. The latter is important because the aircraft models differ in their passenger-carrying capacity and fuel consumption rates, as these factors are both deemed to be important to determining the various outcomes that might ensue from this decision situation.

The relevant state of nature might be (simplistically) described in terms of *demand level* and *fuel prices:*

N_1: High—High
N_2: High—Low
N_3: Low—High
N_4: Low—Low

Thus, N_1 indicates one *possible* state of nature in which both demand and fuel price are high, N_2 is the state in which demand is high and prices are relatively low, etc.

These five strategies and four states of nature could then be arrayed as shown in Table 5-1, and the relevant outcome descriptions entered in the table. These descriptions will normally be stated in terms of measures of the attainment (or degree of attainment) of the decision maker's objectives. The objectives are the decision maker's desire to achieve some state of affairs which is presumably better than that which now exists or that which can be foreseen as resulting from the present course. Most decision problems in business and government involve multiple objectives.

In this business situation, one might suspect that the sole objective is profit and that the outcomes therefore might be described solely in profit terms. However, business firms have become more aware of their social responsibility and might therefore include such objectives in their considerations. For instance, if the various aircraft models were believed to be different with respect to the level of air pollution that they produce, some prediction of this level might be included in each of the outcome descriptions along with profitability.

Although this illustration provides insight into the makeup of the outcome array, it is an easy problem in the sense that relatively good quantitative measures of the degree of attainment of many relevant objectives are available. On the other hand, many other decisions in both government and industry involve situations in which outcomes are not readily measured quantitatively. In choosing among alternative programs in health-depressed areas, for example, the federal government might decide that reasonable objectives are:

1 To prevent maternal deaths
2 To prevent premature births
3 To prevent infants' deaths
4 To prevent mental retardation
5 To prevent physical handicaps
6 To enhance the quality of the environment in which children are raised
7 To provide information about basic rules of sanitation

In practice, of course, the objectives involving prevention would not be feasible in the short run. These goals would be more appropriately stated in terms of reducing the incidence of the various occurrences.

When objectives are stated in these terms, quantitative measurement of the degree of attainment of some goals is made possible. The outcome descriptions in the outcome array would be in terms of such measures. From a practical viewpoint, however, it must be recognized that no valid numerical measures can be found for some objectives. In the case of the sixth objective above, for instance, a numerical measure of quality would probably be difficult to develop. Numerical outcome measures appropriate for the first five objectives (stated in terms of reducing rather than preventing) are rather clear—the number of deaths prevented, the number of premature births averted, etc. If no quantitative measures can be determined or developed, or if it is impossible operationally to measure the values which an existing descriptor assumes, the outcome array may include subjective descriptions such as "quality high" or "corporate image maintained." But, of course, quantitative measures or indexes are always more desirable.

Evaluating and Comparing Outcomes

The solution to a decision problem—or the choice of a best strategy—depends on the outcomes to which each strategy may lead. To compare alternative strategies meaningfully, you must first be able to evaluate and compare the outcomes to which the strategies may lead.

The practical problems arising in the seemingly simple process of comparison of outcomes are immense. Suppose, for instance, that a marketing decision maker has objectives with respect to the market share and profits achie... particular decision problem. Two possible outcomes which might evolve are:

Outcome O_1: Market share = 10%
Profit = $50,000
Outcome O_2: Market share = 15%
Profit = $40,000

There is no clear basis for choice between these two simple outcomes. In O_1, a larger profit is achieved together with a lower market share than in O_2. It is not at all obvious which outcome a particular decision maker or organization might prefer. In fact, it is apparent that two different individuals or organizations might have different preferences regarding these outcomes. The marketing manager associated with a newly introduced product might be very concerned with penetration of the market and might therefore prefer O_2 over O_1, even though less immediate profit is achieved. The manager whose product is mature might prefer O_1.

Economists have long concerned themselves with a concept called *utility,* which may be thought of as the capacity of an event, object, or state of affairs to satisfy human wants. The idea of a basic measure of the degree of human satisfaction which is derived from a state of affairs leads naturally to consideration of using this measure as a way of evaluating and comparing outcomes. If, for example, O_1 is preferred to O_2, O_1 possesses greater utility than O_2. If you had a numerical measure of the utility of each outcome, you could easily select the outcome which had the greatest utility and was therefore to be preferred.

However appealing the concept of utility as indicative of human satisfaction might be, the operational measurement of utility is not well developed.[4] In practice, utility measurements are seldom performed by systems analysts if there is any other reasonable alternative. The reasonable alternative to utility measurement which often suggests itself involves the outcome descriptors themselves.

In general, outcome descriptors—or measures of the degree of attainment of objectives—are of two varieties. The *resource costs* associated with a strategy are an intrinsic part of any outcome description. So, too, are the *benefits* derived from the outcome.

Resource costs are the resources expended in achieving a state of affairs (outcome). Often, it is convenient to think of resource costs as being associated with strategies, e.g., "Spend $1 million of advertising" or "Develop a weapons system at a cost of $1 billion." However, if the strategies are not described in terms of costs, it is often better to think of the costs as being a part of the outcome. Indeed, this sort of thinking is becoming more pervasive as the

[4]See, for instance, P. C. Fishburn, *Utility Theory for Decision Making,* John Wiley & Sons, Inc., New York, 1970, and D. J. White, *Decision Methodology,* John Wiley & Sons, Inc., New York, 1975.

uncertainties involved in complex systems become greater. For example, in a decision problem involving the selection of a weapons system, various alternative designs and contractors are considered. Attached to each proposal is a cost estimate. If we decide to treat this cost as known, we may consider it to be a part of the strategy. For example, one strategy might be "Proceed with development of XYZ Corp. design at estimated cost of $5 million." However, it is probably better to consider the strategy to be "Proceed with development of XYZ Corp. design" and to realize that the actual cost which will result depends on *both* the design and the environmental factors (states of nature) which may ensue. Thus, the cost of $5 million would be attached to one of a set of possible outcomes which might result from the XYZ design.

Benefits are the returns associated with an outcome, in terms of either the resources gained or the psychological, sociological, or other intangible values derived from the state. Examples of such benefits are the profit earned, the time saved, and the number of maternal deaths prevented, as are such intangibles as "greater freedom enjoyed," "freedom from want," and "higher quality of life."

Outcome descriptors—whether of the resource-cost or the benefit variety—are measures of the degree of attainment of objectives, and since most complex decision problems involve multiple objectives, it is common to find each outcome in an outcome array described by a number of measures. This is the curse of multidimensionality in systems analysis, for the comparison of outcomes in terms of multiple descriptors is no simple task.

A key method for obviating the problem of multiple outcome descriptors is to reduce them to an overall single measure which reflects their aggregate worth. This is the essence of the utility concept discussed earlier; however, as already pointed out, the problems involved in utility measurement are so great as to impair the usefulness of this approach. The idea of reducing the description of an outcome to a single dimension is still of great importance. A simple illustration will make this point clear. Consider a decision situation involving two strategies and only a single state of nature. Table 5-2 shows the two outcomes described in terms of revenue (benefit) and cost. Note that these outcomes have the same characteristics as those described previously in terms of market share and profit. The outcome for S_2 is better in terms of one measure—revenue—while the outcome for S_1 is better in terms of the other, since it has the lower cost; yet no

TABLE 5-2
OUTCOME TABLE

	N_1
S_1	Revenue = $100,000 Cost = $40,000
S_2	Revenue = $110,000 Cost = $60,000

TABLE 5-3
OUTCOME TABLE

	N_1
S_1	Market share = 10% Revenue = $100,000 Cost = $50,000
S_2	Market share = 15% Revenue = $110,000 Cost = $70,000

one would hesitate to declare that the outcome associated with S_1 is better.[5] Why? Because the *profit* there is $100,000 minus $40,000, or $60,000, while the profit associated with the outcome for S_2 is only $50,000.

The critical point illustrated by Table 5-2 is that since the resource costs and benefits are expressed in the same terms—dollars—the *difference* between benefits and costs seems to be a valid measure of the aggregate worth of each outcome. Additionally, of course, this difference has an accounting significance which further enhances its intuitive appeal.

Another level of complexity presents itself when we have outcomes such as those discussed previously in terms of market share and profit. Consider the case of Table 5-3, for example. There, we have implicitly superimposed an objective related to market penetration onto those related to dollar costs and revenues.

In Table 5-3, we can utilize the point just demonstrated and make a profit calculation, thus converting the situation to one such as that in Table 5-4. However, after doing so, we have exactly the dilemma discussed earlier, and we cannot readily determine that one or the other of the outcomes is preferable.

If we knew the dollar value which the decision maker places on market share, we could arrive at a conclusion, however. Suppose that each 1 percent of market share above 5 percent is regarded as equivalent to $1,000 in profit. Then, since

[5]This statement implies some assumption about the decision maker's availability of funds, etc. If the alternatives were presented to a large, prosperous corporation, the statements made here would invariably be true.

TABLE 5-4
OUTCOME TABLE

	N_1
S_1	Market share = 10% Profit = $50,000
S_2	Market share = 15% Profit = $40,000

the outcome for S_1 is 5 percent above 5 percent and that for S_2 is 10 percent above 5 percent, the total worth of the upper outcome is $55,000, and that of the lower one is $50,000. Hence, S_1 is the better because it leads to the outcome with the greatest total worth. Of course, the determination of the trade-off between dollars of profit and market share would not be easy to determine in any real situation. Even so, the *method* of developing such trade-offs is important to systems analysis.

Decisions under Certainty

The two cases just discussed involved simple decision problems under certainty. The word "certainty" is applied because we acted as though we were *certain* of the uncontrollable environmental factors which would be imposed on us; i.e., we assumed that only one state of nature needed to be considered.

True decision problems under certainty are a rare occurrence in the real world. Almost all meaningful problem situations involve some degree of uncertainty about the outcomes which may result from a given course of action. However, assuming a certainty situation is often a useful way to formulate and analyze a problem.

Since each strategy leads (or is assumed to lead) to a unique outcome in the case of certainty, the decision problem of choosing among strategies is reduced to one of choosing among outcomes. In simple cases, this may be a rather elementary conceptual process. For instance, in the problem of Table 5-2, after we reduced the resource cost and benefits for each outcome to a single aggregate measure—profit—it was a simple matter to choose the best outcome (the one with the highest profit) and hence the best strategy. The best strategy is simply the one which leads with certainty to the best outcome. On the other hand, we showed in Tables 5-3 and 5-4 that the problem of choice under certainty may not be simple if there is no obvious way of reducing the outcome description to a single measure.

Decision problems under certainty may also be more complex than they seem at first glance, as a result of the size of the problem, i.e., the number of alternative strategies. Consider, for example, the personnel assignment situation discussed in Chapter 4. In that case, the time required by each person to complete each job was assumed to be known with certainty. The outcome array corresponding to Table 4-2 is shown in Table 5-5. There are two alternative strategies:

TABLE 5-5
OUTCOME TABLE

	N_1
S_1	Total person-hours = 5
S_2	Total person-hours = 5

S_1: (Assign person A to job 1 and person B to job 2)
S_2: (Assign person A to job 2 and person B to job 1)

If the only relevant outcome descriptor is "total work-hours to perform both jobs," it is clear from Table 5-5 that there is no basis for preference between the two strategies.

Suppose, however, that the problem involves three people and three jobs rather than two people and two jobs. Such a case is shown in Table 5-6, where each person's performance on each job is still assumed to be known with certainty. The first step in developing an outcome table in such a situation is to enumerate the strategies—the alternative person-job combinations constituting an assignment which will get the three jobs done (one by each person). You are asked to develop this table. You will find that there are six alternative strategies. The best of these is "A to job 1, B to job 3, and C to job 2" since it involves a total of five work-hours. The point is, however, that in going from a two-person–two-job situation to a three-person–three-job one, the number of strategies increases from two to six. In general, this sort of progression holds true, so that for 10 people and 10 jobs, the number of strategies is over 3.6 million. From this, it is easy to see how a seemingly simple decision problem under certainty may indeed be difficult to handle. Consider, for example, the military services assignment of people to jobs. Even though it is done on a batch basis—i.e., each group of enlistees out of basic training is assigned to available spaces—the scope of the problem is immense. Typically, several thousand enlisted men and women must be simultaneously assigned. Hence, the number of alternative strategies is so large as to be incomprehensible.

A Revisit to the Evaluation and Comparison Problem

Having described the simplest formulation of a decision problem—certainty— we can return to the question of evaluating and comparing outcomes. For the moment, we can assume that we are dealing with a problem under certainty and that each outcome is described by one measure of each of the two general varieties—a measure of resource cost and a measure of benefit.

TABLE 5-6
TIME REQUIRED BY THREE PEOPLE TO PERFORM EACH OF THREE JOBS

		Jobs		
		1	2	3
People	A	2	2	4
	B	3	3	2
	C	6	1	5

There are three general situations with which we may be faced, each of which requires a different approach. First, the most general case is the one in which *both resource cost and benefit are variable.* They are variable in the sense that, as we look down the single column of the outcome array, we see various values of both measures. Table 5-7 illustrates this case for some (undefined) measures of resource cost and benefit (where high values of cost are assumed to be bad and high values of benefit are assumed to be good). In such a case, the benefit-cost *difference* or *ratio* may be a meaningful single descriptor of each outcome and, hence, a measure on which to base a choice between outcomes. Profit, for example, is a benefit-cost difference, or a measure of *net* benefit. Return on investment is a ratio measure of benefit to cost which has some accounting significance. Other differences (net benefits) may be appealing as long as the dimensions of benefit and cost are the same, i.e., as long as we are not subtracting apples from oranges. Other ratios may be valid measures even if the dimensions are not the same. For example, "number of infant deaths prevented per dollar spent on health programs" might be a reasonable single aggregate outcome descriptor.

The second general situation which may arise involves *specified benefit levels and variable resource costs.* In this case, outcomes are compared on the basis of the cost required to achieve a given level of benefit. For example, if the objective is to have a mile run as fast as possible, the best alternative is the runner who expends the least resource (time) in doing so.

We should point out here, however, that this general situation—specified benefits and variable costs—is an all-too-easy way of ignoring the significant problems facing an organization. A constant search for cost savings does not necessarily imply good management if benefit levels are ignored. Because of the difficulties in measuring benefits, it is easy to rationalize objectives so that benefit levels appear to be fixed, when really both costs and benefits should be treated as variable. When a physical system is to be selected, it is easy to think of specifying performance levels and then seeking to minimize cost. In practice, however, the small number of alternative designs will vary in performance capability, and variations in both performance (benefit) and cost should be considered in the decision analysis.

The third general situation involves *specified resource costs and variable benefits.* This sort of situation exists in a race in which time is fixed and distance is variable; i.e., all runners will run for, say, two minutes, and the winner will be the one who goes the farthest. The allocation of a fixed budget in an optimal fashion is illustrative of this viewpoint since, presumably, cost is fixed and one wishes to get the greatest total benefit possible.

In the latter two cases, when either resource cost or benefit is fixed and the other is variable, the comparison of outcomes under certainty is rather simple. The single aggregate measure which may be used to describe an outcome is the one which varies. Thus, if cost is variable and benefit fixed, we choose the least-cost outcome and hence, in the case of certainty, the least-cost strategy. If

TABLE 5-7
OUTCOME ARRAY

	N_1
S_1	Cost = 10 Benefit = 20
S_2	Cost = 6 Benefit = 8
S_3	Cost = 8 Benefit = 8

cost is fixed and benefit variable, we choose the benefit-maximizing outcome and (under certainty) the strategy corresponding to it.

However, this does not exhaust the range of difficulties which we may encounter. If we have multiple objectives and multiple benefit or cost measures, we are faced with the dimensional problem discussed previously, and there are no pat answers or formulas which resolve this difficulty. Its resolution depends on the ingenuity of the analyst in using the information at hand to develop benefit measures, cost measures, and aggregate measures which provide meaningful and comprehensive outcome descriptions and at the same time provide a basis for rational choice among alternatives.

Decisions under Uncertainty

A more realistic description of real-world problem situations involves uncertainty. Formally, uncertainty differs from certainty in that the latter involves a specified set of environmental conditions—one state of nature—while uncertainty involves a range of possible sets of environmental conditions which may ensue—more than one state of nature.[6] This is the general situation described by the outcome array of Table 5-1.

Most real-world decision problems look more like problems under uncertainty than like problems under certainty. Of course, since the outcome array form is itself an abstraction of the real world, you might argue that the maze which constitutes a real problem looks nothing like the outcome array of Table 5-1, much less looking like that simpler one-column array describing a problem under

[6]Two problems of semantics arise here. First, a distinction is often made by decision theorists between risk and uncertainty. This distinction is based on a knowledge of the likelihoods associated with the uncontrollable elements. See R. D. Luce and H. Raiffa, *Games and Decisions,* John Wiley & Sons, Inc., New York, 1957. Here, uncertainty is used to describe any problem formulation involving more than one state of nature. Hence, it encompasses both risk and uncertainty as the terms are used by decision theorists. The other problem has to do with the kind of uncertainty. Various authors have distinguished between uncertainty due to different sources. For example, see the early paper by C. H. Hitch, "An Appreciation of Systems Analysis," RAND Corporation P-699, Santa Monica, Calif., Aug. 18, 1955. Here, no such distinction is necessary.

TABLE 5-8
OUTCOME ARRAY IN TERMS OF PROFIT

	N_1	N_2
S_1	\$50	\$100
S_2	\$48	\$96

certainty. This would be a valid comment. The critical point is, however, that it is necessary to think in precise terms such as those of the outcome array in order really to understand most complex problems. Thus, while problems and outcome tables may look quite different, the array is an abstraction which enables us to analyze a problem.

Let us look to the analysis of a well-structured decision problem involving uncertainty as a way of understanding the additional difficulties imposed by this formulation. In doing so, we should remember that the same difficulties inherent in problems involving certainty also occur in those involving uncertainty, so we need not discuss them again here.

Consider the outcome array of Table 5-8, in which the cost and benefit descriptors have already been combined into the aggregate measure profit. With all these simplifying assumptions, we find that it is easy to choose the best strategy in this table. Why? Because, regardless of whether N_1 or N_2 occurs (and recall that only one of them will), S_1 leads to the better outcome. Since S_1 leads to a better outcome than S_2 for every possible contingency, it is clearly the better strategy.

Of course, all we must do to construct a decision situation in which no such simple answer is possible is to use the same profits in a rearranged outcome array such as that in Table 5-9. There, S_1 is better if N_1 occurs, but S_2 is better if N_2 occurs. Which is better—S_1 or S_2? The answer is: "It depends." It depends on the relative likelihoods associated with N_1 and N_2, for one thing. For example, if you knew that N_2 was one thousand times as likely to occur as N_1, you would probably choose S_2 because it is the better if N_2 occurs, which is extremely likely. On the other hand, if the likelihoods are not so extreme—say, there is a 50:50 chance of N_1 or N_2 occurring—no clear best choice is immediately obvious.

The *maximization of expected net benefits* is one way of arriving at a basis for

TABLE 5-9
PROFIT OUTCOME ARRAY

	N_1	N_2
S_1	\$50	\$96
S_2	\$48	\$100

choice in such a situation. The expected net benefit associated with a strategy here is simply the weighted sum of the profit outcomes to which that strategy might lead—the weights being the probabilities for each outcome.[7]

In the illustration of Table 5-9, for example, if it were determined that both states of nature are equally likely to occur, probabilities of one-half would be imputed to them. The expected net benefit (expected profit) associated with strategy S_1 is therefore

$$\tfrac{1}{2}\ (\$50) + \tfrac{1}{2}\ (\$96) = \$73$$

and the expected net benefit for S_2 is

$$\tfrac{1}{2}\ (\$48) + \tfrac{1}{2}\ (\$100) = \$74$$

Therefore, strategy S_2 is presumably the better of the two, since it has the higher expected net benefit.

We use the word "presumably" here since there are a number of deficiencies associated with the maximization of expected net benefit (profit) which restrict its usefulness in systems analysis. Among these are the presumption that an exhaustive list of environmental conditions—states of nature—can be specified and the undue weight that such a presumption puts on extreme outcomes. In practice, you are seldom able to make a meaningful list of *all* possible

[7]The basic ideas relating to probability are familiar to most of us. Any event whose outcome is at least partially determined by chance, such as the flip of a coin, may be described in probabilistic terms.

The probability of an outcome is most easily thought of as the long-term percentage relative frequency which would be realized if the act were repeated again and again. In terms of the coin flip, after a long series of flips you might divide the total number of occurrences of the outcome "heads" by the total number of flips, calling the resulting decimal the *probability of heads;* in this case, you would be likely to determine the probability of heads to be near one-half. Similarly, the probability of a six on a throw of a single die would be about one-sixth.

The use to which the concept of probability may be put is also based upon sequences of events. *The knowledge that the probability of heads is one-half will in no way help you to predict what the outcome of a particular flip of a coin will be,* for it will be either heads or tails, and you are either right or wrong. This knowledge does permit you to predict that in a long sequence of flips, the relative frequency of heads will be close to 50 percent, however.

To apply probabilities to the analysis of decisions, you must recognize that the basic idea is applicable to the uncertain outcomes of events which can influence these decisions—the states of nature. To conclude that the probability is one-third that June rainfall in Cleveland will exceed 2 inches implies that an investigation of weather bureau records for many past Junes has indicated a relative frequency of one-third for such a state. If this is so, and if there is no reason to believe that future weather patterns will be different from those which prevail now or which have existed in the past—i.e., if the process exhibits *stability* over time—you might conclude that the future percentage of occurrence of rainy (over 2 inches) Junes will also be about one-third.

Probability, then, is simply a way of dealing with our uncertainties about the future. In attaching probabilities to states of nature or outcomes, we evaluate the likelihood of their occurring, and we thereby synthesize our information about the future into a single number. The reader who is interested in pursuing basic probability ideas further is referred to William R. King, *Probability for Management Decisions,* John Wiley & Sons, Inc., New York, 1968.

contingencies, and if one outcome is either very much better or very much worse than all others, the maximization of expected net benefit leads to results which are counter to your intuition.[8]

Also, if the states of nature are not simply uncertain environmental factors, but also involve the actions of rational beings whom the decision maker cannot control, the idea of an expected net benefit is only tenuously applicable. Moreover, if you realize that no simple measure of net benefit, such as profit, is usually available in any but the most straightforward decision situations, the maximization of expected net benefit is relegated to a role of minor importance in terms of analytic usefulness. Like the idea of utility, the maximization of expected net benefit is conceptually interesting and potentially useful,[9] but neither is widely applicable in solving meaningful real-world problems.

Other bases for choice under uncertainty have been proposed. In particular, these bases are applicable in the sort of uncertain situation in which nothing is known concerning the likelihoods associated with the states of nature. Such a situation is probably more descriptive of the real world than the restrictive assumptions of an exhaustive list of the states of nature and a knowledge of the numerical probabilities of each which is inherent in a calculation of expected net benefit. Yet, if you have knowledge of the likelihoods, you should obviously make use of it. We shall discuss this further in the next chapter. Here, we conclude our discussion of uncertainty with a brief summary of some of the other bases for decision making which have been proposed. In doing so, we emphasize that they are ways of thinking about problems rather than hard-and-fast rules about how you should make decisions. In particular, since the several bases to be discussed may well be contradictory in any specific decision problem, it is apparent that none is being put forth as *the* way to solve problems involving uncertainty.

The *maximin* approach to problems under uncertainty is based on the argument that, in the face of so large a degree of uncertainty, the decision maker might choose to act pessimistically by deciding to maximize the *security level.* You might reason that if nature is going to be perverse and attempt to give you as small a payoff as possible, you should in turn act to maximize the return consistent with this malevolent intention of nature. In effect, you would identify the worst outcome which could possibly result from each strategy and choose the strategy that will give you the best of these worst outcomes.

In the problem of Table 5-10, in which we have assumed a single (undefined) measure of net benefit, the decision maker finds that if S_1 is chosen, the worst

[8]Consider the situation which many people face when their alma mater asks for conributions, saying that the average contribution last year was $1,000. The alumnus who is considering giving $50 is awed until he realizes that several people have given very large sums, while most have given $100 or less. The weighted sum of the contributions, called an *average,* is grossly unrepresentative of one's intuitive idea of an average. The expected net benefit discussed here is similar to an average and has the same deficiency.

[9]In particular, the combination of these two ideas—maximization of expected utility—is the theoretical basis for choosing among alternative strategies. However, in practice, it is almost impossible to measure utilities meaningfully. Hence, the concept is not readily implemented.

TABLE 5-10
PAYOFF TABLE

	States of nature			
	N_1	N_2	N_3	N_4
S_1	50	30	30	5
S_2	80	40	10	15
S_3	120	50	5	0

which can happen is a net benefit of 5. Similarly, under S_2 it is 10, and under S_3 it is 0. These worst outcomes are security levels for each of the strategies in that the decision maker can be confident that no worse can happen under that strategy.

To make the best of the solution, the decision maker may act to ensure the best of this set of worst outcomes. You can do this by choosing S_2. The worst which can happen under S_2 is a net benefit of 10, while lower net benefits could result under both S_1 and S_3. These 10 net-benefit units represent the *maximum security level* which you can obtain. By choosing S_2, you can be certain of getting no less than 10 units, and you cannot be certain of as much as 10 under any other strategy.

This idea of a best strategy is such a case of uncertainty, the one which maximizes the minimum net benefit which can be achieved, is appropriately termed a *maximin strategy*. To select a maximin strategy, determine the worst outcome for each strategy and then select the strategy which has the largest value of this minimum. The decision maker is guaranteed at least the maximin payoff, since whatever state of nature is actually realized, no lesser payoffs can be obtained using the maximin strategy.

Operationally, you determine the maximin strategy simply by finding the lowest payoff for each strategy (the smallest entry in each row of the payoff table) and then choosing the largest of these lowest payoffs. The strategy associated with this "largest of the smallests" is the maximin strategy.[10]

If you, the decision maker, choose to be completely optimistic rather than pessimistic, you might use a *maximax* criterion; i.e., choose the strategy which makes the best of the best which can occur. In this case, you would choose the largest net benefit for each strategy and then choose the strategy which is associated with the largest of these—in effect, simply determine the greatest payoff in the table and select the strategy which *may* result in this greatest payoff. An optimistic decision maker in this case decides to "go for broke."

[10]If you are dealing with the minimization of resource costs for a given benefit level, the equivalent to maximum is *minimax,* i.e., the minimization of maximum loss. This is so because the worst outcomes have the smallest numerical values in such cases.

To determine a minimax strategy, you simply choose the largest number in each row and then the smallest of the largest. The associated strategy is a minimax strategy. All other arguments are the same as for the maximin criterion. You must remember only that to ensure the best security level, you play *maximin when dealing with gains* and *minimax when dealing with losses.*

In the example of Table 5-10, the largest payoff is the 120 units associated with O_{31} (involving strategy S_3 and state of nature N_1); hence, the maximax strategy is S_3.

It should be noted that this criterion completely ignores the relative sizes of the payoffs for each strategy. The choice of S_3 would still be dictated by the maximax criterion if the least of the other payoffs under S_3 were 119, 0, or minus 1 billion. The maximin criterion has the same failing, since it completely ignores the larger payoffs associated with a strategy.

Savage[11] has proposed an alternative criterion for decision making under uncertainty which is based upon a common psychological quirk. Most of us, when confronted with the outcome of a choice situation, apply our 20/20 hindsight vision and view the outcome which we *could* have obtained had we known in advance that the realized state of nature would occur. Savage used the term "regret" to describe the dissatisfaction associated with not having fared as well as one would have if the state of nature had been known. In effect, the decision maker views the past and regrets having chosen the alternative that was chosen.

A measure of this *regret* for any outcome might be the difference between the payoff for the outcome and the largest payoff which could have been obtained under the corresponding state of nature. In the situation of Table 5-10, the regret associated with outcome O_{11} is 70—the difference between the 50 net-benefit units realized and the 120 which could have been realized had you known in advance that N_1 would occur. The remaining regrets are given in the regret Table 5-11. It should be noted that there is at least one outcome which has no regret for each state of nature—the outcome which has the highest payoff for that state of nature. If you choose the strategy which leads to the highest payoff for the state of nature which is actually realized, you experience no regret at not having chosen a better strategy.

To utilize Table 5-11 in a decision criterion, you must recognize that the regrets are of the nature of losses. Hence, the pessimistic decision maker might decide to use a strategy that would minimize the maximum regret—a *minimax regret* criterion. In this case, the maximum regret for each strategy would be 70, 40, and 25, respectively. The strategy which minimizes maximum regret is S_3, with a security level of regret of 25.

[11]L. J. Savage, *The Foundations of Statistics,* John Wiley & Sons, Inc., New York, 1954.

TABLE 5-11
REGRET TABLE

	States of nature			
	N_1	N_2	N_3	N_4
S_1	70	20	0	10
S_2	40	10	20	0
S_3	0	0	25	15

The appeal of this criterion is diminished for many by a logical flaw which appears to be inherent in it. If we add a fourth strategy (S_4) which has net benefits of 60, 40, 30, and 60, respectively, for the four states of nature, the regret table becomes that shown in Table 5-12. The minimax regret payoff here is 45, and the minimax regret strategy is S_2.

Consider now what has resulted from the introduction of S_4 into the decision problem. Although S_4 is not chosen as best, its consideration has shifted the preference from S_3 to S_2 *within the set of previously considered strategies.* That this is illogical is perhaps best illustrated by a simple example. If you are offered a choice of an apple or an orange and choose the apple, it is assumed that in a choice among an apple, an orange, and a peach, you will choose either the apple or the peach—but not the orange, since you have already expressed a preference for the apple over the orange. The introduction of a new alternative—the peach—should not cause a shift in preference between the apple and the orange. This is exactly what has occurred here. The introduction of S_4 has caused a shift in preference from S_3 to S_2. Logically, if S_3 were the best strategy from the set consisting of S_1, S_2, and S_3, the introduction of S_4 should not result in a switch in preference from S_3 to S_2.

This logical deficiency—termed the *independence of irrelevant alternatives*—forms the basis for much of the criticism of the regret idea. To some, however, the cases in which the invalidity of the criterion can be demonstrated on these grounds seem to be pathological. To them, the criterion is a good one for most practical purposes.

SUMMARY

The conceptual framework for systems analysis discussed in this chapter is not a panacea which will solve all decision problems. On the contrary, it sometimes raises more questions than it solves; yet we shall argue that these questions are more apt to be the right ones than those which might be raised without the approach.

The first aspect of the conceptual framework for systems analysis is problem formulation. Before a problem can be solved, or a decision made, it must first be precisely defined (formulated). Proper attention to formulation ensures that the "right" problem is being addressed.

TABLE 5-12
REVISED REGRET TABLE

	States of nature			
	N_1	N_2	N_3	N_4
S_1	70	20	0	55
S_2	40	10	20	45
S_3	0	0	25	60
S_4	60	10	0	0

The basic tool of the conceptual framework is the outcome array. Conceptually, it represents the array of all possible circumstances which might ensue after any of the specified actions are taken and environmental conditions are realized. In any real problem, you may not actually construct an outcome array. Indeed, the problem may be too large to permit you to do so. The value of the outcome array is that it requires the decision analyst to think in terms of the wide scope of consequences (outcomes) which can result from an action. In doing so, the decision maker is precluded from focusing undue attention on what is thought most likely to occur. Of course, the decision maker should give attention to this, but should also carefully consider the other possible outcomes.

This simple table has many other values aside from its formal use as an enumeration device. In attempting to describe outcomes meaningfully, you are forced to consider resource costs, benefits, and their relationship. More important, this often leads to a reconsideration and questioning of objectives, which produce valuable insight into the problem.

Decision problems in the real world do not exist in terms of outcome arrays. Even if they can be formulated as such, they do not normally fall neatly into the certainty-uncertainty dichotomy discussed here. Indeed, the majority of problems are most naturally formulated in terms of uncertainty, since the very nature of meaningful problems is such that the potential consequences are uncertain. However, as we shall see in the next chapter, the certainty-uncertainty taxonomy is useful for purposes of analysis. For example, sometimes the analyst can analyze a complex problem as though the outcomes were certain and learn enough in doing so to reformulate the problem in a more meaningful way under uncertainty.

The essential appeal of the conceptual framework is that it is descriptive of the way in which we all know we should, or think that we do, make decisions. Its value is that it forces us to give proper attention to enumerating the alternative strategies, the entire range of contingencies, and the scope of possible outcomes in a fashion which would be impractical if the problem were approached at an intuitive or informal level. Moreover, the outcome descriptions force the decision makers to spell out clearly just what it is that they are trying to do—the objectives—and to develop reasonably precise measures of the degree of attainment of the objectives. In most instances, this process leads to a degree of comprehension of the salient features of a decision maker's responsibilities which is superior to any insights and understanding that might otherwise be gained. Frequently, this is itself sufficient justification for a formal problem analysis, and the better strategy which is selected as a result of analysis is something of a by-product.

DISCUSSION QUESTIONS

1 In carrying out systems analysis, a theoretical framework for analysis must be used. What is the justification for such a framework?
2 The first step in solving a decision problem is the formulation of the problem. What is

involved in the formulation of a problem? What is the difference between a problem and the symptoms of the problem? Why is it important to differentiate?

3 One might say that it is human nature to jump to the identification of a problem and begin to solve that problem before alternative ways of viewing the problem have been carried out. Do you agree with this? What might be done to reduce the chances of selecting the decision problem too quickly?

4 What is the process that we go through, either explicitly or implicitly, in solving a decision problem?

5 Decision problem analysis involves the ways in which alternative strategies interact with the general state of things. What is meant by this statement?

6 How might the outcome array be used in solving decision problems? How might this be applied to some of the key personal problems that an individual encounters in a lifetime?

7 If profit were the sole objective of the business firm, then decision problems involving the firm would be easier. Defend or refute this statement.

8 Most real-world problems involve varying degrees of uncertainty. How would the decision problem analyst deal with uncertainty?

9 Distinguish between decisions under certainty and uncertainty. Illustrate the difference in the form of the outcome table for these two cases.

10 What are some general philosophies that can be used to guide the system analyst in dealing with the process of decision problem formulation?

11 What is meant by the controllables and uncontrollables in a decision problem? Give examples.

12 What role do contraints play in dealing with decision problems? Give some examples.

13 How is effectiveness to be measured in a decision problem? What are some examples?

14 The range of outcomes which may result from a particular decision situation is often quite wide. How might this be related to the notion of second-order consequences mentioned earlier in the text?

15 To compare alternative strategies meaningfully, we must first be able to evaluate and compare the outcomes to which the strategies may lead. How is this best done?

RECOMMENDED READINGS

Ackoff, R. L.: *The Art of Problem Solving,* Wiley-Interscience, New York, 1978.

Bender, E. A.: *An Introduction to Mathematical Modeling,* Wiley-Interscience, New York, 1978.

Beltrami, E. J.: *Models for Public Systems Analysis,* Academic Press, Inc., New York, 1977.

Checkland, P. B.: "Formulation of Problems for Systems Analysis," in Majone, G., and E. S. Quade (eds.), *Pitfalls of Analysis,* John Wiley & Sons, Inc., New York, 1980, chap. 3.

Fishburn, P. C.: *Decision and Value Theory,* John Wiley & Sons, Inc., New York, 1964.

———: *Utility Theory for Decision Making,* John Wiley & Sons, Inc., New York, 1970.

Hayes, R. H.: "Qualitative Insights from Quantitative Methods," *Harvard Business Review,* July 1969.

Herbert, T. T., and R. W. Estes: "Improving Executive Decisions by Formalizing Dissent: The Corporate Devil's Advocate," *Academy of Management Review,* October 1977.

Hitch, C. H.: "An Appreciation of Systems Analysis," RAND Corporation, P-699, Santa Monica, Calif, August 18, 1955.
Howard, N.: "The Analysis of Options in Business Problem" *INFOR: The Canadian Journal of Operational Research and Information Processing,* February 1975.
King, William R.: "Human Judgment and Management Decision Analysis," *Journal of Industrial Engineering,* December 1967.
——— and B. Dutta: "Metagame Analysis of Competitive Strategy," *Strategic Management Journal,* vol. 1, no. 4, October–December, 1980, pp. 357–370.
McKenna, Christopher K.: *Quantitative Methods for Public Decision Making,* McGraw-Hill Book Company, New York, 1980.
Quade, E. S.: "Pitfalls in Formulation and Modeling," in Majone, G., and E. S. Quade (eds.), *Pitfalls of Analysis,* John Wiley & Sons, Inc., New York, 1980, chap. 3.
———, *Analysis for Public Decision,* 2nd ed. North Holland, New York, 1982.
Rivett, P.: *Model Building for Decision Analysis,* John Wiley & Sons, Inc. New York, 1980.
Sassone, P. G., and W. A. Schaffer: *Cost-Benefit Analysis,* Academic Press, Inc., New York, 1978.
Savage, L. J.: *The Foundations of Statistics,* John Wiley & Sons, Inc., New York, 1954.
Shell, R. L., and D. F. Stelzer: "Systems Analysis: Aid to Decision Making," *Business Horizons,* December 1971.
Vickers, Geoffrey: *The Art of Judgment,* Basic Books, Inc., New York, 1965.
White, D. J.: *Decision Methodology,* John Wiley & Sons, Inc., New York, 1975.

CASE 5-1: Developing Outcome Descriptors for a Bank

The following is a statement of the objectives and strategies of a bank. Your project team has interviewed various people in the bank, perused the strategic plan, and established that these are the "true" objectives and strategies and not just public relations statements. Now you must develop outcome descriptors that can be used to evaluate various computer-based systems that have been proposed to support the bank's efforts to implement these strategies and to achieve the objectives.

Statement of Bank Objectives and Strategies[1]

Objectives for individual and commercial customers and prospects:

a Accessible, flexible facilities for deposit and receipt of cash checks, bonds, drafts, and other negotiable documents
b Interest-paying system to encourage time deposits
c Safekeeping facilities for valuable records
d Personal, confidential, knowledgeable consultation on all financial matters

Objectives for loan customers and prospects:

[1]Adapted with permission from IBM Study Organization Plan: Documentation Techniques, Manual C20-8075, 1961.

a Facilities and experienced personnel available for consultation and financial advice on all loan matters

b Readily accessible facilities for the closing of (and payment of) personal, commercial, or mortgage loans

c Extensive advertising program to attract loan prospects to the bank for consultation

d Specialist available with a broad knowledge of income-producing investments

e Specialists available having detailed information on the financial status of local individuals and businesses

f Analysts available who are well informed on relative valuations of all types of property

g Flexible interest-charging structure to encourage large loans and rewards for those who pay when due

Planned strategies to meet objectives:

a Expand advertising program to reach more potential customers.

b Enlarge number of drive-up banking facilities.

c Increase emphasis on installment-type loans.

d Modernize and recognize physical and personnel facilities as necessary for most efficient operation.

e Develop services and systems to facilitate achievement of objectives with view in mind of offering these services to local businesses on a fee basis.

Work as a team to formulate a set of outcome descriptors that will enable you to assess alternative systems in terms of the degree to which they support these objectives and strategies. Consider how each potential descriptor relates to each objective and strategy, as well as to more global objectives that are not included in the above list. Also consider how each proposed descriptor variable could be readily measured and predicted.

CHAPTER 6

METHODS OF SYSTEM ANALYSIS

We have no simple problems or easy decisions after kindergarten.[1]

The conceptual framework for systems analysis which was spelled out in the previous chapter does not represent a practical operating procedure. There, we illustrated the way in which the analysis *should* be done and enumerated some of the related problems. In this chapter we shall discuss how it *is* done, and as we shall see, although there are real problems involved in applying the framework, the two are not very different.

MODELS IN SYSTEMS ANALYSIS

The basic operating device used by systems analysts is a model. The layperson's understanding of models is simultaneously helpful and injurious to understanding of the scientific use of the term. Discounting the fashion variety, if the proverbial man in the street were asked to react to the word "model," he would be likely to respond with some familiar example such as a child's model airplane. Indeed, this kind of model is familiar to all of us.

Such a model is simply a scaled-down *representation* of a real-world system—the full-size airplane. Each exterior dimension of the real airplane is accurately represented in miniature on the model. In addition, many features of the real airplane are completely excluded from the model; e.g., the model is often of solid construction, while the airplane's interior is a maze of electronic gears and

[1]John W. Turk, as quoted in *Forbes,* Jan. 22, 1979, p. 104.

cables. This feature of the child's model is an intrinsic part of the scientific model; that is, in both kinds of models, some aspects of the real system are included (such as exterior dimensions, color, and markings), and some are excluded (such as interior configuration, and materials). This is consistent with the scientist's idea of a model as an *abstraction of reality*.

Other models of the same system might abstract different elements of the system. For example, the ground training devices used for pilots are little more than movable enclosures which incorporate all the interior makeup of the cockpit, but little else. The pilot is seated in an exact representation of a portion of an airplane's interior and closed off from the surroundings. By reference to the instruments in the cockpit, the trainee proceeds to "fly" the training device while sitting in a room with other such devices. The device responds to the pilot's actions in a manner which is quite similar to that in which the real airplane would respond in actual flight.

Such training devices are also models of airplanes. The elements of the real system which they incorporate are nevertheless very different from those included in the child's model. In a training device, no attention is paid to exterior detail and dimensions since the pilot cannot see outside. The interior instruments and dimensions are faithful duplicates of those of the real system, however, because the pilot is expected to carry over what is learned in the training device to the operation of the real airplane.

These two kinds of airplane model illustrate, in their similarity, the applicability of the layperson's view of a model as a "representation of something else" to the scientific concept of a model. Although they are different models of the same system, their similarity lies in their inclusion of important aspects of the real system and exclusion of unimportant aspects of the real system. The determination of which factors are important and which are unimportant clearly depends on the use to which the model is put. In the case of the model to be used to decorate a child's room, the exterior configuration, color, and markings are important, and the portions which are not visible are unimportant. In the case of the training device, the aesthetic value is insignificant, and the interior design and control responses are of utmost importance.

However, since both of these models involve changes in physical structure, their applicability to the more abstract scientific use of the term "model" is limited.

A model, in the scientific sense, is a representation of a system which is used to predict the effect of changes in certain aspects of the system on the performance of the system.

The applicability of the first part of this definition—"a representation of a system"—to the layperson's use of the term "model" has already been illustrated. The essential distinction lies in the next part—"used to predict the effect of changes in certain aspects of the system on the performance of the system." Clearly, most commonly known models are not used in this particular way.

In terms of the conceptual framework, a model is a set of relationships among the objectives, strategies, and states of nature. Looking back to the outcome arrays of Chapter 5, we see that this is precisely what an outcome array represents—a relationship among strategies, states of nature, and objectives (since it will be recalled that each outcome is described in terms related to objectives). Thus, the outcome array may be thought of as a model, although it is better to think of it as presenting the *results* of a model. In this case, the model is the set of calculational *predictive* relationships used to determine the array for a particular decision problem. This conceptualization reflects the time dimension inherent in most decision problems. In general, the point in time at which a strategy is selected by the decision maker precedes that at which the outcome is realized. This implies that at the time the alternatives are being evaluated, the outcome descriptors in the outcome array are predictions of future states which will result from the various strategy–state of nature interactions.

We have defined scientific models as representations of systems which may be used in predicting performance. If we recognize that the outcome descriptors which we must choose are of the system-performance variety, one possible use for models becomes apparent. We need to have outcome (performance) predictions in order to apply the conceptual framework, and models may be used as devices to predict performance. Hence, models may be used to generate the predicted states of affairs associated with each strategy–state of nature combination in the outcome array. This is one simple way of viewing the role of models in the conceptual framework—as devices for predicting the performance level associated with each outcome. These performance levels will normally involve both resource cost and benefit dimensions, and perhaps a number of each.

Models are widely varied in form. A profit and loss statement is a calculational model of a system which can be used for predictive purposes. A model may also be graphical in nature; for example, Weimer[2] uses the model shown in Figure 6-1 to depict the flow of cash through a business firm. Of course, such a model is limited in the predictions which can be made from it.[3] However, gross qualitative predictions can be made from the model more easily than they could be made without it. Moreover, such a model contributes to the ease with which one can comprehend the intricacies of cash flows through an organization.

The models used to represent systems for purposes of decision analysis are often symbolic in nature. Such models may take the form of mathematical equations, inequations, or other logical relations. Even in the simplest such instance, there is usually no single equation which constitutes the model; rather, it is usually made up of a number of components. For example, systems analysts frequently use *cost models*—devices for predicting costs—in conjunction with similar mathematical devices for predicting benefits—*benefit models*. Alternatively, a decision model might well be in the form of a flow diagram, a graph, or a physical model (such as the "toy" models used by industrial engineers to try out various locations for machines in industrial plants).

[2]A. M. Weimer, *Business Administration: An Introductory Management Approach*, Richard D. Irwin, Inc., Homewood, Ill., 1966, p. 426. Chart developed by Robert R. Milroy and Donald H. Soner. Reproduced with permission.

[3]Most such models are used in textbooks for teaching purposes.

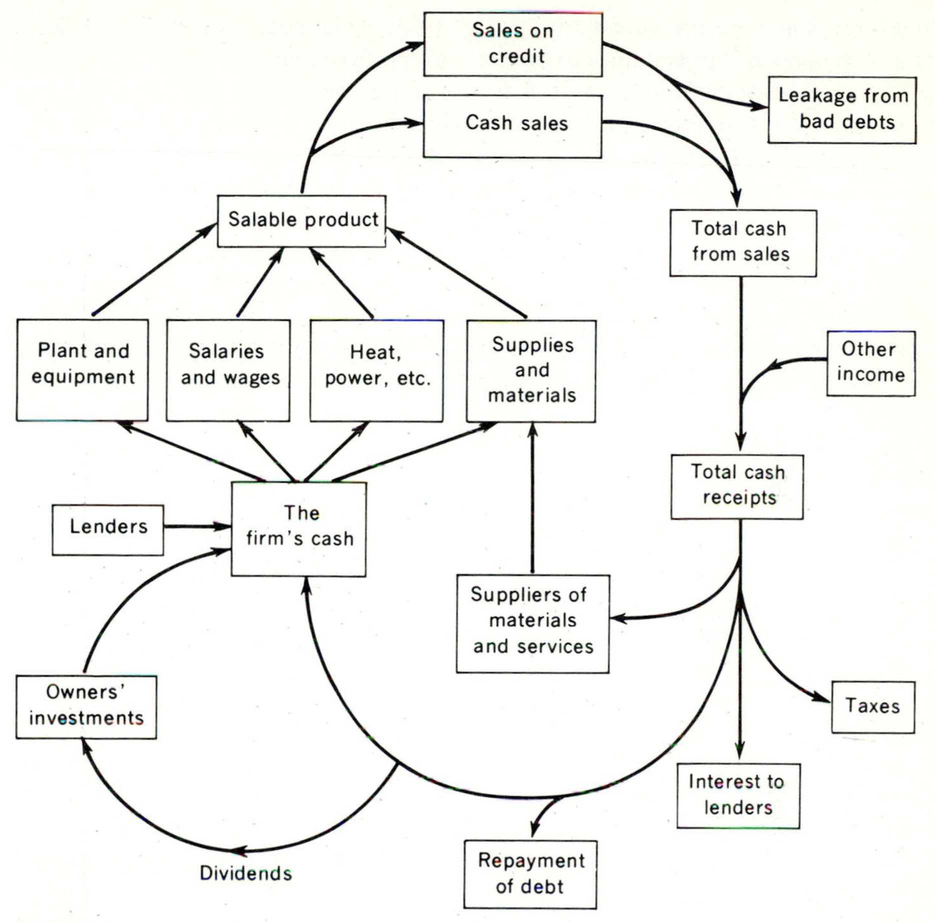

FIGURE 6-1
Pictorial model of flow of cash through a business firm.

Basically, *the value of a model lies in its substitutability for the real system for achieving an intended purpose.* The physical scientist might proceed toward the solution of a decision problem by experimentation, i.e., trying various combinations of controllables in a planned fashion and observing the results. In most problems of business and government, this procedure is either impossible or impractical. Who, for example, would propose that we simply "try" a number of new products? To do so, even on a restricted pilot-study basis, would obviously be too costly. Thus, the need in decision-problem analysis is for a representation of the system which can be used in place of the real system. A model is such a device. In using a model, the analyst makes assertions which express the relationship of various elements of the system with one another and, in turn, their effect on the performance of the system. In doing so, the analyst creates an entity—the model—which can be used in lieu of the actual system. He or she can then *experiment on the model* and on this basis make predictions of the effects which changes in the system will have on its performance.

A model airplane might be used in the same way. If we were to place the model in a wind tunnel and vary the angle of sweep of the wings in order to predict the effect of various angles on the performance of the airplane, we would have a representation of the airplane which was being experimented on. Hence, the model airplane would be a model in the scientific sense. The key distinction between the common, garden-variety model and a scientific model is *the use to which the model is put.*

In designing a new airplane, for example, Boeing engineers and designers might develop a cost model on the basis of experience with aircraft that have been developed in the past. The form of this model might be

$$TC = aW + bN + c$$

where TC is total cost, W is the weight of the aircraft, N is the passenger capacity, and a, b, and c are parameters which are estimated from past data. Of course, this is an overly simple illustration, but it shows the basic form of a statistically developed cost model.

In the illustration involving programs in health-depressed areas, one of the objectives was related to a reduction in the incidence of infant mortality. A little research reveals that there is a high correlation between the weight of a newborn baby and its chances of survival. This suggests that a program to provide expectant mothers with proper nutrition would result in their giving birth to larger babies, and hence would further the objective of reducing the incidence of infant mortality. This is a verbal model which is gross but useful. It is predictive in the sense that it spells out observable (and probably causal) relationships between alternatives and objectives.

To perform an analysis, one would need to quantify some of these relationships, but even this qualitative set of relations is useful. Such a gross model is deficient, of course, in some obvious ways. Yet, even with its deficiencies, it may be a useful predictive device, which should be supplemented by a systematic study before large-scale expenditures were recommended.

GENERAL COST-BENEFIT MODELS

The models which are most frequently used in systems analysis are those general economic models involving mixes of resources. This is the case because the nature of many strategic decisions is economic. However, we should point out that these models are not representations of all strategic planning problems. Rather, they represent approaches to particular classes of subproblems which are frequently important parts of overall strategic planning. In a subsequent section, we shall deal with the question of putting these pieces together to form a unified whole.

If we think back to an earlier discussion of strategies which may be composed of combinations of other strategies, we get the flavor of resource-mix situations.

For example, one resource to use in support of ground troops is artillery; another is missiles. However, some combination of both may be superior to either alone under some circumstances. In fact, some combination is likely to be best to achieve an overall capability for a wide variety of possible circumstances.

Generally, there are a number of submodels used in any analysis involving costs and benefits. Here, we shall illustrate a general economic model and method of approach which itself involves the use of a number of submodels. The context for the illustration will be a resource mix in a manufacturing enterprise.[4] An industrial firm can utilize various combinations of workers and machinery to produce goods, just as the military can use either artillery or missiles in support of ground troops. If we use "quantity of goods produced" (simply a physical count) as our measure of benefit, we can think of a number of strategies, ranging from virtually complete automation (with few human workers) to a situation involving much hand labor. In general, some combination of workers and machinery might be best, and it is the industrial manager's problem to determine which.

The two resources—workers and machines—interact with each other to produce benefits. Often they reinforce each other's effectiveness, and sometimes they detract from each other. To begin, let us think of a certainty situation in which each strategy (combination of workers and machines) is represented by a point on a set of axes. Figure 6-2 shows that the axes are scaled in terms of number of workers and number of machines. The point in the upper right involves the strategy "100 workers and 10 machines," and the assumed benefit is 1,100 output units of the product. This is indicated by the number attached to

[4]Parts of the following illustration are adapted from an excellent paper by R. N. Grosse, *An Introduction to Cost-Effectiveness Analysis,* Research Analysis Corporation, McLean, Va., RAC Paper RAC-P-5, July 1965.

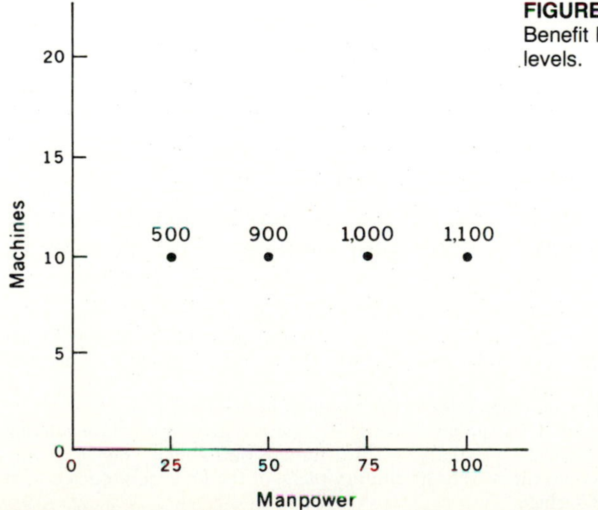

FIGURE 6-2
Benefit levels for four worker-machine levels.

the point. If the number of machines is held constant at 10 and fewer workers are used—say, 75, 50, or 25 workers—output would be less. This is indicated by the series of points showing outputs (benefit) of 1,000, 900, and 500 units, respectively, across the top of Figure 6-2. Thus, 10 machines in conjunction with 75 workers can produce 1,000 units of output, but 10 machines and 50 workers can produce only 900. And if 10 machines are used but the number of workers is reduced to 25, only 500 units of output will result (since presumably this few workers will not be able to operate the machines efficiently).

If we were to determine the output associated with each of the feasible points on the graph in Figure 6-2, we would be generating the benefit portion of the outcome description for each of the feasible strategies.[5] To do so would require a *benefit model*—a way of representing the production system which permits us to express benefit in terms of resources (workers and machines) and to predict the benefit levels that will result from various combinations of resources[6] In this case, the resource mixes are the controllables which make up the strategies.

One way of concisely depicting many benefit levels is through the use of equal-benefit lines.[7] Hypothetical lines of equal benefit are shown in Figure 6-3. There, benefit (output) of 900 units is shown to be attained for various combinations of labor and machinery. The line connecting all points for which the benefit level is 900 units is the 900-unit equal-benefit line. Other equal-benefit lines can be similarly interpreted. The four strategies shown in Figure 6-2 would be represented as points on their appropriate benefit lines in Figure 6-3.

To arrive at a mix decision, we need a *cost model* to complement this simple benefit model. In this case, we shall assume that the unit costs of the worker and machinery resources are known, so that lines of equal cost may be drawn on the worker-machinery graph as shown in Figure 6-4. These lines assume that each unit of labor costs $200 and that each machine costs $1,000. Thus, an expenditure of $10,000 on machines alone would result in 10 machines being used. This is indicated by the circled point at the upper left. A similar expenditure on labor alone would result in 50 workers being used, as represented by the circled point at the lower right. All other combinations of workers and machines which would result in a total expenditure of $10,000 lie on the $10,000 equal-cost line, all mixes involving a $15,000 total expenditure lie on the $15,000 equal-cost line, etc. In effect, these equal-cost lines are *budget lines.* A given budget level fixes the line, and the points on the line represent all possible mixes of resources which equal that level of total expenditures.

If we superimpose the benefit model's lines of equal benefit and the equal-cost lines, we obtain Figure 6-5. This information is useful for analysis,

[5]Note that the delineation of feasible strategies is rather simple in such cases. All those strategies involving a requirement for more workers or machines than it would be possible to add to the existing organization during the planning period might well serve to define the infeasible ones, for example.

[6]The economist refers to this particular submodel as the *production function.*

[7]In economics jargon, these are called "isoquants." In reality, the production function relating output to the two resource inputs could be plotted in a third dimension "above" Fig. 6-2. The isoquants would then be projections onto the worker-machinery plane of the lines of intersection of equal-output planes with the output surface.

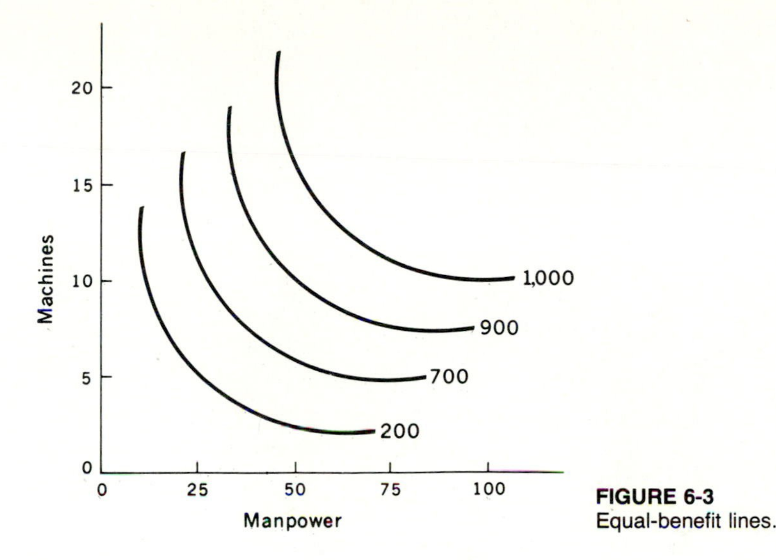

FIGURE 6-3
Equal-benefit lines.

since it enables us to answer such questions as: "What is the *best* mix of workers and machinery for a budget level of \$15,000?" Referring to Figure 6-5, we can see that the 700-unit equal-benefit line just barely touches (is tangent to) the \$15,000 equal-cost line. No equal-benefit line involving more than 700 output units touches this equal-cost line, but many such lines involving *less* than 700 output units do so. For example, the 200-unit equal-benefit line intersects the \$15,000 equal-cost line, as do all those equal-benefit lines between 200 and 700 which are not shown on the graph. Thus, the point at which the 700-unit equal-benefit line touches the \$15,000 equal-cost line gives the answer to our

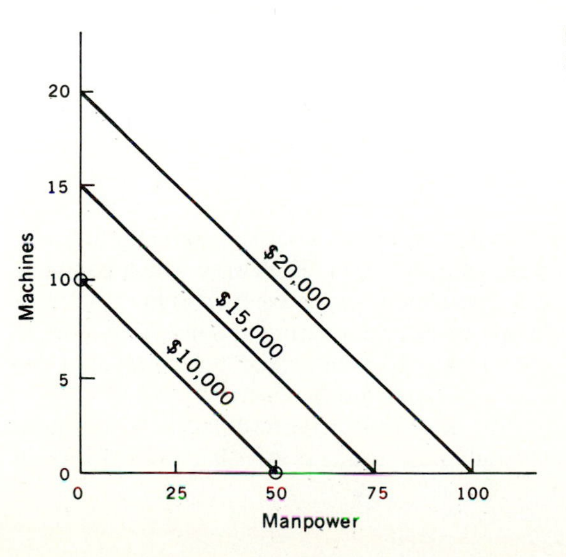

FIGURE 6-4
Equal-cost lines.

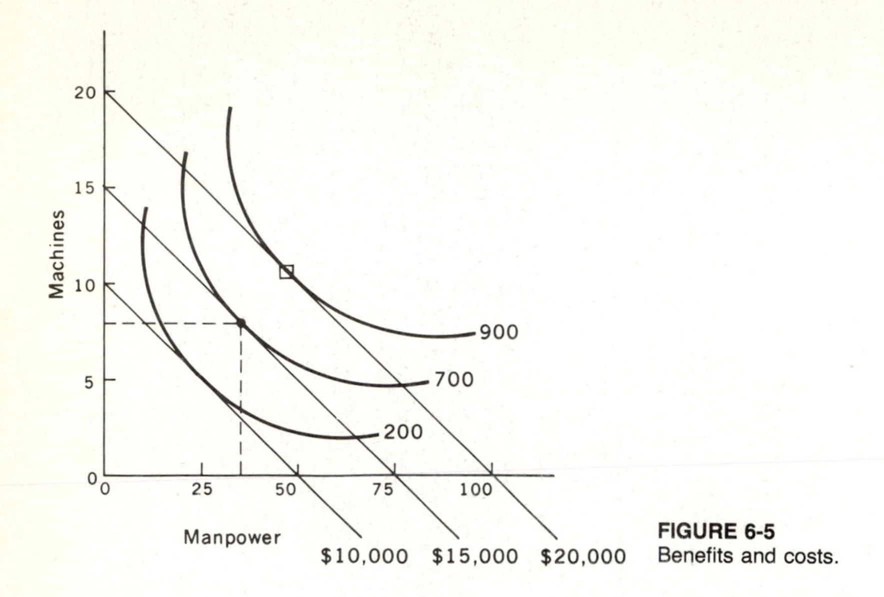

FIGURE 6-5
Benefits and costs.

question. This point is shown on the graph of Figure 6-5. The dashed lines to the worker and machinery axes indicate that the strategy which results in the greatest benefit at a $15,000 expenditure level is the one involving about 8 machines and 35 workers.

So, too, will the information provided by Figure 6-5 provide the answer to questions like: "What is the least-cost resource mix which will enable us to achieve an output of 700 units?" The answer is the same. At the resource mix indicated by the point of tangency of the 700-unit equal-benefit line and the $15,000 equal-cost line, 700 units of benefit are achieved at a cost of $15,000. This level of benefit cannot be achieved with any other mix at a lesser cost. This is so because the 700-unit equal-benefit line crosses only other equal-cost lines involving a greater expenditure than $15,000.

If we are not attempting either to maximize the benefit achieved for a fixed expenditure or to minimize the cost associated with achieving a required benefit level, we need further information to determine the best resource-mix strategy. The cost model provides one of the necessary elements of information. From Figure 6-5, we can determine the lowest-cost mix to achieve each benefit level, or the greatest benefit from a given expenditure level. The curve which connects these series of points is plotted (on a cost-benefit set of axes) as Figure 6-6. The point determined above (representing a strategy consisting of 8 machines and 35 workers) is the one which achieves the lowest cost for a fixed benefit level of 700 units. Hence, it appears on the curve of Figure 6-6 as shown.

This curve is a *total-cost curve* in the sense that it represents the cost-benefit combinations which the rational person would use, i.e., the least cost for each specified benefit level.

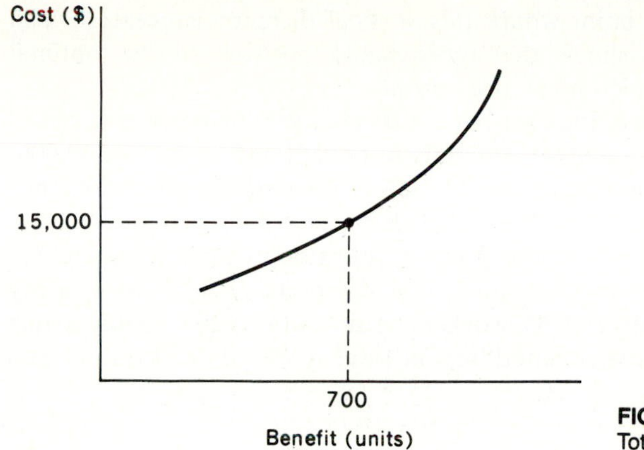

FIGURE 6-6
Total-cost curve.

The other necessary element of information to determine the best overall resource mix is the revenue which will be derived from each level of output. If the output is priced at $30 per unit and the firm's sole objective is to maximize profit, the best benefit level can be determined from Figure 6-7. There, the revenue function is plotted along with the total-cost curve. The revenue curve is simply $30 times the number of units of output (benefit)—a straight line. The vertical distance between the revenue and total-cost curves represents the organization's profit.

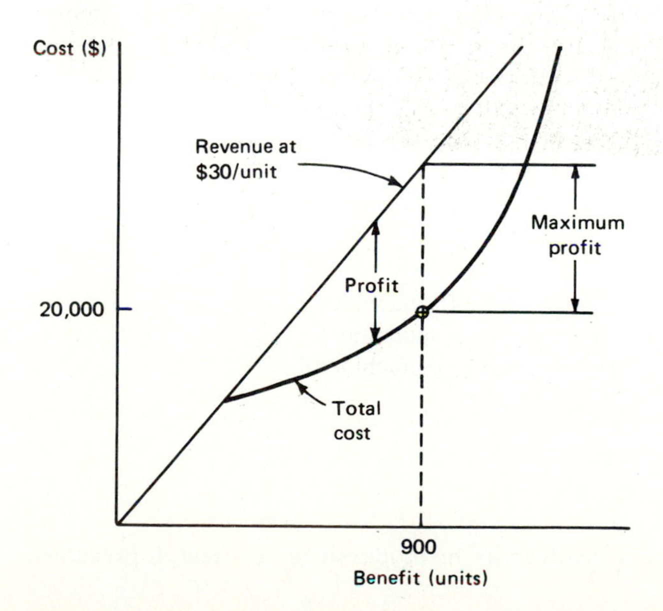

FIGURE 6-7
Total-cost and revenue curves.

Profit is maximized at a point where this vertical distance is greatest. The dashed line in Figure 6-7 shows the approximate position of the optimal benefit—the output level which maximizes profit.

The circled point represents the cost-benefit combination which is associated with the maximum profit. It involves 900 units of benefit and a cost of $20,000. Since this is a point on the total-cost curve, it is also one of the points of tangency between an equal-cost ($20,000) and equal-benefit (900) line in Figure 6-5. By referring back to Figure 6-5, where this point is indicated by a small square, we can determine the resource mix associated with this point. This is the strategy involving about 11 machines and 45 workers. Thus, this is the resource mix which maximizes profit. The associated benefit level is 900 units of output at a cost of $20,000.

We can also determine the optimal profit by recognizing that the product sells for $30 per unit. Hence, 900 units will produce $27,000 in revenue. Since the total cost is $20,000, the gross profit is $7,000.

If such a model were available, the profit-maximizing firm would have no difficulty in determining the best mix of resources to use. Of course, if the revenue curve of Figure 6-7 were unavailable, as it is in most social-policy decision contexts, this would not be the case. Since many meaningful problems in a variety of contexts—both governmental and industrial—have this characteristic, we shall discuss it extensively. First, however, we should place the general cost-benefit model into the context of the conceptual framework which we developed in the preceding chapter.

Interpreting the Model in Terms of the Conceptual Framework

The conceptual framework for systems analysis which was developed in Chapter 5 may appear to be quite different from the general cost-benefit model just described. In fact, however, the model is simply an implementation of the conceptual framework in terms of graphs rather than in tabular form.

Let us consider first the salient elements of the conceptual framework—strategies, states of nature, and outcomes. Each strategy is a particular combination of workers and machines. Thus, an infinity of available strategies might be enumerated as:

S_1: (50 workers, 15 machines)
S_2: (35 workers, 8 machines)
S_3: (45 workers, 11 machines)
.
.
.
etc.

Only one state of nature is involved in the model since, in effect, it presumes that each strategy leads with certainty to a single known outcome. This is made more clear when one considers the form of the outcome array as shown in Table 6-1. There, the state of nature N_0 is not explicitly defined, but the form of the table shows that the outcome array is equivalent to the graph of Figure 6-5; i.e., each point on the graph represents a particular outcome described in terms of cost (dollars) and benefit (units of output). The equal-cost and equal-benefit lines in Figure 6-5 have no significance other than the obvious one of facilitating interpretation of the general relationship between the strategies (resource mixes) and outcomes (described in terms of costs and benefits) which they permit. Also, of course, these lines enable us easily to recognize the strategies which lead to minimum cost for a fixed benefit level and to maximum benefit level for a fixed cost. In effect, by drawing equal-cost lines, we are defining a subset of all the possible outcomes which can be compared on the basis of the benefit outcome descriptor. In doing so, we are thinking in the terms described in Chapter 5 as "specified resource costs and variable benefits." Of course, we are applying this concept only to a subset of the possible outcomes rather than to all possible outcomes. This would be descriptive of a real-world situation if a fixed budget were to be spent, for example. In such a case, the best strategy (of the subset involving exactly the budgeted expenditure level) is the one which achieves maximum benefit for a fixed cost level.

The equal-benefit lines of Figure 6-5 have a similar significance. Each of these lines defines a subset of outcomes having specified benefit levels and variable costs. To choose the best strategy from among the subset, one merely chooses that which leads to an outcome having minimum cost for the fixed benefit level.

Of course, since a revenue curve of the form of that shown in Figure 6-7 is presumed to be known in this case, the array of outcomes—which in its entirety involves, in the language of the previous chapter, outcomes described in terms of variable resource costs and variable benefits—can be described using the benefit minus cost aggregate measure (profit), and the strategy can be chosen (from the entire range of possible outcomes) which maximizes profit. So, the selection of the profit-maximizing strategy in Figure 6-7 is conceptually equivalent to the choice of a strategy from an outcome array involving an aggregate cost-benefit

TABLE 6-1
OUTCOME ARRAY CORRESPONDING TO FIGURE 6-5

		N_0
S_1 (50 workers, 15 machines)	Cost =	$25,000
	Benefit =	1,000 units
S_2 (35 workers, 8 machines)	Cost =	$15,000
	Benefit =	700 units
S_3 (45 workers, 11 machines)	Cost =	$20,000
	Benefit =	900 units

measure as described in Chapter 5. Table 6-2 shows such an outcome array. Thus, the conceptual framework and the general cost-benefit model are indeed equivalent. The graphical form of the model as outlined in this chapter is merely a clear and efficient way of *implementing* the conceptual framework. Indeed, since there are an infinity of available strategies here, it would not be possible to apply the conceptual framework without some variety of graphical or mathematical simplification.[8]

OPERATIONAL SYSTEMS-ANALYSIS MODELS

Models such as the general cost-benefit models are sometimes directly applied to decision problems in business and government. Frequently, however, theory and practice are somewhat different because of the difficulties encountered in the real world. In this section we shall discuss some of these practical difficulties and their implications for systems analysis. In doing so, we shall be contrasting the real world with two things simultaneously—the conceptual framework of Chapter 5 and models such as the general cost-benefit model just discussed. In effect, these form the theory, and it is the practice on which we shall focus attention here. The difficulties in applying the conceptual framework, and the way in which analysts have surmounted them, form the practice of systems analysis.

Measuring Costs and Benefits

The first difficulty which arises in the application of a cost-benefit model involves measurement. In the general cost-benefit model, it was assumed that the unit costs of the resource inputs were known and fixed. Of course, this is not normally the case. Most frequently, the unit cost of the resource inputs must be estimated in advance—frequently in the absence of much "hard" information. For example, how is one to predict the costs of future transportation systems that exist only in the mind? Such costs incorporate costs of research and development, initial investment, operation and service and maintenance; thus, they are not simple entities that are easy to predict.

However, it is the measurement of benefit which causes the most practical difficulty. In determining costs, the measure is clearly defined in terms of dollars; in measuring benefit, this is frequently not the case. The benefit concept is directly related to the objectives—to what we are trying to achieve. Often it is difficult to determine measures of the degree of attainment of objectives. For

[8]In fact, this is the role played by many of the sophisticated mathematical techniques and algorithms which are commonly associated with systems analysis, and it is the explanation for the disclaimer which systems analysts frequently make that, although the use of mathematics often provides a useful and efficient way of solving problems, there is no intrinsic connection between sophisticated mathematics and systems analysis. If you have doubts about this, you should note that little or no mathematics has been introduced in this text.

TABLE 6-2
PROFIT OUTCOME ARRAY CORRESPONDING TO FIGURE 6-6

	N_0
S_1	Profit = $5,000
S_2	Profit = $6,000
S_3	Profit = $7,000
.	.
.	.
.	.

example, what if the morale of our troops is an important concept related to the outcomes of a decision problem—i.e., what if one of our objectives in a military decision problem has to do with morale? As Charles J. Hitch has put it:[9]

> How do you quantitatively distinguish between men who are highly motivated, and those who are demoralized? In fact, how do you quantitatively predict what it is that motivates or discourages a man? And which man? The fact that we simply cannot quantitize such things (and there are many other similar examples) does not mean that they have no effect on the outcome of a[n] . . . endeavor—it simply means that our analytical techniques cannot answer every question.

In making this evaluation, Hitch is restating the idea which was put forth in Chapter 4 concerning the role of systems analysis. There, we argued that objective analysis cannot treat all aspects of a problem and that it is complementary to, and not duplicative of, qualitative analysis. Here, in the context of measuring the benefits to be associated with the outcomes of a military decision problem, we find a good illustration of an aspect of a problem which cannot be easily handled quantitatively—morale. Hitch recognized this when he said, in the same speech, that this difficulty is

> . . . widely recognized, particularly by those . . . who have had to live with these realities. But does that mean that all analysis becomes meaningless? I think not. Every bit of the total problem which can be confidently analyzed removes one more bit of uncertainty from our process of making a choice.

The second aspect of benefit measurement which causes difficulty is the *comprehensiveness* of the measure. Often, it is possible to define a benefit concept and a measurable quantity which is related to this concept but is not a comprehensive indicator of it. Consider the concept of marketing performance, for example. If we wish to measure this, we might choose the rather comprehensive measure "profit" (which obviously encompasses both costs and benefits), as

[9]Charles J. Hitch, address before the U.S. Army Operations Research Symposium, Duke University, Durham, N.C., Mar. 26, 1962.

we did in the general model in the previous section. If the sole objective of the firm is to maximize profit, this may be adequate. However, suppose we are engaged in introductory sales of a new product. Is profit a comprehensive measure of the degree of attainment of corporate sales-performance objectives? The answer is probably no because marketing performance, in this context, means something more. One of the additional things that it means has to do with expectations; i.e., how well was the product expected to do during its introductory period? Most frequently, new products are not expected to be immediately profitable, and a small loss might well have positive implications. Also, there is the market-penetration aspect of performance. What share of the market has been attained? What proportion of the total consuming public has tried the product at least once? What proportion has purchased it more than once? Then, there is the trend aspect of performance, which is particularly important with new products and which is not measured by profit. A market in which the trend of sales is highly positive is better than one in which trends are stable or negative, even though the total profit may be the same. Another aspect is the time dimension. Is the profit to be maximized in the short run or the long run; e.g., would present profits be sacrificed in the hope of reaping future profit rewards? Thus, while profit is presumably a good cost-benefit measure for the short-run-profit-maximizing firm, few firms have such simple objectives. Most have objectives related to overall marketing performance, but for most there is no simple numerical measure which comprehensively describes the degree of attainment of this concept.

A similar difficulty in benefit measurement involves both the comprehensiveness of the measure and the validity of a predictive model involving the measure. Consider, for example, that we are using a resource-mix model of the kind described previously to determine the best mix of fixed-wing aircraft and helicopters, both of which are to be used for observational support of ground troops. What benefit measure can be used? The basic question is: "What are our objectives?" To answer this requires a definition of "observational support of ground troops" and a further question: "What is the objective of the military operation?"

In these days of limited war, this question is difficult enough to answer. Is our objective to win the war or to gain territory from the enemy? Let us suppose that the objective is the latter—to gain ground—since it is easily measurable. Now, if we wish to support ground troops, this should be our objective in the observational aircraft problem too. But how is it possible to relate the number of helicopters and airplanes to ground gained, i.e., to develop a model which will enable us to predict the ground gained as a result of various mixes of the two varieties of observational aircraft? Of course, it is very difficult to do so, and because of our lack of understanding of the relationships involved, we would probably choose a less comprehensive but more easily predicted benefit measure, say, the number of satisfactory missions flown. This could be used as the benefit measure on a graph such as that in Figure 6-3, and we could perform an analysis such as was described for the general cost-benefit model.

In such instances, the benefit measures used are usually directionally related to the benefit concept in the objectives; e.g., we believe that more missions are directly related to more ground gained, or at least that more missions do not cause less ground to be gained.

As another example, suppose we wish to develop a military establishment which provides maximum deterrence. What is deterrence? In fact, it is something that exists only in the minds of our potential adversaries, so the systems analyst will choose a benefit measure which is measurable, predictable, and directionally related to deterrence, such as the number of mortalities inflicted or the number of buildings destroyed.[10] In doing so, the analyst is optimizing with respect to a *proxy* benefit measure—one used as a substitute for the benefit concept which defines our objectives.

Incommensurate Benefit Measures

Incommensurate benefit measures are measures which have different dimensions. Thus, if we are setting the federal budget and we must compare a program involving aid to health-depressed areas with a military expenditure for helicopters, how do we compare the benefits? Suppose we have thoroughly analyzed both problems independently and have used "number of deaths averted" as a benefit measure in the one case and "number of successful missions flown" in the other? How do we compare the two?

Of course, you may question the need to make the comparison at this level. However, the same problem exists at much lower levels. In the military case, suppose one analyst applied the general cost-benefit model to the mix of helicopters and fixed-wing aircraft and used "number of successful missions flown" as the benefit measure, while another analyst modeled the mix of artillery versus missiles using another benefit measure. How could these two be compared in trying to analyze strategic problems of force composition? As you might surmise, there is no simple answer to such a question. The answer revolves about our knowledge of a trade-off between the two benefit measures; i.e., how much is one unit of one measure worth in terms of the other? As we have illustrated in Chapter 5, such trade-offs are difficult to determine in most real problem situations.

However, such difficulties do not prevent us from performing a systems analysis or even from arriving at a definite conclusion from the analysis. To illustrate another of the many ways in which you may use common sense and judgment to complement formal analysis, consider a situation involving the comparison of benefits from strategic and tactical weapons. General cost-benefit curves could be developed for strategic weapons and tactical weapons, but the

[10]Deterrence is such a complex concept that it may well be argued that these benefit measures are not even directionally related to it, i.e., that an ability to inflict more damage does not necessarily involve increased deterrence. Also, even if a casualty-producing capability is directionally related to deterrence, it is not our assessment to which it is related but, rather, our enemy's; i.e., how many casualties does the *enemy* think we can produce?

benefit (military effectiveness) measures are incommensurate. Suppose that these curves were those shown in Figure 6-8*a* and *b*. Although the analyst is unable to compare the benefit units symbolized as B_1 for strategic weapons with those symbolized as B_2 for tactical weapons, other useful information can be obtained to help the decision maker. For instance, the analyst could try to determine the *current operating level* on both graphs.

Suppose that the current operating levels are those described by the circled points and dashed lines in Figure 6-8*a* and *b*. What would this tell the decision maker? Rigorously, it would tell little, since the benefit measures are incommensurate and he or she does not rigorously know how to compare units of B_1 with units of B_2, but practically speaking, the decision maker's intuitive grasp of the relative worth of these two varieties of weapon might lead to the reasoning that (1) the current operating point in strategic weapons indicates that additional expenditures would buy little additional benefit B_1 and (2) the current operating point in tactical weapons indicates that additional expenditures would buy relatively large increments of B_2. This is so because as more money is spent for strategic weapons, we move up the curve of Figure 6-8*a*. Since we are currently operating on a steep portion of that curve, little additional benefit is gained. Consider, for example, the point marked with a cross in the figure. It involves only slightly more than benefit B_1 at x dollars greater cost than the current operating situation does—as indicated by the circled point. Similarly, a small move down the strategic-weapons curve involves only slightly less benefit at significantly less cost.

On the tactical-weapons curve of Figure 6-8*b*, the situation is quite different. An expenditure of x dollars more above the current operating level will involve a move up the curve to the point indicated by the cross. Significantly greater benefit B_2 is obtained at that point than at the current operating level.

Thus, although such an analysis would not provide a decision maker with a rigorously determined best solution, it would give a sound basis for using a

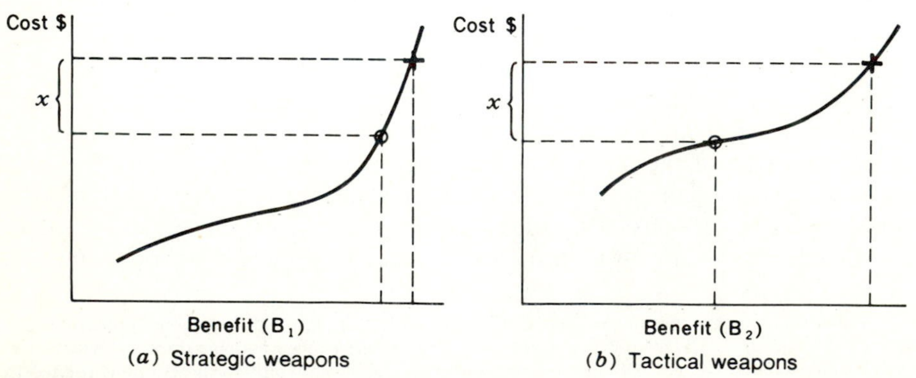

FIGURE 6-8
Cost-benefit curves for strategic and tactical weapons.

"feel" for the situation, and the decision maker might well choose to spend additional moneys for tactical weapons.

Suboptimization

Suboptimization is both a difficulty associated with systems analysis and a method of attacking real-world problems in a way which integrates the conceptual framework with the practicalities of the real world. *Suboptimization* means, in simple terms, *choosing an alternative to be "best" in the context of a subsystem of the total system.*

In business and government, the natural tendency of the manager to focus attention on a particular specialized function, rather than on the complex system composed of the business organization and its environment, has often led to an avoidance of true system solutions. Even though problems which involve complex systems are usually difficult to solve and, in fact, often difficult to formulate in understandable form, they are nonetheless the truly important problems which organizations face. The enterprise will not progress if actions are constantly taken which enhance the performance of one department at the expense of another. Such actions are called *suboptimum.* It is the desire of the systems-analysis approach to develop optimum solutions to problems—those which are best, or at least good, for the organization as a whole, rather than for a single portion of it. The systems approach to management is therefore incompatible with the idea of operating a functional unit just as you would operate the family grocery store, for there may be occasions when a small decrease in the apparent performance of one functional unit results in a large increase in the performance of the total organization.

In the production-marketing context, for example, the production of a wide variety of products with high machine teardown and setup costs between each product may result in increased total sales, since no customers who desire quick delivery will be turned away because of low inventories of the product they desire. These increased sales may more than offset the increased production setup costs, resulting in a higher profit for the enterprise. In this case, the apparent high costs incurred by the production department are not at all indicative of poor performance, but rather are indicative of the systems approach to the solution of decision problems.

In public decisions, when you choose, for example, the highway-safety program which is predicted to save the greatest number of lives at a given cost, you are suboptimizing in the sense that other ways to save lives (such as by providing nutrition to pregnant women) are not considered and compared. If the objective is simply the saving of lives, government expenditures in many programs in all departments are presumably aimed at the same end, and the federal government decision system involves both highway-safety and health programs.

So, too, when a military planner determines the best mix of helicopters and fixed-wing aircraft and another independently determines the best mix of artillery and missiles, neither has answered the question of whether it would be best to forgo some artillery pieces in order to buy additional helicopters. Thus suboptimization is first an analytic "error" to be avoided by the systems analyst, for, indeed, the word "systems" implies an overall, and not a piecemeal, approach to decision problems.

To develop optimum solutions to managerial problems, it is necessary to view the organization in as large a context as possible. In solving a marketing problem, for example, you must consider the effect on production, finance, quality control, and other relevant organizational entities, while simultaneously weighing the relative influence of competitors, the government, suppliers, and other elements of the system. To do this in its entirety, of course, might require infinite analytic resources. In practice, you account for limited analytic capability by abstracting out much of the real-world complexity of the problem and considering systems which may involve less than the total organization. At all times, however, *the effect of making these approximations should be evaluated.* In effect, the scope and complexity of many problems, the limited analytic resources which are usually available, and the demand for results and not just studies often force the systems analyst to begin analysis at a low level and to build toward the solution of the strategic problem. In doing so, the analyst finds that the answers to all the strategic questions are not simply the aggregate of the answers to smaller questions. What is best for the production department, the sales department, and the accounting department do not necessarily combine to constitute what is best for the organization as a whole. In fact, these suboptimum answers are frequently incompatible with one another.

Thus, a prudent systems analyst utilizes suboptimization because many real systems are too complex to be viewed in the whole. Yet, when putting together suboptimum solutions, the analyst recognizes that the best alternative for the organization as a whole may not be simply the sum of the alternatives chosen at each stage. He or she recognizes that while it may not be possible to assess these interactions with perfect accuracy, it is necessary at least to consider their implications. The difficulties involved in doing this should not be discounted. Quade illustrates these difficulties as follows:[11]

> Suppose a family has decided to buy a television set. Not only is their objective fairly clear, but, if they have paid due attention to the advertisements, their alternatives are well-defined. The situation is then one for cost-effectiveness analysis, narrowly defined. The only significant questions the family need answer concern the differences among the available sets in both performance and cost. With a little care, making proper allowance for financing, depreciation, and maintenance, they can estimate, say, the five year procurement and operating cost of any particular set and do so with a feeling that they are well inside the ball park. They will discover, of course, that finding a standard for measuring the performance of the various sets is somewhat more difficult. For one thing, the problem is multidimensional—they must consider color quality, the option for remote control, portability, screen size, and so forth. But,

[11]E. S. Quade, *Cost-Effectiveness: An Introduction and Overview,* RAND Corporation, P-3134, Santa Monica, Calif., May 1965, pp. 2–3.

ordinarily, one consideration—perhaps color—dominates. On this basis, they can go look at some color sets, compare costs against color quality, and finally determine a best buy.

Now suppose the family finds they have more money to spend and thus decide to increase their standard of living—a decision similar to one to strengthen the U.S. defense posture by increasing the military budget. This is a situation for a broader analysis. They first need to investigate their goals or objectives and look into the full range of alternatives—a third car, a piano, a country club membership. They then need to find ways to measure the effectiveness of these alternatives and establish criteria for choice among them. Here, because the alternatives are so dissimilar, determining what they want to do is the major problem; how to do it and how to determine what it costs may become a comparatively minor one.

However, the analyst who is able to determine these interactions by specifying overall objectives, determining satisfactory ways of measuring joint performance, etc., is better able to develop an optimal solution. Of course, such analysts are seldom certain that they can do so, but they do know that a bad solution has not been used because of a failure to consider the effect of trade-offs and interdependencies between various subproblems.

Certainty versus Uncertainty

In each of the illustrations of this chapter, we have dwelled on decision problems formulated as involving certainty—one state of nature or set of environmental conditions. In practice, most real-world problems involve great uncertainty as to the state of nature which will occur. By analyzing a decision problem under certainty, we are again reducing the scope of the problem so that it can be managed in the time and with the resources allocated to analysis. In studying the certainty case, the analyst learns about the problem to a degree which is usually impractical without the searching questioning of objectives, alternatives, costs, and benefits that is an intrinsic part of system analysis. Having done so, the analyst is better prepared to analyze a more realistic problem model involving more than one state of nature.

Thus, in this context we view the problem-solving process as sequential in nature. First, the problem is formulated as though a specified state of nature were certain to occur. Then, after the system is better understood and some tentative results are achieved in the certainty case, the analysis is expanded to include other states of nature which may occur. In military planning in the Department of Defense, analysts might assume a given state of nature: "Our enemy will be country X; the war will be of the guerrilla variety; it will be fought in the desert of North Africa; only conventional weapons will be used by our enemy; etc." A force structure would then be developed which would best meet this contingency. In answering this restricted question, the analyst will have asked the right questions; e.g., "What controllables significantly affect the battle's outcome?" and "How are they related to one another?" Then, when these questions are answered and the best force structure determined for this contingency, the analyst is better able to ask the right questions about another enemy fighting in another part of the world with different weapons, etc.

Of course, as was illustrated in the discussion of uncertainty in the previous chapter, the determination of a best overall strategy is not a direct extension of our knowledge concerning the best strategy for each state of nature. In general, the strategy which leads to the best outcome for one state of nature will not necessarily lead to the best outcome for another. Hence, problems under uncertainty, as we discussed in the previous chapter, are complex.

This is particularly true in the context of military force-structure decisions. It is relatively easy to design a force structure which would be best to fight a particular enemy in a particular location, but it is much more difficult to design a force structure to meet a wide variety of possible contingencies, say, war in the jungles of Asia, the deserts of North Africa, or the polar region. Similarly, it is relatively easy to determine the best structure if our competitors proceed with a given course of action, but it is rather difficult to determine a structure which is good for a wide range of competitive actions.[12]

Values and Limitations

In discussing operational systems-analysis models, we have emphasized the discrepancies between theory and practice. We would be less than fair to systems analysis if we did not point out that the difficulties are in the problems—not in the analysis. If we look at each of the areas in which some practical limitations need to be placed on the conceptual framework and general economic cost-benefit models, we quickly realize that the same difficulties are inherent in any other approach to solving problems. Indeed, these difficulties are characteristic of the problem rather than of the analysis, and *any* other approach, whether analytic or not, either encounters the same difficulties or unknowingly avoids them altogether.

One of the great virtues of the systems-analysis approach has to do with the question of avoidance. Most, if not all, alternative problem-solving approaches "assume away" or ignore the difficulties with which systems analysis explicitly grapples, however unsuccessfully. It is a matter of reasoned faith to the scientist that it is better to have considered an insurmountable problem than to have completely neglected it. This is exactly what systems analysis does—grapples with the imponderables—rather than providing answers which either are not really answers at all or are answers to the wrong questions. If, by using a systems-analysis approach, the decision maker is led to ask the right questions about the salient features of the problem, he or she has gotten the greatest benefit from analysis, even though it has provided no quick answers. If at the same time some insight is obtained into the determination of the strategy which is best in the real world, the decision maker is doubly blessed.

[12]We do not mean to imply that such situations cannot be handled; some of the ways in which they can be approached are discussed in Chap. 5 under "Decisions Under Uncertainty."

However, all these arguments are theoretical. The really significant issues dealing with the values and limitations of systems analysis revolve around its practical use. Alice Rivlin has dealt directly with these issues in terms of four critical kinds of questions which a systems-oriented decision maker dealing with social programs would like to have answered.[13]

1 How do we define the problems and how are they [the problems][14] distributed? Who is poor or sick or inadequately educated?

2 Who would be helped by specific social action programs, and how much?

3 What would do the most good? How do the benefits of different kinds of programs compare?

4 How can particular kinds of social services be produced most effectively?

In reviewing progress in these four areas within the context of social action programs, Rivlin concluded that systems analysis had made great strides in developing answers to the first two questions. For instance, systems analysis debunked the then popular myth that most of the poor were black women with many children who lived in the city. Most, in fact, were white, and over half of all poor families had male heads.[15]

Systems analysis also has served to provide new insights into what is wrong with the welfare system and has provided new alternatives, supported by cost and benefit data, for consideration. These cost and benefit data are in terms of the various clienteles (interest groups) who have a stake in the welfare system. Thus, the question of "who wins and who loses," which is so much a systems consideration, is paramount in evaluating alternative programs of this variety.

With regard to the latter two questions, Rivlin concluded that "little progress has been made in comparing the benefits of different social action programs," and that "little is known about how to produce more effective health education and other social services."[16]

We have already discussed the former problem—that of incommensurate benefit measures. However, in social programs you face an additional complication beyond that faced in the context in which we described the problem. That additional consideration is the systems recognition that *different programs benefit different clientele groups.* Therefore, cross-program comparisons are difficult not only because the natural benefit measures are different in each, but also because *the benefits accrue to different people.*

Of course, systems analysts can, and do, play the role of indicating the consequences, in cost-benefit terms, of the various strategic alternatives which are available. In doing so, they act as a part of the decision maker's information system and not as a decision maker. Although their search for "good" answers often puts them in the position of being viewed as decision makers, this role as

[13]Alice M. Rivlin, *Systematic Thinking for Social Action,* The Brookings Institution, Washington, D.C., 1971, pp. 6–7.

[14]Authors' insertion.

[15]Ibid., p. 12.

[16]Ibid., p. 7.

an element of the decision maker's information system is, in fact, the proper one. Whether the systems analyst recommends a particular course of action or simply points out the consequences of the alternatives which are available, the analyst's role always should be that of supplying information to provide a basis for better-informed decision making by those who have the ultimate responsibility for decisions.

With regard to the question of producing better social services, the negative conclusion of Rivlin is based on the fact that many government programs have not been set up in a manner which facilitates the evaluation of their effectiveness. Thus, programs have been implemented on the basis of predicted effects, without the incorporation of methodologies for comparing actual and predicted results. Although this is a serious problem, it is now well recognized and is being corrected through the incorporation of specific evaluative mechanisms in government programs. We shall have more to say about this point in Part Four, where we deal with the *control* aspect of projects and programs.

The Emerging "Modeling Support" Technology

Despite the many difficulties that exist in modeling ill-structured problems, there has emerged a computer-based technology that makes the work of the modelers easier. It does not address these basic structural problems, but it takes some of the difficulty out of the process of developing, testing, and using a model.

The emerging technology is represented by a wide variety of commercially available high-level computer languages.[17] They are referred to in a number of different ways, such as "user friendly" languages, modeling languages, and financial planning languages.[18] However, the "user friendly" description is probably most descriptive.

These languages permit an individual to develop, test, and use a formal model while sitting at a computer terminal and interacting with a computer language that is relatively nonprocedural. People who are not familiar with the intricacies of computer languages such as FORTRAN and COBOL can readily use these languages.[19]

As perceived by the modeler who may not be a proficient computer programmer, these languages require much less knowledge of the "rules," are much more forgiving of error, and do not require extensive formatting of outputs.

Therefore, the modeler can build a model within the computer rather than first doing it outside and then relying on a computer programmer to interpret it

[17]Such as IFPS, SIMPLAN, and EMPIRE. See R. Klein, "An Examination of the Corporate Use of Computer-Based Financial Modeling," *Journal of Systems Management,* June 1982.

[18]Sometimes they are (inaccurately) referred to as "decision support systems." See Chapter 16 and R. H. Bonczek, "The Evolving Roles of Models in Decision Support Systems," *Decision Sciences,* vol. 4, no. 2, 1980.

[19]E. B. McClean, and G. L. Neale, "Computer-Based Models Come of Age," *Harvard Business Review,* July–August 1980.

and to program it correctly. The modeler can experiment with different versions of a model and, in sophisticated systems, can access existing data bases in order to exercise the model.

While these languages do not make modeling itself any easier, they do serve to minimize the burdens associated with the mechanics of modeling. As such, they are positive motivators of more and better systems analyses.

SUMMARY

In this chapter, we have tried to emphasize that the practice of systems analysis is neither the dogmatic application of a set of rules to a situation which may not be susceptible to rules nor the ceding of decision-making authority to a mystical set of mathematical equations or to a computer.

At this stage of development, the practice of systems analysis is largely an art. Indeed, the foundations of the art evolve from basic scientific and logical precepts, and the conceptual framework is a rigorous basis for analysis, but in practice, human judgment and intuition play an overwhelmingly significant role in the decision itself, in the analysis which is made of the decision problem, and in the decisions involved in "deciding how to decide," i.e., in constructing the analysis, the measures to be used, etc.

Here, we have focused attention on the role of models in systems analysis and on the practicalities of the real-world application of models. As we have shown, explicit systems-analysis models—whether mathematical, graphical, or physical —are not really much different from the mental models which everyone constructs in solving any problem. The primary difference is that systems-analysis models are explicit, and thus they can be manipulated more easily and constructed to be a more comprehensive description of the real world than the subjective models which most people use to solve problems.

DISCUSSION QUESTIONS

1 An interrelationship exists between all elements and constituents of a society. What is the significance of this statement with respect to the methods and process of systems analysis?
2 Models are used in carrying out systems analysis. Both quantitative and qualitative models can be used. Give some examples. What are some of the models that we use implicitly in decision problems in our personal lives?
3 The most useful models are the simple ones. Defend or refute this statement. The value of a model lies in its relationship to the real world. Do you agree?
4 The traditional organizational chart is both useful and not useful as a model for organizational analysis. For what purposes is it useful and for what purposes is it of limited use?
5 The models which are frequently used in systems analysis are those general economic models involving mixes of resources. Give some examples.
6 What are the salient elements of the conceptual framework of systems analysis? Give some examples.

7 General cost-benefit models are not always applied to decision problems in the real world. Why is this so?
8 What are commensurate benefit measures? Give some examples.
9 What is meant by suboptimization? Which managers in a business firm might have the greatest predisposition toward suboptimization in their decision process? What can be done to reduce the probability of this occurring?
10 It is better to have considered an insurmountable problem than to have completely neglected it. What is meant by this statement? Give some examples of real-world situations where this statement is appropriate.
11 What role do the personal values of the decision maker play in the systems-analysis process?
12 One of the fears that a manager has who is not conversant with systems-analysis methodology is the relegation of the decisions to a computer, to a mathematical model, or to a specialized staff analyst. A general manager might claim that the analyst has no real appreciation of the general nature of the forces which the manager must consider in making decisions. Do you agree? What can be done to reassure the manager in this regard?
13 Systems analysis is as much art as science. Would you agree with this statement? Why or why not?
14 Within the spirit of systems analysis, most of the affairs of modern organizations can be counted, measured, and evaluated. Would you agree? Is this a useful attitude for the general manager to have?
15 In the final analysis, nothing replaces the intuition, judgment, and experience of the decision maker. Do you agree with this statement? If so, then it follows that the successful decision maker has no need for formal systems analysis. Do you agree?

RECOMMENDED READINGS

Ackoff, R. L.: *The Art of Problem Solving,* John Wiley & Sons, Inc., New York, 1978.

Argyris, Chris: "How Tomorrow's Executives Will Make Decisions," *Think,* November–December, 1967.

Asnes, M., and A. King: "Prototying: A Low Risk Approach to Developing Complex Systems," *Business Quarterly,* vol. 46, August 1981.

Bell, D. E.; R. L. Keeney, and H. Raiffa (eds.): *Conflicting Objectives in Decisions,* John Wiley & Sons, Inc., New York, 1977.

Beltrami, E. J.: *Models for Public Systems Analysis,* Academic Press, Inc., New York, 1977.

Bender, E. A.: *An Introduction to Mathematical Modeling,* Wiley-Interscience, New York, 1978.

Bennett, P. G., and M. R. Dando: "Complex Strategic Analysis; a Hyper-game Study of the Fall of France," *Journal of the Operational Research Society,* vol. 30, 1979.

Braun, T. H.: "The History, Evolution and Future of Financial Planning Languages," *ICP Interface,* spring 1980.

Checkland, P. B.: "Toward a Systems Based Approach for Real World Problem Solving," *Journal of Systems Engineering,* vol. 3, 1972.

Eckstein, O.: "Decision Support Systems for Corporate Planning," Data Resources Review, February 1981.

Green, M., and N. Waitzman: "Cost Analysis Needs Analyzing," *The New York Times,* February 8, 1981, p. F3.

Grosse, R. N.: *An Introduction to Cost-Effectiveness Analysis,* Research Analysis Corporation, McLean, Va. RAC Paper RAC-P-5, July 1965.
Keen, P. G. W.: "Information Systems and Organizational Change," *Communications of the ACM,* vol. 24, no. 1, 1981.
——— and M. S. Scott Morton: *Decision Support Systems: An Organizational Perspective,* Addison-Wesley Publishing Company, Inc., Reading, Mass., 1978.
——— and G. R. Wagner: "DSS: An Executive Mind Support System," *Datamation,* November 1979.
Klein, R.: "An Examination of the Corporate Use of Computer-Based Financial Modeling," *Journal of Systems Management,* June 1982.
Lave, C. A., and J. G. March: *An Introduction to Models in the Social Sciences,* Harper & Row, New York, 1975.
Lyneis, James M.: *Corporate Planning and Policy Design,* The M.I.T. Press, Cambridge, Mass., 1980.
Naylor, T. H.: *Corporate Planning Models,* Addison-Wesley Publishing Company, Inc., Reading, Mass., 1979.
Pidd, M.: "Systems Approaches and OR, " *European Journal of Operations Research,* vol. 3, 1979.
Quade, E. S.: *Cost-Effectiveness: An Introduction and Overview,* RAND Corporation, P-3134, Santa Monica, Calif, May 1965.
———: *Analysis for Public Decisions.* 2nd ed., North Holland, New York, 1982.
Radford, K. J.: "Decision Making in a Turbulent Environment," *Operational Research Quarterly,* vol. 29, July 1978.
Rivlin, Alice M.: *Systematic Thinking for Social Action,* The Brookings Institution, Washington, D.C., 1971.
Sprague, R. H., and R. L. Olson: "The Financial Planning System at Louisiana National Bank," *MIS Quarterly,* 1979, vol. 3, no. 3, pp. 35–46.

CASE 6-1: Service to the Nation

In the early 1970s, the United States went to an "all volunteer force" (AVF) military personnel procurement system after having long relied on a draft to ensure that military forces would be adequately manned.

A study of the volunteer military[1] summarized the following "National Goals" of the United States as prescribed in 1960 by the President's Commission on National Goals:

> *The Individual* The status of the individual must remain our primary concern. All our institutions—political, social, and economic—must further enhance the dignity of the citizen, promote the maximum development of his capabilities, stimulate their responsible exercise, and widen the range and effectiveness of opportunities for individual choice.
>
> *Equality* Every man and woman must have equal rights before the law, and an

[1]W. R. King, "Achieving America's Goals: National Service or the All Volunteer Armed Force?" Study prepared for Committee on Armed Services, U.S. Senate, U.S. Government Printing Office, Washington, D.C., February 1977.

> equal opportunity to vote and hold office, to be educated, to get a job and to be promoted when qualified, to buy a home, to participate fully in community affairs.
>
> *The Democratic Process* The degree of effective liberty available to its people should be the ultimate test of any nation. Democracy is the only means so far devised by which a nation can meet this test. To preserve and perfect the democratic process in the United States is therefore a primary goal in this as in every decade.
>
> *Education* The development of the individual and the nation demands that education at every level and in every discipline be strengthened and its effectiveness enhanced. This is at once an investment in the individual, in the democratic process, in the growth of the economy, and in the stature of the United States.
>
> The basic foreign policy goal of the United States should be the preservation of its independence and free institutions.[2]

The AVF study argued that the AVF, the draft, and other alternatives, such as national service, should be evaluated in terms of how well they serve to further these goals.

Your team should consider these three alternatives: the draft, the AVF, and a "national service" program under which all youth would be required to devote at least a year to some form of service to the nation—either military or civilian depending on individual preferences. Outline how each alternative contributes to national goals and how this degree of contribution might be measured. Evaluate and compare the three alternatives in terms such as those described in this chapter and the preceding ones.

[2]Excerpted from "Report of the President's Commission on National Goals," November 1960.

CHAPTER 7

PLANNING AND SYSTEMS ANALYSIS IN ORGANIZATIONS

Striking a balance between formal planning tools and creative thinking may be the answer to the strategic planning riddle.[1]

Systems analysis is a planning "decision aiding" or problem-solving approach that is in wide use in a variety of organizations. As noted in Chapter 4, it is somewhat unfortunate that it operates under different names in various contexts. In this chapter, we review some of the "organizational manifestations" of systems analysis.

Before doing so, however, it is useful to introduce some of the basic concepts of planning at a more detailed level than has been done previously. These concepts are more mundane than are the ideas of strategic systems planning discussed in Chapter 2, but they are essential to developing a full understanding of the role of systems analysis in planning.

BASIC PLANNING CONCEPTS

There are a variety of planning concepts that serve to define the variety of planning that an organization performs. These may be thought of in terms of the

[1]Harold W. Fox, "The Frontiers of Strategic Planning: Intuition or Formal Models," *Management Review,* April 1981, pp. 8–12.

substance of planning—the variety of plan that is produced—the time dimension of planning, and the values to be anticipated from planning. One notion—that of the "program" plan—will also be given attention, since it is basic to planning in some organizations.

A Classic Taxonomy of Plans

There is a wide variation in the planning processes used in various organizations and by managers at various levels within organizations. One useful way of gaining an appreciation of the magnitude and variety of the planning actions is to review the various types of plans that may be a product of the planning process.

Objectives Objectives are the end results to which the organizational activity is directed. The decision involved in setting necessary objectives is a fundamental responsibility of top-level management. Objectives are hierarchical in nature in the sense that the general overall objectives are formulated by top management after economic, social, legal, technological, and political forces affecting the company have been appraised. These overall objectives provide guidelines for subobjectives throughout the organization and become the standard against which progress can be measured.

Policies Policies are general statements of intended behavior of the organization; they provide guidance for thinking and decision making within the framework of existing or anticipated resources. Policies tend to limit the scope within which decisions must be made and to ensure that necessary decisions will contribute to the accomplishment of overall objectives.

Budgets Budgeting is the process of defining anticipated circumstances involving the funds of an organization. A business budget is a plan covering the funding implications for all phases of operations for a definitive period in the future.

Program/Project Plans Program/project plans are a combination of objectives, policies, procedures, budgets, and other elements necessary to carry out a predetermined objective. Project plans are the basic building elements of the organization's system of plans. Program plans encompass a broader scope of resources and objectives.[2]

The Time Dimension of Planning

One of the important aspects of strategic planning is that it tends to focus the attention of decision makers on the longer-range future, whereas many other

[2]Program plans are treated in more detail later in this chapter.

planning approaches, such as "management by objectives,"[3] tend to emphasize the nearer-term (usually one year hence). The question of *the* appropriate time dimension for planning has been much discussed in the classic literature of planning. For instance, most literature cites three to five years as the most common long-range planning term. However, consider the problem of determining a corporate long-range planning period. The span of an overall planning period could be determined as any of the following:

1 The average planning period for the various functional areas of effort

2 The longest single period of functional-area long-range planning

3 The time period required to provide for the amassing of necessary resources

4 An arbitrary period which in the judgment of the executive group best fits the long-range objectives of the organization

5 A period which encompasses the most critical areas of long-range planning within the organization

6 A period which provides for the best market advantage in terms of economic cycles and long-term growth

The above list suggests a multidimensional horizon of planning. The precise period of time is less important than the determination of *the ability of the organization to realize a return on the resources that have been committed in the planning process.* According to Koontz and O'Donnell:[4]

> There should be some logic in selecting the right time range for company planning. In general, since planning and forecasting that underlie it are costly, a company should probably not plan for a longer period than is economically justifiable; yet it is risky to plan for a shorter period. The logical answer as to the right planning period seems to lie in the "commitment principle," that planning should encompass the period of time necessary to foresee (through a series of actions) the fulfillment of commitments involved in a decision.

For example, pulp, paper, and lumber companies are reported to plan in terms of a forty-year horizon, whereas a cosmetics manufacturer would have no need for such a long time frame.

Values of Long-Range Planning

E. Kirby Warren[5] has delineated four "realistic expectations" for long-range planning which, in effect, represent the values of a long-range planning system:

[3]The MBO idea was introduced by Peter Drucker in *The Practice of Management,* Harper & Row, Publishers, Incorporated, New York, 1954. It has now been extended and widely applied at operational planning levels in industry. MBO is discussed in some detail later in this chapter.

[4]Harold Koontz and Cyril J. O'Donnell, *Principles of Management,* 3d ed., McGraw-Hill Book Company, New York, 1964, p. 87.

[5]E. Kirby Warren, *Long-Range Planning: The Executive Viewpoint,* Prentice-Hall, Inc., Englewood Cliffs, N.J., 1966, pp. 29–31.

1 Clearer understanding of likely future impact of present decisions
2 Anticipating areas requiring future decisions
3 Increasing the speed of the flow of relevant information
4 Providing for faster and less disruptive implementation of future decisions

The values of long-range planning, then, have to do with better current decisions and better and more efficient future decisions. Hence, strategic decisions, long-range planning, and the systems-analysis approach, which in previous chapters we have argued is the best basis for good decisions, are intrinsically intertwined. In effect, long-range planning is the overall objective of systems analysis, for one can apply analysis to a particular decision and "solve" a problem without effectively integrating that solution into an overall plan for the future. Thus, while strategic decisions and the analysis of those decisions are a necessary part of long-range planning, they are not synonymous with it.

PROGRAMS

One level of "plan" which has been added to the classic taxonomy is that of a *program* (and hence, a *program plan*). The idea of a program is not precisely defined, because it must be given a somewhat different interpretation in different organizations. Programs are closely related to the objectives of the organization. In fact, it is in part because the objectives of large organizations are often difficult to define that no single definition of a program is satisfactory. In practice, a number of criteria may be applied in the definition of a program.

One essential feature of a program is clear; it is *output-oriented.* In other words, programs are defined first in terms of what the organization is trying to achieve, rather than in terms of the resources which the organization can bring to bear (inputs). The Bureau of the Budget[6] used the following principles to guide the initial development of appropriate output categories:[7]

> 1 *Program categories* are groupings of agency programs (or activities or operations) which serve the same broad objective (or mission) or which have generally similar objectives. Succinct captions or headings describing the objective should be applied to each such grouping. Obviously, each program category will contain programs which are complementary or are close substitutes in relation to the objectives to be attained. For example, a broad program objective is improvement of higher education. This could be a *program category,* and as such would contain Federal programs aiding undergraduate, graduate and vocational education, including

[6]Now part of the Office of Management and Budget (OMB).
[7]U.S. Bureau of the Budget Bulletin 66-3, Oct. 12, 1965, p. 4. More recent documents such as Bulletins nos. 68-2 (July 18, 1967) and 68-9 (April 12, 1968) use somewhat different descriptions, but those given above are more illustrative of the points to be made here.

construction of facilities, as well as such auxiliary Federal activities as library support and relevant research programs.

2 *Program subcategories* are subdivisions which should be established within each program category, combining agency programs (or activities or operations) on the basis of narrower objectives contributing directly to the broad objectives for the program category as a whole. Thus, in the example given above, improvement of engineering and science and of language training could be two program subcategories within the program category of improvement of higher education.

3 *Program elements* are usually subdivisions of program subcategories and comprise the specific products (i.e., the goods and services) that contribute to the agency's objectives. Each program element is an integrated activity which combines personnel, other services, equipment and facilities. An example of a program element expressed in terms of the objectives served would be the number of teachers to be trained in using new mathematics.

For instance, a program structure for the United States Forest Service might be made up of such categories as timber production, outdoor recreation, and natural beauty. The Coast Guard's structure might include search and rescue, aids to navigation, and law enforcement. A university might use science and humanities categories, involving teaching and research subcategories. In business, natural programs are products or product lines. The obvious differences in these program structures imply that the designation of appropriate structures is not subject to any inviolable rules.

Objectives and Programs

The basic difficulty involved in defining appropriate programs has to do with the nebulous objectives of most large organizations. Even businesses, where profit maximization may seem to be the sole objective, in reality have much more complex objectives. This is evidenced by their charitable contributions and the periodic objections of some stockholders at annual meetings to their non-profit-oriented activities. In government and educational organizations, the objectives are not usually operationally defined, i.e., defined in a way in which attainment can be readily measured. For instance, the typical educational institution's objectives are, at best, vaguely stated as "preparing students to be good citizens" or "advancing our understanding of our environment."

Because objectives are often vaguely stated, the relationships between the organization's activities and its objectives are seldom precisely understood. However, most frequently the organization knows, or believes that it knows, those areas in which achievement will lead to the attainment of its goals. Such a nonrigorous relationship between outputs and goals is often an adequate basis for the definition of a meaningful program structure for an organization. The examples of potential program structures for the Forest Service, the Coast Guard, and a university which were discussed above show this to be true, even though comprehensive goals for those organizations on which all branches of the federal government would agree might not be easy to determine.

Determining a Program Structure

Since there are no hard-and-fast rules involved, the determination of a program structure is a pragmatic undertaking, involving alternative structures and a recognition that the structure chosen can itself have an important impact on the decisions which are to be made.

An important criterion to be used in this determination is that *the program structure should permit comparison of alternative methods of achieving objectives,* however vaguely defined these objectives may be. However, programs may also consist of a number of interactive components—the effectiveness of each of which depends on the others. Thus, in the Defense Department, Strategic Retaliatory Forces might be a reasonable program because it relates to a national defense objective and can be broken down into elements—manned bombers, ICBMs, submarine-launched missiles, etc.—which are to some extent substitutes for one another and which involve questions of resource mix such as were treated in the previous chapter. On the other hand, the elements of a program may be complementary to one another, as in the case of the research and teaching elements of a university program.

Programs should emphasize extended planning horizons, say, five or ten years. This enhances the value of programs as bases for long-range planning.

Other practicalities and peculiarities of particular organizations enter into the definition of programs. Frequently, for instance, the time period over which the goal will be pursued is a natural criterion. Pure research is the best illustration of an activity which is not easily related to organizational outputs. Thus, its long-range nature naturally defines it as a distinct program of many organizations. However, applied research and development is most often related to specific organizational objectives, and thus it should be included in the same program category as the other activities related to the same objective.

It is sometimes necessary to have programs defined in terms of intermediate outputs rather than final outputs. Thus, while Strategic Retaliatory Forces, Continental Defense, and General Purpose Forces represent Defense Department programs related to final outputs (national objectives), Airlift and Sealift might be a program which clearly relates only to an intermediate output. However, since it lends itself nicely to the comparison of alternatives (mixes of aircraft and sealift, for example), it is a reasonable program.

A General Support program in the Department of Defense's program structure included items which could not be identified with another specific program but for which an accounting must obviously have been made. In most organizations, it will be necessary to have some sort of catchall program as this. It is axiomatic to say that such catchalls are potential devices for hiding activities which are in fact closely identifiable with objectives and for which alternatives are conceivably available for consideration. Care should be taken at all levels to avoid having this happen, either purposefully or accidentally, since such a perversion of the program structure can easily negate the greatest value of the concept—that of comparing alternatives.

The program structure of an organization should not be allowed to become either (1) a reflection of the organization's administrative structure or (2) a way of putting new labels on old budget activities. The program structure need not reflect the organization's structure. It is often desirable to have basic program categories which cut across organizational lines to facilitate the comparison of alternative elements which are potential substitutes for one another. So, too, should a relabeling of old budget activities be avoided. As we shall see, there is nothing incompatible about a traditional budgeting process such as that which has been used by the federal government and the program budget. The former can be developed from the latter. To do this meaningfully, however, requires that planning and programming be done in the light of objectives and not in terms of inflexible budget categories.

METHODS OF ORGANIZING PLANNING AND SYSTEMS ANALYSIS

Every organization that plans also does systems analysis, although many do not use the terminology. Most organizations use some planning framework within which systems analysis is conducted. Among the frameworks are "special study" projects, program budgeting, *planning-programming-budgeting systems* (PPBS), *zero-base budgeting* (ZBB), and *management by objectives* (MBO).

Systems-Analysis "Special Study" Projects

Systems-analysis special projects are conducted in virtually every modern organization. Sometimes they are conducted by a staff department called "operations research" or "management science"; sometimes they are conducted as a part of the "computerization" analyses that are continuously under way; sometimes they are conducted by a "planning" group that is studying new product or process opportunities.

These staff groups apply the framework and methods of systems analysis to "special projects"—the analysis of specific complex problems of choice within the organization. Under whatever title, systems analyses are being conducted on an everyday basis in most organizations. This is because the body of knowledge that has been developed is so useful in addressing complex decisions and problems and because such situations are ever more prevalent.

Across all kinds of organizations, systems-analysis "projects"—the application of systems-analytic methods to specific one-time decisions or problems—will probably always be the most common variety of application. This is because many questions or issues arise only once, and when they are satisfactorily addressed, they do not arise (at least in the identical form) again. This is true, for example, of strategic opportunities to acquire a business firm, to place a new product on the market, or to develop a new social program.

Such opportunities are unique and nonrecurring. Those that do seem to recur, such as the opportunity to purchase a similar business, or the same business one year later, or to introduce a social program that has previously been considered and rejected, actually are *different* decisions or problems when they recur because *the circumstances have changed.*

In such situations, the best way to apply systems analysis is to have the organizational resources, in the form of trained analysts and their appurtenant computers, to address these problems when they occur, or when interest in them becomes sufficiently high to warrant the allocation of costly resources to their solution.

"Project basis" systems analysis is straightforward in concept but not in practice. There are difficulties in maintaining a high-quality staff resource whose contributions are not always easy to measure during periods of austerity in an organization. There are difficulties in determining the "best" problems to work on, because "best" may be defined in various ways, such as "highest potential payoff," "highest likelihood of identifying a system improvement," or "highest likelihood of having the results accepted and implemented."

Such special studies also form the basis for many of the comparative analyses that are intrinsic to other more formal planning frameworks, such as planning-programming-budgeting systems and zero-base budgeting.

Program Budgeting

One of the most common "systems" for planning is that which revolves around the *program budget.* In the most theoretical framework, a program budget is merely one part of a broader planning system. In that context, it is the financial expression of the underlying program plan, i.e., the program plan translated into financial terms. In fact, some organizations base their planning on the program budget, in which case it represents a framework for planning.

An output-oriented program budget is best understood by comparing it with the input-oriented budgets that are traditionally used in many organizations, including the federal government.[8]

Traditionally, the budget of the federal government has had two important characteristics which its critics regard as limiting the effectiveness of government expenditures. First, it is an annual budget, and thus it has limited usefulness as a basis for comprehensive long-range planning. Second, it is broken down into functional or object-class categories such as "pay and allowances" and "con-

[8]A. Wildavsky, "A Budget for All Seasons?: Why the Traditional Budget Lasts," *Public Administration Review,* November–December 1978, pp. 501–509.

struction," rather than into categories which are related to governmental *objectives*. Thus, it is an input-oriented budget.

In the context of defense planning, it became apparent that a budget of this variety was inadequate. The short-range nature of the annual budget precluded its use as a long-range planning document. Using the budget information, it was extremely difficult to gather together all the costs which would eventually result from a decision to undertake a program or to purchase a weapons system. Since the costs involved were scattered among a variety of appropriations and the budget projected requirements for only a year, the United States government found itself in the position of having purchased and operated many weapons systems and programs with little regard to, or knowledge of, the total cost. For example, operating costs, which may often be more significant than development and procurement costs, were not made apparent in the budgeting process.

Clearly, this confusion concerning the amount and timing of costs did not discourage various governmental units from proposing projects that could not be funded to completion. The visible evidence of this was the large number of weapons-system development projects which were canceled, at least partly because adequate advance planning was not required by the budgeting process.

Some important weapons systems had been specially "costed" under the traditional federal government budgeting process for purposes of decision making, but such a procedure was not a natural product of the budget. The institution of a program budget, as a complement to the traditional budget, provided a basis for costing weapons systems and output-oriented programs of the government. Such costings are essential to the comparison and evaluation of alternatives, which are, in turn, a vital part of the conceptual framework for analysis of decision problems. Thus, the concept of a program and the program budget provide the basis for utilizing objective scientific analysis in strategic decision making.

The interaction of the programs, operating units, and staff functions of an organization in program budgeting is illustrated in Figure 7-1, which gives a hypothetical description of the U.S. Department of Defense. Each of the Armed Forces and each staff function conceivably cuts across each major program. Thus, by budgeting on a program basis, duplications can be eliminated and valid requirements for the accomplishment of objectives can be determined. For example, both the Navy and the Air Force contribute toward the strategic retaliatory mission via submarine-based missiles and long-range bombers. Also, all staff functions must exert efforts toward accomplishing these objectives.

In the civilian sector, we may construct a hypothetical corporation to illustrate a similar interaction. Figure 7-2 depicts the interaction of operating units, staff functions, and programs for a corporation in the same fashion that Figure 7-1 does for the U.S. Department of Defense. One major program of the fictitious organization is plant nutrition. The corporation's chemical products division is obviously involved, as is the agricultural marketing division (which may sell bulk fertilizers to farmers) and the consumer products division (which

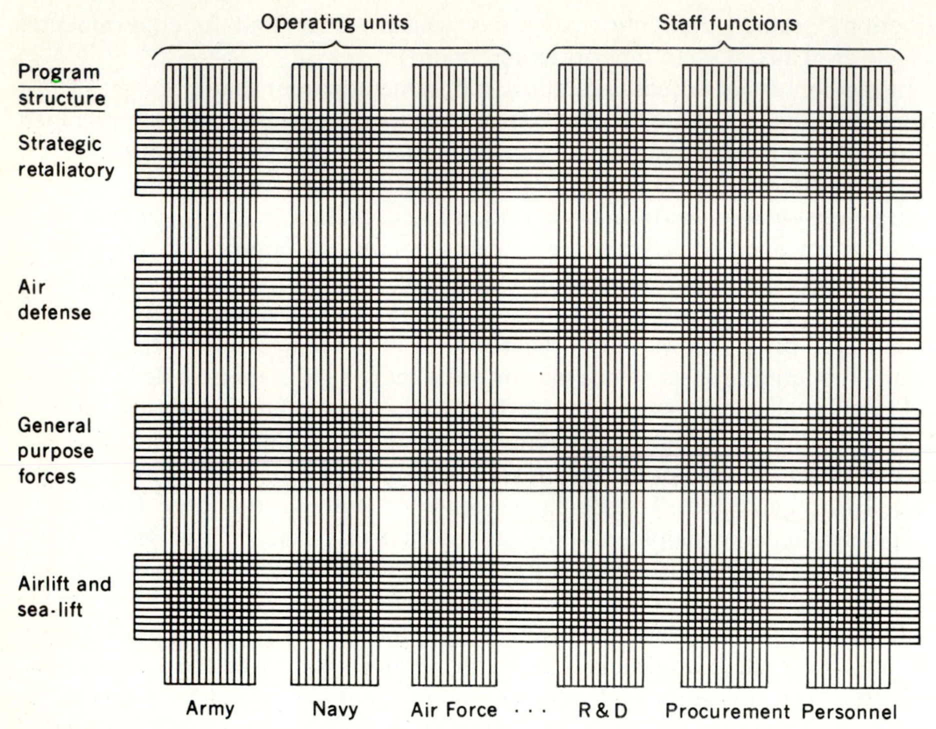

FIGURE 7-1
Programs and functions in the Department of Defense.

may sell both lawn-care chemicals and equipment to individual consumers). Similarly, in the animal nutrition program, the chemical products and agricultural marketing divisions have interests in farm animal foods, just as the chemical and consumer products division might be concerned with pet food products.

Planning-Programming-Budgeting Systems

In the early 1960s, the Kennedy administration formally introduced some of the processes of systems analysis into the U.S. Department of Defense through a system of planning, programming, and budgeting. In the mid-1960s, this system was formally spread throughout the federal government's executive branch and to many state and local governmental levels.

Planning, Programming, and Budgeting Systems (PPBS) have since been formally dropped by the U.S. federal government,[9] as has the related zero-base

[9]A. Schick, "A Death in the Bureaucracy: The Demise of Federal PPB," *Public Administration Review,* March–April 1973, pp. 146–156.

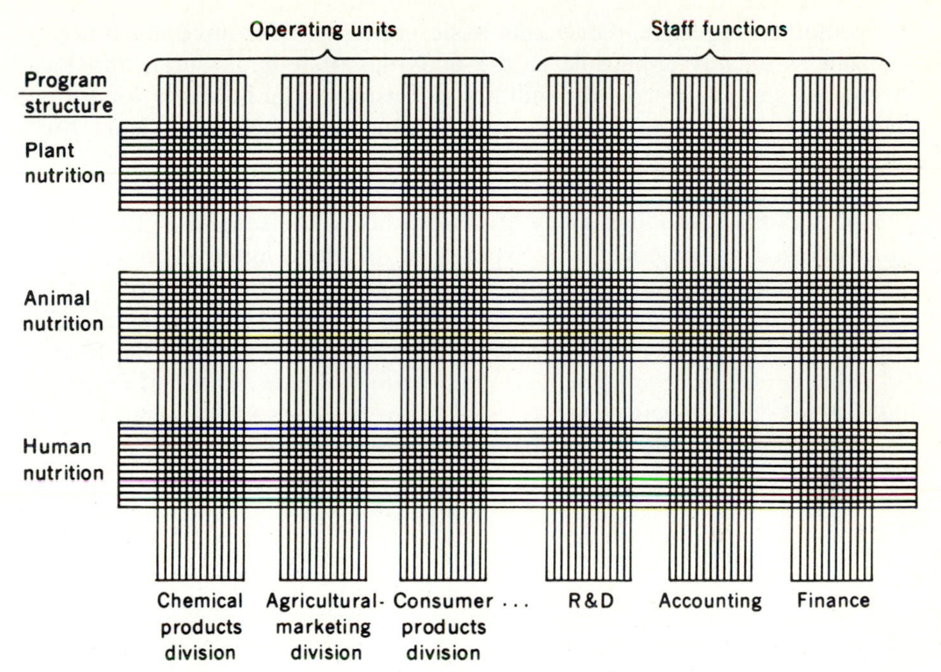

FIGURE 7-2
Programs and functions in a corporation.

budgeting framework that was subsequently introduced by the Carter administration. However, the impact on the thinking of managers and the formal use of PPBS in other organizations remain real and significant.

The Need for PPBS The need for a formal comprehensive system of planning such as PPBS arises from the scarcity of resources that is endemic to organizations. At the level of the federal government, the organization's objectives are clearly stated in the preamble to the United States Constitution—to "provide for the common defence" and to "promote the general welfare." In modern terms, this is interpreted to mean defense; the maintenance of order; the promotion of health, education, and welfare; economic development; and the conduct of essential services such as the postal service. Each modern government does all these things to one degree or another, but no one attains each of these objectives to the ultimate.

No one would claim, for instance, that the United States defenses are perfect, or perhaps even as good as those of some other nations of the world. On the other hand, the educational level of the United States population is higher than that of most other countries. Thus, the attainment of government objectives is a process involving the *allocation of resources*—of compromises between money spent in one way to better achieve one objective and that spent in another way to achieve another objective. Once this basic concept of resource allocation is recognized as equally applicable to the activities of federal, state, and local governments and to those of the military or business, the necessity for formal planning and analysis to achieve "best" allocations becomes clear. Such planning and analysis is the goal of PPBS.

A Model PPBS Although many of the details of the federal PPBS are no longer followed in practice, the experience of the United States federal government in implementing planning, programming, and budgeting has served as a model for the development of similar systems at various governmental levels and in a wide variety of other agencies. Therefore, an analysis of the federal PPBS is important to developing an understanding of many different organizational systems. The objectives of the federal government's PPBS were stated in the original Bureau of the Budget bulletin as follows:[10]

> The overall system is designed to enable each agency to:
>
> **1** Make available to top management more concrete and specific data relevant to broad decisions;
> **2** Spell out more concretely the objectives of Government programs;
> **3** Analyze systematically and present for agency head and Presidential review and decision possible alternative objectives and alternative programs to meet those objectives;
> **4** Evaluate thoroughly and compare the benefits and cost of programs;
> **5** Produce total rather than partial cost estimates of programs;
> **6** Present on a multi-year basis the prospective costs and accomplishments of programs;
> **7** Review objectives and conduct program analyses on a continuing, year-round basis, instead of on a crowded schedule to meet budget deadlines.

The accomplishment of these objectives in the federal government was based on a recognition of the three-dimensional nature of planning, programming, and budgeting. The first of these dimensions is universally recognized—resource inputs. These are the people, equipment, and organizational units to be allocated to various activities. The second dimension involves the variety of missions and goals to which the organization is directed—the output. Third is the time dimension, since few complex organizations pursue programs which are entirely short-range in nature.

These three dimensions are illustrated in Figure 7-3. Each element of that figure represents an expenditure of a given resource on a specific program during a given year. For example, the darkened element represents R&D allocations to program P_4 during the current year, and the row of elements behind the darkened one represents future R&D allocations to P_4.

Three components of the federal "model" PPBS serve to demonstrate how PPBS operates as a planning framework as well as the role of systems analysis

[10]U.S. Bureau of the Budget Bulletin 66-3, p. 3.

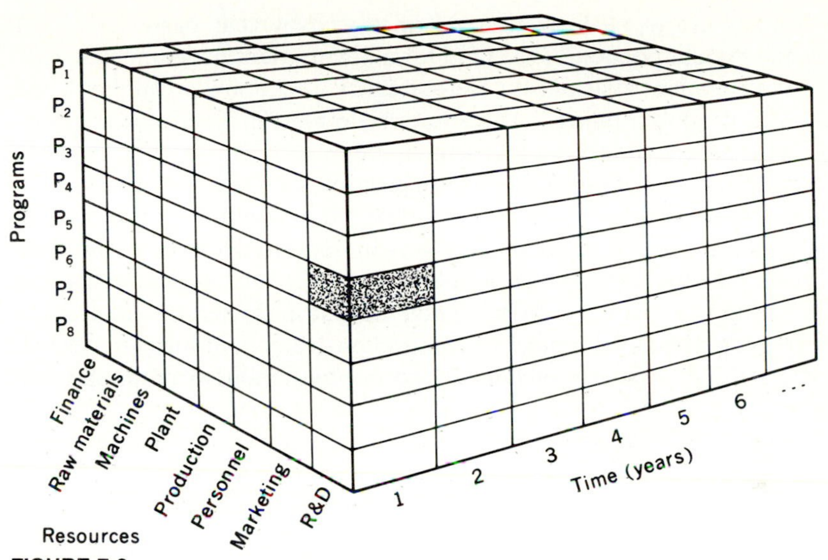

FIGURE 7-3
The three-dimensional nature of PPBS.

"special studies" in PPBS. These components are the multiyear Program and Financial Plan (PFP), Program Memoranda (PM), and Special Analytic Studies (SAS).[11]

The Program and Financial Plan A multiyear program and financial plan is a translation of concretely specified organizational objectives into combinations of activities and operations designed to achieve objectives within specified time periods.

The PFP is stated in terms of the program structure and covers, on a year-by-year basis, a period of years as determined by the nature of the organization's objectives and operations. In the federal government, this period is frequently of five years' duration, but for many projects, longer durations may well be appropriate; e.g., for timber production, the appropriate time period may be much longer than five years.

The PFP should reflect a plan for the future. It is not constrained because authorizations for programs are expiring or because appropriate approvals have not been obtained. Rather, it should show programs at the levels which are deemed best for the duration of the multiyear plan.

The form of the PFP is essentially that of an outcome array under certainty such as was discussed in Chapter 5. The multiyear nature of the plan implies that

[11]For illustrative purposes, the following describes elements of both the "old" PPBS as outlined in the U.S. Bureau of the Budget Bulletin 66-3 and "newer" versions outlined in subsequent versions (68-2 and 68-9). None is currently followed by the federal government.

costs and benefits are to be broken down on a year-by-year basis. Table 7-1 summarizes the benefit portion of the plan for a government program category, "manpower development assistance"; a subcategory, "manpower training"; and an element, "on-the-job training." The appropriate benefit descriptor might be the number of workers trained. Other similar tables can also be presented as part of the PFP for other important benefit measures. The reason for determining these outcome descriptors of benefit is to permit the assessment of the degree of accomplishment of organizational objectives in quantitative terms. Doing so permits comparisons to be made on an objective basis. Of course, it will not always be possible to do this. As we have pointed out in discussing the practice of systems analysis, the development of comprehensive benefit measures is one of the greatest difficulties in applying the conceptual framework in the real world.

Two bases for comparison often prove useful in illustrating program benefits. One of these is historical. Note that Table 7-1 includes columns headed "last year" and "this year." Since this is a planning document for subsequent years (beginning with "next year"), historical benefit data on continuing programs are provided as a point of reference. Also, relative measures should be included wherever possible. By "relative measures," we mean those which relate the benefits to a standard such as the total possible benefit which could potentially be achieved. In using a benefit measure involving the number of workers trained, for example, one could relate the absolute number to the entire population of workers desiring or requiring training.

Other comparative bases are often included in PFPs as "special tabulations." For example, federal agencies might tabulate the known programs of state and local governments or private companies which are directed toward similar objectives.

To complement these tabular presentations of the benefit portions of outcomes the cost portion should be presented in a form similar to that of Table 7-1. In doing so, the systems costs should be used, i.e., capital outlay, research, development, operating, maintenance, etc.

TABLE 7-1
BENEFIT PORTION OF OUTCOME ARRAY FOR PFP

	Last year	This year	Next year	Two years hence	Three years hence	Four years hence	Five years hence
I. Manpower development assistance							
A.							
B. Manpower training							
1.							
2. On-the-job training (No. of workers trained—000)	xx	xx	xx	xx	xx	xx	xx

Thus, the basic elements of the PFP are simply a sequence of tables which present the benefits and costs associated with various program categories, subcategories, and elements of an organization. It represents, in concise summary form, a plan of recommended levels and mixes of programs for an organization. The word "recommended" should be emphasized since the PFP represents the results of analyses involving choice among alternatives and the assessment of priorities.

The PFP is reviewed, reappraised, and updated by the organization annually. Of course, the column headings in the tables are indexed by one year in doing so. In the federal government, these reviewed PFPs after approval and modification by the agency head, budgeting experts, and the President, formed the basis for the agency's annual budget request.

Program Memorandums A program memorandum (PM) for each of the program categories in PFP *supports the conclusions and recommendations presented in the PFP.* The PM should present the results and a summary of methodology used in analyzing and comparing alternative programs.

As outlined in the basic Bureau of the Budget bulletin on PPBS, the PM should:[12]

1 Spell out the specific programs recommended by the agency head for the multi-year time period being considered, show how these programs meet the needs of the American people in this area, show the total costs of recommended programs, and show the specific ways in which they differ from current programs and those of the past several years.

2 Describe program objectives and expected concrete accomplishments and costs for several years into the future.

3 Describe program objectives insofar as possible in quantitative physical terms.

4 Compare the effectiveness and the cost of alternative objectives, of alternative *types* of programs designed to meet the same or comparable objectives, and of different *levels* within any given program category. This comparison should identify past experience, the alternatives which are believed worthy of consideration, earlier differing recommendations, earlier cost and performance estimates, and the reasons for change in these estimates.

5 Make explicit the assumptions and criteria which support recommended programs.

6 Identify and analyze the main uncertainties in the assumptions and in estimated program effectiveness or costs, and show the sensitivity of recommendations to these uncertainties.

In effect, the PM objectively summarizes the systems analyses which have been performed and presents the results in the form of recommendations. The PM serves as a basic planning document throughout the organization, since it represents in summary form the analytic statement of, and justification for, the organization's current objectives and programs.

[12]U.S. Bureau of the Budget Bulletin 66-3, p. 8.

One of the basic virtues of objective analysis is made clear by the PM. This feature—the reproducibility of analysis—means that a succinct statement of assumptions, objectives, and methodology should be an adequate basis for a check on the analysis performed. If, for example, other alternatives are brought to light after the PM is made up, it should be possible to reproduce the analysis and thereby compare the new alternatives with those which had previously been considered.

Special Analytic Studies The role of "special analytic studies" in the federal PPBS further serves to clarify the inherent relationship between systems analysis and PPBS.

> The Special Analytic Studies provide the analytic groundwork for the decisions reflected in the PMs. Studies are of two types, both of which are essential to effective operation of an agency PPB system and to annual budget review.
>
> Some SASs will be performed in order to better resolve an issue in the budget year. These studies will be initiated and completed during the year and their results will be shown in the PM submitted in support of the budget request.
>
> The second type involves studies which continue beyond the budget year. A continuing study will develop on a longer-run basis the conceptual understanding necessary to improve the data available, to evaluate the implications of agency objectives, and to provide an analytic basis for deciding future Major Program Issues[13]

The *major program issues* have been specifically identified as a device for relating the more mundane resource-allocation aspect of planning to the higher-level policy choices. A major program issue is a "question requiring decision in the current budget cycle, with major implications in terms of either present or future costs, the direction of a program or group of programs, or a policy choice."

Zero-Base Budgeting

Another planning framework which may be used in organizations is that termed "zero-base budgeting." Despite ZBB's identification with the Carter administration and its subsequent demise under the Reagan administration,[14] ZBB, like PPBS, represents a viable framework within which systems analysis may be applied to organizational planning.

The Objective of ZBB Peter Pyhrr, one of the originators of ZBB, defined it as:

> . . . an operating, planning, and budgeting process which requires each manager to justify his entire budget request in detail from scratch and shifts the burden of proof to each manager to justify why he should spend any money at all. This approach requires

[13]U.S. Bureau of the Budget Bulletin 68-9, p. 3.
[14]"Zero-Base Budgeting Is Abandoned by Reagan," *The Wall Street Journal,* August 10, 1981.

that all activities be identified in "decision packages" which shall be evaluated by systematic analysis in rank order of importance.[15]

The essence of ZBB lies in the notion of "decision packages," in which discrete activities, both existing and proposed, are collected together, and the subsequent evaluation of decision packages through the development of a ranking, or prioritization. This ranking is based on systems analyses of the alternative "packages."

ZBB requires managers to justify their entire budget "from scratch" (hence, "zero-base"). They are required to propose and assess alternatives to accomplish objectives. Just as in PPBS, this process may be conducted through cost/benefit or subjective systems analyses as part of the overall planning process and prior to the formal ZBB portion of the process, but just as in PPBS, alternatives and their analysis are intrinsic to ZBB.

A Model ZBB System Just as the federal PPBS provides a model system that can be used to describe the underlying principles, so, too, does the federal ZBB system. As outlined by the federal Office of Management and Budget, the objectives of the federal ZBB were to:

1 Involve managers at all levels in the budget process;
2 Justify the resource requirements for existing activities as well as for new activities;
3 Focus the justification on the evaluation of discrete programs or activities of each decision unit;
4 Establish, for all managerial levels in an agency, objectives against which accomplishments can be identified and measured;
5 Assess alternative methods of accomplishing objectives;
6 Analyze the probable effects of different budget amounts or performance levels on the achievement of objectives; and
7 Provide a credible rationale for reallocating resources especially from old activities to new activities.[16]

The similarity between these objectives and those for the federal PPBS are apparent.

The federal ZBB process involved a series of steps:

1 Identification of objectives
2 Identification of decision units
3 Preparation of decision packages
4 Ranking of decision packages
5 Higher-level review

Identification of objectives The first step in the process is the identification of output-oriented objectives of a program or organization. Along with objectives, key indications that will be used to measure results should also be indicated.

[15]P. A. Pyhrr, *Zero-Base Budgeting,* John Wiley & Sons, Inc., New York, 1973.
[16]U.S. Office of Management and Budget Bulletin 77-9, April 19, 1977, p. 2.

Identification of Decision Units The entities in the program or organization structure whose managers will develop decision packages should then be identified. These units should be at a level at which the manager makes decisions on the amount of spending and the scope, direction, or quality of work to be performed.

Preparation of Decision Packages Decision packages are then prepared in a fashion that is meant to suggest:

a Where funding reduction may be feasible
b The increased benefits that can be achieved through additional spending
c The effects of such funding changes.

This is done by each manager first identifying alternative ways of accomplishing major objectives and then selecting the best alternative for further analysis involving alternative funding levels. A series of decision packages is prepared for all activities where there is discretion as to the level of activity and funding levels.

The decision package set normally includes the specification of:

a A minimum level (normally below the current level unless it is infeasible to operate below the current level)
b The current level
c A level specified that is between the minimum and current level
d Any increment desired above the current level

Then the decision unit manager prepares a decision package set that includes packages which cumulatively represent the total budget request of the decision unit. Generally, a series of packages will be prepared to show the effect of various funding levels on performance or alternative ways of achieving objectives.

Ranking of Decision Packages The decision packages are then ranked by the decision unit manager. The minimum level for a decision unit is always ranked higher than an incremental level for the same decision unit, since the minimum level represents that below which the activity cannot be conducted. However, the minimum level for one decision unit need not be ranked above an incremental level for another unit. In such a case, if funding does not encompass the lower-ranked minimum level, the activity would not be pursued.

Each review level prepares a ranking sheet for decision packages to submit to the next higher level. The ranking sheet includes summary data on budget authorizations and outlays for each ranked package.

Higher-Level Review Once a decision unit manager has submitted the various decision packages, they are reviewed at higher organizational levels. This review process may include revision, deletion, or addition of decision packages and revision of rankings. In some cases, management by exception may be practiced in that only lower-ranked packages are examined in depth, since they are the ones that would be directly affected by incremental increases or decreases in budgets.

Consolidation of decision packages may be done at various review levels to reduce detailed paper work. Normally, this consolidation is done on the basis of natural groupings of decision units.

A minimum-level consolidated decision package is prepared that may or may not include each of the minimum-level packages from the decision package sets being consolidated.

Figure 7-4 shows a decision unit overview that provides information necessary

FIGURE 7-4
Decision unit overview. (Office of Management and Budget Bulletin 77-9, April 19, 1977.)

DECISION UNIT OVERVIEW
Department of Health, Education, and Welfare
Mental Health Administration
Federal Support of Community Mental Health Services
Mental Health: 75-0001-0-1-550

Goal.
To ensure needy citizens access to community-based mental health services, regardless of ability to pay. Services should be of high quality, provided in the least restrictive environment, and in a manner assuring patients' rights and dignity.

Major objective.
To assist in the establishment and operation of a nationwide network of 1,200 qualified community mental health centers (CMHCs) by 1984 to ensure availability and accessibility of services to residents of each mental health catchment area.

Current method of accomplishing the major objectives.
Grants are made to public and nonprofit entities to plan and operate community mental health center programs. The planning grants are one-time grants, not to exceed $75,000 each. The operating grants are for eight-year periods with a declining Federal matching rate.

Alternatives.
1. Consolidate Federal funding for community mental health services and other categorical health service programs into a single formula grant to the States.
2. Consolidate Federal funding for community mental health services and other community-based inpatient and outpatient services—as well as institutionally based short-term acute and long-term care services—for the mentally ill and mentally retarded.

These alternatives are not being pursued because the States thus far have not been able to ensure that funds will be targeted into high priority areas. The Secretary believes the Federal Government must have the ability to control the funding.

3. Provide for mental health services coverages through the national health insurance proposal. This alternative is not presently viable because passage of the national health insurance act is not near. Intensive study is now being directed toward this alternative for possible consideration next year.

Accomplishments.
Since the establishment of the CMHC program in the mid-1960s, 670 CMHCs have received Federal funding of nearly $2.0 billion. In 1977, nearly 600 centers were operational, covering 45% of the population (90 million people), and providing treatment services to 2 million individuals annually.

In 1977, 450 centers received Federal grant support and 100 centers completed the eight-year Federal grant cycle. To qualify for an operational grant, P.L. 100-63, requires centers to provide the following services on a 24 hour a day, seven day a week basis:

1. Inpatient hospitalization;
2. Outpatient treatment and counseling;
3. Partial hospitalization as an alternative to full-time hospitalization;
4. 24-hour emergency services by telephone or on a walk-in basis;
5. Consultation and education services;
6. Services to children;
7. Services to the elderly;
8. Screening services to the courts and other agencies;
9. Follow-up care for former full-time patients from a mental health facility;
10. Transitional services for same;
11. Alcoholism and alcohol abuse program and drug addiction and abuse program.

for higher-level management to evaluate the choices made on each of the decision packages.

Figure 7-5 illustrates the overview in Figure 7-4. The "other information" category normally includes explanations of any legislative or policy changes that are needed, the predicted impact or consequences of not approving the package, the effect of a zero funding (for minimum-level packages), an explanation of what will be sacrificed (for packages below the current level), and a description of the relationship of the decision unit to other decision units.

Figure 7-6 shows an illustration of the decision package ranking and consoli-

FIGURE 7-5
Decision package. (Office of Management and Budget Bulletin 77-9, April 19, 1977.)

DECISION PACKAGE
Department of Health, Education, and Welfare
Mental Health Administration
Federal Support of Community Mental Health Services
Mental Health: 75-0001-0-1-550

Activity Description:

Fund 50% more newly qualifying CMHCs. That is, for every two CMHCs whose eight-year eligibility period ends, fund three newly qualifying CMHCs.

Resource Requirements: (Dollars in thousands)

	1977	1978	1979 This Package	1979 Cumulative Total
Planning grants ($)	1,000	1,000	0	0
Operating grants ($)	97,000	147,000	10,000	150,000
Total obligations	98,000	148,000	10,000	150,000
Budget authority	98,000	148,000	10,000	150,000
Outlays	97,000	145,000	10,000	148,000

Five year estimates	1979	1980	1981	1982	1983
Budget authority	150,000	162,000	172,000	183,000	194,000
Outlays	148,000	161,000	171,000	182,000	193,000

Short-term Objective:

To ensure in 1979 access to qualified comprehensive mental health services to 51% of the population (this results in treatment of about 2.2 million patients).

Impact on Major Objectives

	1977	1978	This Pkg.	1979 Cum.	1980	1981	1982	1983	1984
Number of public and non-profit CMHCs	700	710	25	775	850	925	1,000	1,075	1,150
Number of CMHCs providing comprehensive services, as now defined	550	600	25	675	750	825	900	975	1,050
Number of CMHCs receiving grants	400	450	25	475	500	525	550	575	600
Percent of population covered	43	45	6	51	65	75	80	85	90
Percent of probable patients covered	45	50	6	56	66	77	83	87	90

Other Information

By 1982 95% of the high priority catchment areas will have a qualified CMHC. If stretched out from 1984 to 1986, total program costs for establishing 1200 CMHCs will increase from $3.6 billion to about $3.8 billion.

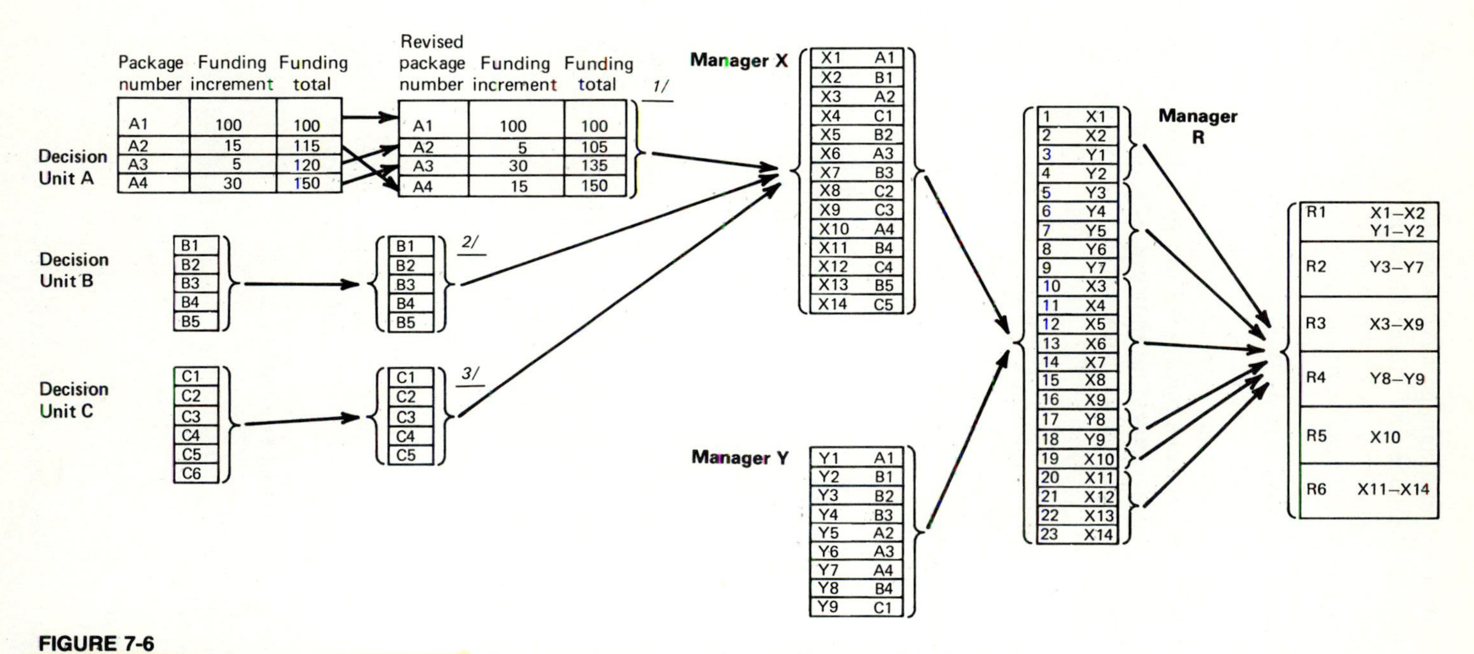

FIGURE 7-6
Decision ranking and consolidation process. (Office of Management and Budget Bulletin 77-9, April 19, 1977.)

dation process in which three decision unit managers A, B, and C rank packages for their units and submit them to Manager X. Here, Manager X revises the rankings and funding levels for unit A, accepts the rankings and funding levels of Unit B, and accepts the rankings of unit C but decides not to propose funding of the lowest-priority package, C_6.

Manager X then ranks the packages for units A, B, and C against each other, as shown to the right of the designation "Manager X" in Figure 7-6. He then submits these revisions to Manager R, who must consolidate them with similar packages submitted by Manager Y. She ranks packages from X and Y against each other and develops consolidated packages, as shown at the right of the figure (e.g., Packages X_1, X_2, Y_1, and Y_2 are consolidated into R_1, Y_3–Y_7 into R_2, etc.).

Management by Objectives

Another planning framework that is in common use in organizations is "management by objectives" (MBO). MBO is a "bottom up" planning technique that focuses on "self-management"—the delineation of objectives by individuals, groups, departments, etc. The generic MBO process is one in which superior and subordinate confer to gain agreement on what is to be done in the planning period that lies ahead. The superior and subordinate both outline their expectations, with the objective of clarifying the role of the individual in the objectives of the organization.

The two parties agree on specific objectives to be accomplished during the planning period. This is the most important and difficult step in the MBO process, since both must generally modify their prior expectations. Also, specific objectives and ways of assessing their degree of fulfillment must be carefully defined, so that it is clear whether or not they have been attained.

The subordinate is then allowed to allocate resources and effort to fulfilling the agreed-upon objectives. Both superior and subordinate participate in an appraisal after the planning period is completed. The degree of attainment of the objective becomes the input to the next planning period.

Clearly, this simple description of MBO is oriented to the planning for, and assessment of, individual performances. To integrate this well-accepted process into a strategic planning framework for an organization has proved to be extremely difficult.[17]

Carroll and Tosi[18] have concisely summarized MBO as:

> Although opinions vary on how to use the MBO approach and some persons disagree about its purposes, most authorities in the field agree that this approach involves the

[17]M. W. Dirsmith, S. F. Jablonsky, and A. D. Luzi, "Planning and Control in the U.S. Federal Government: A Critical Analysis of PPB, MBO, and ZBB," *Strategic Management Journal,* vol. 1, pp. 303–329, 1980.

[18]S. J. Carroll, Jr., and H. L. Tosi, Jr., *Management by Objectives: Applications and Research,* The Macmillan Company, New York, 1973. p. 3.

establishment and communication of organizational goals, the setting of individual objectives pursuant to the organizational goals, and the periodic and then final review of performance as it relates to the objectives. In addition, agreement would be likely on the following elements as necessary to an effective MBO program:

1 Effective goal setting and planning by top levels of the managerial hierarchy.
2 Organizational commitment to this approach.
3 Mutual goal setting.
4 Frequent performance review.
5 Some degree of freedom in developing means for the achievement of objectives.

Organizations that have implemented these ideas refer to their programs variously as "management by results," "goals management," "work planning and review," "goals and controls," and so on. However, all these programs are similar, despite the differences in terminology.

MBO, as most frequently used, is less involved with formal systems analysis than is either PPBS or ZBB. While it is an appropriate planning framework for use in some organizations, its emphasis on participatory management and self-management rather than on formal comparison and analysis of alternatives means that it is less appropriate for the broad, complex organizational systems planning that is the focus of this book.

SUMMARY

A variety of formal planning frameworks have been used in various organizations. This chapter discusses some basic planning concepts and the relationship of systems analysis to these frameworks.

Particular attention is devoted to planning-programming-budgeting systems (PPBS) and zero-base budgeting (ZBB), with some attention to management by objectives (MBO). Each of these approaches has been tried, without great success, in the United States federal government. However, each of these applications at that broad and complex level has had lasting impact on other organizations at lower levels in the federal government, at state and local levels, and in business. So the ideas, despite the fact that various administrations have been identified with particular viewpoints, and consequently with the rejection of others, have had lasting impact. In other organizational contexts, their potential is great.

DISCUSSION QUESTIONS

1 In this chapter, we have discussed the organizational manifestations of systems analysis. What is meant by this? Give some examples.
2 What are some of the traditional or classical products of the planning process? How should these products be defined? What is the usefulness of these products in supporting the organizational mission?
3 Project plans are the basic building blocks of an organization's strategy. What is

meant by this? If project planning has not been successful in the organization, one might reasonably expect that other planning will not be carried out effectively. Discuss.

4 The time dimension for planning is an important consideration. How might these time dimensions differ with respect to the type of plan in the organization? How might the dimension differ between companies in the same industry? In different segments of the same industry? In different industries?

5 If a manager does not do effective long-term planning, someone else will do it for the organization. What is meant by this? Who might that "someone else" be?

6 Give a definition of a program. Define a program plan. Why is it useful to consider the concept of a program in the development of strategies for an organization?

7 Most organizations use some planning framework within which systems analysis can be conducted. What is the systems-analysis context of such planning?

8 What is program budgeting? How does the program budget fit into higher levels of activity in the organizational planning system?

9 Although planning-programming-budgeting systems (PPBS) and zero-base budgeting have been dropped by the U.S. federal government, the impact on the thinking of today's managers from these techniques is very real. What is the nature of this impact?

10 What is the objective of zero-base budgeting? How might the technique be used in the management of a business organization? In a local government, such as that of a township or a county?

11 The federal zero-base budgeting process involves a series of steps. Could this process be applied to a university? What might be some of the problems and constraints in applying the concept to a university?

12 The concept of management by objectives has been with us for a long time. What is the process? What are its advantages? What are its disadvantages?

13 If a superior and a subordinate jointly develop objectives, their interdependent roles can be better understood. Would you agree? Why or why not?

14 The successful operation of both a strategic-planning system and a project-management system in a business organization depends on the continued profitability of that organization. Do you agree? Why or why not?

RECOMMENDED READINGS

Anderson, D. N.: "Zero-Base Budgeting: How to Get Rid of Corporate Crabgrass," *Management Review,* October 1976.

Austin, L. A.: "Zero-Base Budgeting: Organizational Impact and Effects," *AMACOM,* 1978.

Beckhard, R., and R. T. Harris: *Organizational Transitions: Managing Complex Change,* Addison-Wesley Publishing Company, Inc., Reading, Mass, 1977.

Carroll, S. J., Jr., and H. L. Tosi, Jr.: *Management by Objectives: Applications and Research,* The Macmillan Company, New York, 1973.

Conference Board, Inc.: "Zero-Base Budgeting Revisited," FB-79-2, 1978.

Dirsmith, M. W.; S. F. Jablonsky, and A. D. Luzi: "Planning and Control in the U.S. Federal Government: A Critical Analysis of PPB, MBO, and ZBB," *Strategic Management Journal,* vol. 1, 1980.

Dudick, T. S.: "Zero-Base Budgeting in Industry," *Management Accounting,* May 1978.

Fox, Harold W.: "The Frontiers of Strategic Planning: Intuition or Formal Models," *Management Review,* April 1981.
Francl, T. J.; W. T. Lin, and M. A. Vasarheli: "ZBB Fits DD to a Tee," *Datamation,* September 1980.
Herzlinger, R. E.: "Zero-Base Budgeting in the Federal Government: A Case Study," *Sloan Management Review,* Winter 1979.
Hobbs, J. M., and D. F. Heany: "Coupling Strategy to Operating Plans," *Harvard Business Review,* May–June 1977.
Howell, R. A.: "Management by Objectives: A Three Stage System," *Business Horizons,* February 1970.
Kahalas, H.: "A Look at Major Planning Methods: Development, Implementation, Strengths and Limitations," *Long Range Planning,* August 1978.
Kolodny, Harvey F.: "Matrix Organization Designs and New Product Success," *Research Management,* September 1980.
McCorkney, D. D.: "MBO: Twenty Years Later, Where Do We Stand?" *Business Horizons,* August 1973.
Novick, D.: "Long Range Planning Through Program Budgeting," *Business Horizons,* February 1969.
Olm, Kenneth W.; F. J. Brewerton; Susan R. Whisnant, and Francis J. Bridges: *Management Decisions and Organizational Policy,* 3d ed., Allyn and Bacon, Inc., Boston, 1981.
Pyhrr, P. A.: *Zero-Base Budgeting,* John Wiley & Sons, Inc., New York, 1973.
Schick, A.: "A Death in the Bureaucracy: The Demise of Federal PPB," *Public Administration Review,* March–April, 1973.
Sherlekar, V. S., and B. V. Dean: "An Evaluation of the Initial Year of Zero-Base Budgeting in the Federal Government," *Management Science,* August 1980.
Slocum, John W., Jr., and Don Hellriegel: "Using Organizational Designs to Cope with Change," *Business Horizons,* December 1979.
Stonich, P. J.: *Zero-Based Planning and Budgeting,* Dow Jones-Irwin, Inc., Homewood, Ill., 1977.
Suver, J. D., and R. L. Brown: "Where Does Zero-Base Budgeting Work?," *Harvard Business Review,* November–December 1977.
Wildavsky, A.: "A Budget for All Seasons?: Why the Traditional Budget Lasts," *Public Administration Review,* November–December 1978.
Office of Management and Budget Bulletin 77-9, April 19, 1977.
"Planning-Programming-Budgeting (PPB) System," *Bulletin No. 68-9,* Bureau of the Budget, Washington, D.C., April 23, 1968.
U.S. Bureau of the Budget Bulletin 66-3, October 12, 1965.
"Zero-Base Budgeting Is Abandoned by Reagan," *The Wall Street Journal,* August 10, 1981.

CASE 7-1: Planning Concepts: Their Strengths and Weaknesses

Have each member of your team take responsibility for one or more of the approaches to organizing planning and systems analysis—"special study" project, program budgeting, PPBS, ZBB, and MBO—and outline its strengths and weaknesses. Then, in a group meeting, try to assess which of the approaches might be best suited to each of the following situations:

a Top-level federal government policy making
b Planning within a department of the executive branch of the government
c A regional office of a federal department
d The U.S. Congress
e A committee of the U.S. Congress
f The office of a U.S. senator

PART FOUR

PROJECT MANAGEMENT

CHAPTER **8**

THE PROJECT ENVIRONMENT

Projectization has evolved . . . to the point of being a management revolution. It carries to its most elaborate development the task force concept now becoming the fashion in management doctrine.[1]

The idea of a *project—a combination of human and nonhuman resources pulled together in a "temporary" organization to achieve a specified purpose*— reflects the systems approach to the implementation function of management. This chapter introduces this major section of the book, which deals with the implementation activity of management. The particular philosophy espoused throughout this section is the application of systems ideas to the implementation function. This *systems approach to implementation* is most often spoken of as *project management, program management,* or *matrix management,* but we shall try to show that it has applications and ramifications which go far beyond most parochial definitions of these terms.

In Chapter 1, we argued that *the solutions to complex problems are themselves usually complex systems.* Thus, the transportation problems of a city will be solved only through the development of a complex transportation *system*—one which integrates various modes and which is designed to take account of the interests of myriad clientele groups. So, too, will most other varieties of complex problems require systems solutions of an equal degree of complexity and sophistication.

The previously considered planning process provides a framework for the

[1]"The Unexpected Payoff of Project Apollo," *Fortune,* July 1969.

consideration of alternative systems solutions and alternative systems designs. However, once these strategic choices are made, the implementation function of management becomes paramount; i.e., the actions or programs which have been decided upon must be put into action and evaluated for effectiveness.

PROJECT PROGRAMS AND PROJECT MANAGEMENT

The American Heritage Dictionary of the English Language defines a *project* as "an undertaking requiring concentrated effort." *Project* management has been described thus:

> Project management occurs when management gives emphasis and special attention to the conduct of non-repetitive activities for the purpose of meeting a single set of goals.[2]

Projects vary in size and complexity. One extreme, a project to build a nuclear generating station, involves massive engineering, architectural, and management effort.[3] At the other extreme is a personal project, such as the writing of a term paper. While the two are quite different in complexity, size, and amount of resources involved, they both have the characteristics of a project (since even the simpler one may involve many different research, organizing, and writing activities) and both can be "managed" in the same general way.

Clearly, the size and complexity of a power plant project demands a more disciplined approach to the planning, organization, execution, and control of the resources to support the effort. In this case, elaborate cost and schedule systems with resource allocation capability are required. Many thousands of engineering and construction activities must be planned at different levels of detail. Special computer programs are needed to perform design studies. Large one-time inventory-control systems are necessary to track and account for the hundreds of thousands of material line items that will be involved. Detailed project information has to be summarized and retrieved and presented in a meaningful progress-report form for the use of the project manager. A full-time project manager with a large support staff and with strong decision-making authority is necessary to lead the large project team of selected experts from the disciplines that are involved.

An example of a small organizational (rather than "personal") project is the development effort undertaken by a transportation company which was interested in improving the effectiveness of its cross-country shipments. In investigating alternative forms for containerization, this company identified containers which had been used in military airlift operations. A project manager was appointed and a team provided to support him to design, build prototypes, and test the new

[2]Quoted in James J. O'Brien, "Project Management: An Overview," *Project Management Quarterly,* vol. 3, no. 3, September 1977, p. 27.

[3]For instance, see Barry M. Miller and Charles D. Williams, "Management Action through Effective Project Controls: A Case Study of a Nuclear Power Plant Project," 1978 Proceedings of the Project Management Institute, Los Angeles, Oct. 8–11, 1978.

container on high-volume freight routes. This project manager developed a plan and a schedule to fabricate and test a dozen containers. The project manager, formerly a lead engineer, was given $100,000 to do this work. The work was accomplished as a "special project" within the existing structure of the transportation company. The result was a design for these containers which can be seen today in modular truck container equipment operations.

At the far extreme in size and complexity, we find the project for the construction of the Alaskan oil pipeline. Other types of projects commonly undertaken by organizations include:

Design, development, and fabrication of a weapons system or new product
Reorganization of a company
Project team to assemble autos
Project team to do design engineering on common items of equipment
Applied research project to improve production processes

The manner in which organizational resources are aligned to support a project takes many different forms. In one form, an individual project is established as a distinct organizational element. Instead of one functional group, smaller functional counterparts are established within each project. Thus, instead of one production organization for an entire plant, groups of production people are identified to support specific projects. Their functional units participate by the provision of members from the parent functional department to serve on the project teams. The focal point for all the activity on a project going on within an organization is an individual designated as a project manager. The task of this individual is to achieve the objective of the single project which is assigned to him or her.

From this brief description of basic project ideas, you can see that the systems approach to implementation is as sophisticated as the systems approach to planning. In this chapter, we shall attempt to explain why this is necessarily the case and, in doing so, to develop an understanding of the contexts in which project management is appropriate.

Projects and Programs

The notion of a "program" was introduced in the planning context in Chapter 7. We frequently see the terms "project" and "program" used interchangeably, so the relationship of projects and programs needs to be clarified.

Some organizations indeed use the two words to mean the same thing. However, more generically, projects and programs are distinguished by their size and duration. For instance, one company makes the following distinction:

> A project is a narrowly defined activity which is planned for a finite duration, with a specific goal to be achieved (such as a new product introduction). When a project is successfully completed, the project will be terminated. A program is defined as a

functional or multi-functional broadly defined business activity, which is planned to be of continuing duration and which will provide leadership influencing group strategies.[4]

Another way of differentiating projects and programs is from the perspective of the manager. Often, project managers will not have full-time administrative responsibility for the individuals assigned to their project. On the other hand, because of the continuing nature of programs, a program manager is more likely to have direct administrative responsibility for those individuals who are necessary to support the program.

Because of the implied scale of programs, program managers typically are assigned to manage only one continuing program on a full-time basis, whereas a project manager may be assigned to manage one or several projects. The assignment of a project to a project manager may be on a full-time basis or in addition to that individual's full-time permanent job responsibilities within the organization.

The Genesis of Project Management in the Organization

There are certain environmental conditions that stimulate the need for the use of project management. An increasing fluidity in the operations of an organization as it tries to accommodate to technological change, to introduce new products, to address multiple markets, and to deal with new and different customer needs leads to the realization that the conventional hierarchical mode of organizing management systems may be inadequate to accomplish the desired results.

Sometimes temporary teams are formed, composed of people drawn from the various parts of the organization, brought together to do a task, and then returned to their permanent functional home.

As the need to identify an individual as the "general manager" of such an ad hoc activity emerges, a project manager comes into being. Often one or more functional managers are identified to provide resources to a project manager who, in turn, integrates the resources to obtain results. The use of two managers who work together cooperatively to achieve project results provides an approach to deal with the complexity and ambiguity of dynamic multiproduct and multimarket situations. Project management results in a mutual concession between functional and product departments; a *matrix* organization and a matrix management system may eventually grow out of this mutual concession.[5] The matrix organization is built around the tasks or activities to be completed. Obviously, this is a form of management that departs from traditional bureaucratic models. A review of the characteristics of the project management context should make this clear.

[4]"Guideline for Use of Program/Project Management in Major Appliance Business Group," General Electric Corporation, Mar. 14, 1977. Reprinted with permission of the GE Company, Louisville, Kentucky.

[5]The matrix organization is an advanced form of project-management system. It will be treated in more detail in Chapter 10.

THE PROJECT CONTEXT

Although project-management ideas are applied under a variety of di[illegible]nt titles in different organizations, there are a number of salient factors which characterize most applications:

1 The manager (project manager, task-force leader) operates independently of the organization's normal chain of command—a "horizontal hierarchy" comes into being, reflecting an amalgamation of interfunctional resources directed toward a specific goal having time, cost, and technical performance parameters.

2 The project manager negotiates directly for support from the functional elements; normal line and staff relationships give way to a "web of relationships" directed to the beginning and completion of specific undertakings within the organization.

3 While the role of the project manager may vary widely from one of a coordinating nature to "general manager" function, this manager is the single focal point of contact for bringing together organizational effort toward a single project objective.

4 The organizational life of the project tends to be finite in nature. A particular project ends but will be replaced by another which involves a different product mix, advanced technology, or slightly different objective. In the ongoing and healthy organization, there will be a continuous flow of projects which represent the basic building blocks of the organization's business.

5 A deliberate conflict exists between the project and the functional purposes of the organization. The functional elements are charged with maintaining a pool of resources to support the organizational effort, whereas the project is directed to delivering a product or service on time, within budget, and in satisfaction of the technical performance requirements. Open hostility can break out between the project manager and functional department heads over the competition for the time and talent of department personnel.

6 Each project involves more than one subdivision of the organization; often the project has companywide application.

7 Projects can originate from anywhere in the organization. Projects having as their purpose the application of technology will usually originate on the R&D side of the house; those primarily of a marketing nature (such as a product modification) will usually emerge in marketing.

8 There is a dual chain of command in project management: functional and project. Some people report to two supervisors rather than to the traditional single supervisor.

9 There is a sharing of decisions, results, rewards, and accountability.

10 The people on a project come from different disciplines; integration of these disciplines to support project objectives becomes a responsibility of the project team under the leadership of the project manager.

11 Peer evaluation comes into play among the members of the team.

12 The introduction of project management sets in motion the development of supporting systems such as personnel evaluation, finance, accounting, and information.

Evolution of Project Management

As would be expected, the current state-of-the-art in project management is the result of evolutionary changes over the years. One of the earliest expositions on how modern project management evolved is contained in the research conducted by Davis in 1962.[6]

Davis studied manufacturing firms in the western United States to analyze the types of organization which these firms were using that were sometimes designated as a "project management" type of organization. He distinguished four types of project-management organization. The first was a "project expediter." Such an individual did not perform primary management functions, but rather performed two essential activities: expedition of the work, and serving as a center of communication to be able to report to general management on the totality of the project. The project expediter accomplished unity of. He served also as an interpreter and translator of complex scientific concepts into the cost, market, and other business interests which general management has. In some cases, his expediting role dealt only with technical aspects. In other cases, he was responsible for expediting the entire budget.

The second type of project organization was called a "project coordinator." Such a coordinator had independent authority to act and was held responsible for his action, but he did not direct the work of others. This project coordinator was more of a staff leader, exercising leadership through procedural decisions and personal interaction rather than through any line authority.

The third type of project organization was headed by a manager who actually performed a full range of management functions: planning, organizing, motivating, direction, and control. Hence, such an individual could properly be called a manager. Davis called this third type of organization a "project confederation." In addition to the unity of command and of control found in the roles of project expediter and project coordinator, the project-confederation approach achieves unity of direction.

In the last type of project organization, an individual—a project manager—exercised unitary command. In this arrangement, people were temporarily withdrawn from their functional homes and assigned to the project under the project manager, who became their "chain of command" manager until the individuals went back to their functional homes. This fourth type was called "project general management." The project manager virtually directed the complete project as if he or she were a profit-center manager. Davis noted the

[6]Keith Davis, "The Role of Project Management in Scientific Manufacturing," *IEE Transactions on Engineering Management,* vol. 9, no. 3, pp. 109–113, 1962.

role adaptability of the project manager, who maintained a wide range of contacts, from general managers to vendors, and from the research to the sales department. As the project manager actively managed the project, he visited with customers, subcontractors, the home office, and the general public. Davis also noted a key activity of the project manager when he stated:

> . . . the function of project manager requires a balancing of technical solutions with time, cost, resource and human factors. The project manager is an *integrator and a generalist* rather than a technical specialist.[7]

Some organizations evolve through various forms of project management as they grow more sophisticated and as their management problems become more complex. However, now that project management has reached a level of maturity, this does not mean that its simpler forms have been discarded. For instance, according to *Business Week,* the expediter role is on the rise:

> . . . as the size and cost of plants and many industrial manufacturing projects increase and as the lead time for delivery of vital supplies grows longer, the demand for effective, top level trouble-shooters has started to soar.[8]

Perhaps the best overall indicator of the growth of interest in and the application of these various forms of project management is the growth of membership in the Project Management Institute (PMI), a nonprofit professional organization dedicated to advancing the state-of-the-art in project management. This institute was started in 1969 with only a small cadre of members. By 1981, it had 3,500 active members. PMI membership includes a broad cross section of individuals from industrial and manufacturing companies, engineering-design and architectural firms, construction companies, utilities, educational institutions, pharmaceutical companies, aerospace companies, consulting firms, and all levels of government. The membership is widely dispersed throughout North, Central, and South America, Asia, and Europe.

More broadly, however, the use of projects and programs in modern organizations highlights the growing cultural changes that are affecting the way we manage contemporary organizations. The introduction of a new integrating component such as project management into a traditional organizational structure has important implications for organizational style and behavior. It also involves significant departures from prevailing patterns of management practice. Heretofore, the bureaucratic form of organization has prevailed as a model to assemble and manage resources. Toffler recognized how contemporary cultural changes were swaying organizational styles:

> We are witnessing the breakdown of bureaucracy. We are, in fact, witnessing the arrival of a new organizational system that will increasingly challenge and ultimately supplement bureaucracy. This is the organization of the future. . . . The high rate of

[7]Ibid., p. 112 (emphasis added).

[8]"As Costs Climb, a Rising Demand for Expediters," *Business Week,* Feb. 23, 1981, pp. 132J–132N. Reprinted with permission. © Copyright by McGraw-Hill, New York. All rights reserved.

> turn-over (in organizational relationships) is most dramatically symbolized by the rapid rise of . . . project or task force management. . . . Indeed, project management has in itself become recognized as a specialized executive art. . . .[9]

Anyone who has worked in a bureaucracy or a rigid hierarchy realizes that success depends on effective interaction with other people, including line superiors, subordinates, peers, and a host of others who play some role or who can provide information and resources. Everyone seems to talk to everyone else. Vertical contact occurs in the "chain of command"; horizontal contact is carried out in functional departments, with geographical units and product segments. In some cases "dotted line" relationships are used to establish some form of formal interaction. But, more often than not, informal contact or "underground" communication helps to pull things together. The perceptive individual finds that people coordinate their work on their own without having to work through a supervisor. For many managers, an adopted behavior characteristic of a formal matrix management system emerges—even if the name is not applied to it.

THE HORIZONTAL CULTURE

Because project management superimposes a horizontal dimension onto the traditional vertical organization, many have come to think of project management as "horizontal management." Indeed, in some organizations, project ideas are so well accepted that a *horizontal culture* exists. Such a culture is based on the premise that an organization has great difficulty in efficiently and effectively accomplishing more than one major venture if it is organized totally along traditional vertical lines.

The number of applications of project management in new organizations and in new kinds of project contexts continues to grow. For instance, the applications of horizontal organizations to strategic planning are among the newest such innovative uses of the concepts.

One of the best examples of this is reflected in the way the General Electric Co. has organized for strategic planning, namely, through the use of "strategic business units," or SBUs, as they are commonly called. An SBU is an organizational structure for strategic planning which is superimposed over the operating structure of the company. These SBUs may involve one or more profit centers so long as they are unique from a planning point of view. Different priorities are assigned to different SBUs, forming a matrix type of enterprise within the General Electric Co.[10]

In some organizations, the horizontal culture manifested in the use of project management is commonplace. From a parochial beginning in the aerospace and construction industries, project management has proliferated in multinational

[9]Alvin Toffler, *Future Shock,* Random House, Inc., New York, 1970.

[10]See "Management: GE's Search for Synergy," *The New York Times,* Apr. 16, 1978, for further information on GE's matrix-based SBU Strategic Planning System.

companies, service institutions, health-care agencies, government units, educational organizations, and most large businesses and industries. The use of a permanent matrix management system, while a fairly recent phenomenon, continues to grow. A horizontal culture is truly emerging in the management of contemporary organizations.

We may pause to reflect on the significance and implications of the horizontal culture. What, for instance, will it mean to future managers? Will they need to be trained differently—e.g., to be given a "horizontal bias"? To explore these implications, let us examine the underlying bases for the emergence of horizontal management concepts.

The Role of Managers

Managers appear to have entered an era of identity crisis in the last decade—gradually relinquishing an autonomous role in organizations to adopt a new one of developing bridges to link the organization to its environment. This move from the traditional hierarchical position of downward supervision to a facilitating and coordinating responsibility within complex economic, social, political, and technological systems calls for a much needed look at the *structure and process* makeup of the contemporary organization. While management theory has made great progress in the development of scientific tools and methodology for solving complex resource allocation problems, the organizational interfaces that affect the outcome of the application of these resources are not fully understood.

Organizations emerge to satisfy the need for pooling the efforts of many individuals. The development of organizational theory took its modern shape in response to the requirement for an interface between two isolated segments of society—one represented by society's need for products and the other by the availability of scientific management methods and discoveries. Based on this, one might hypothesize that the "business" that management is in is the interaction between these two segments. Today, when a subsistence level of product demand has been largely satisfied in our society, and when people are playing a more aggressive role in controlling their organizational destinies, the manager emerges in a somewhat different role.

Contemporary managers play *three key roles* in their implementation function. *First,* they are responsible for a given organization—their *direct supervisory duties. Second,* today's managers also act in a *facilitating capacity*— bringing about effective organizational interfacing with all aspects of technical, legal, social, economic, and political systems. *Third,* they provide leadership in the effective meshing of organizational and product (or service) interfaces. It is this new role which project management directly addresses.

Skill in horizontal relations is required. Conflict must be dealt with while preserving positive interpersonal skills and attitudes. Influence and power will be based more on expertise than on formal authority. For many managers, this requires an adjustment of their values and expectations.

Personal Values and Expectations

We are at a time in history when many values of the past generations are being questioned, past methods are being challenged, and goals and objectives are being reevaluated. There is a rapid and continuous change taking place in public priorities, with emphasis shifting from an economically dominated set of values to socially conscious, intangible, and aesthetically oriented standards. Although there has always been a series of changes in society, the pace is getting much faster now. Since any industry has to abide by the rules of the society it is serving, management theory should adjust itself to the changes in public attitudes.

The reassessment of management philosophies is in no small part due to the attitudes of young people—the managers-to-be whose value systems and attitudes reflect a family culture far different from that of previous times. According to Lee:

> There can be little doubt that the family is the most powerful single environmental force in shaping the personality structure, social response patterns, values, and attitudes toward work, authority, and autonomy. The power relationships between government, business, and labor are mirror images of the changes that have taken place in the individual family.[11]

Family influences provide the basis from which cultural attitudes are built and personal expectations emerge.

Of course, this new set of values may well be temporary. In the last thirty years, we have seen several types of cultural attitude reflected by managers in our society. In the 1950 time period the *organization man* prevailed. He was devoted primarily to the service of the organization. He found little difficulty in reconciling his personal goals with those of the organization. If a real conflict developed, the cultural influences of the organization strongly conditioned him to subordinate his goals to the purposes and well being of the organization. Typically, the validity of organizational goals was accepted—certainly not openly questioned. To do so would have been disloyal, if not subversive.

The organization man was followed by the *scientific manager,* who was more deliberate, platonic, yet zealous, and skilled at problem solving; he was dedicated to a scientific methodology of problem solving rather than to an organization. Organizational goals could be questioned but only through the medium of decision models and within the context of evaluating risk and uncertainty factors in the objective analysis. These young managers were exemplified by the "whiz kids" of the Department of Defense who have been charged with "abandoning the durable lessons of experience for the frail guide of logic." Nevertheless, these managers made their influence felt.

Today's young managers were the concerned youth of the early 1970s. They

[11]James A. Lee, "Behavioral Theory versus Reality," *Harvard Business Review,* March–April 1971, p. 25.

are influencing, and will in the future influence, modern organizations. A review of the concerns expressed by these young people in the early 1970s may help us to understand the present and the future. They were concerned about identity, stimulation, openness, concern, change, and questioning the existing order of affairs. These people are now managers who are determined to be heard—to participate in designing the strategies of the organizations to which they belong. A look at the personal values of the business student during the early 1970s when these cultural changes were under way illustrates the point. According to Miner, they included the following:

> Less trusting, especially of people in positions of authority.
>
> Increased feelings of being controlled by external forces and events, and thus believing that they cannot control their own destinies. This is a kind of change that makes for less initiation of one's own activities and a greater likelihood of responding in terms of external pressures. There is a sense of powerlessness, although not necessarily a decreased desire for power.
>
> Less authoritarian and more negative attitudes toward persons holding positions of power.
>
> More independent, often to the point of rebelliousness and defiance.
>
> More free and uncontrolled in expressing feelings, impulses, and emotions.
>
> More inclined to live in the present and to let the future take care of itself.
>
> More self-indulgent.
>
> Moral values that are more relative to the situation, less absolute and less tied to formal religion.
>
> A strong and increasing identification with their peer and age groups, with the youth culture.
>
> Greater social concern and greater desire to help the less fortunate.
>
> More negative toward business, the management role in particular. A professional position is clearly preferred to managing.
>
> A desire to contribute less to an employing organization and to receive more from the organization.[12]

Today, many sense that something is wrong with management in America. Inflation, low productivity, unemployment, lack of product quality, the growing threat of foreign competition, the threat of energy crises, all serve to remind today's managers-to-be that America's managers can no longer manage effectively without some form of help from those who are the object of management. These young people know that they are maturing in a democracy, yet too few of them see any real democratic practices at their place of work.

The impact of these managers-to-be of the late 1960s, 1970s, and 1980s is being felt as they take on added responsibility in their organizations. Older managers are having to deal with new lifestyles and demands for increased personal attention, intensive career planning, openness with information. As these young managers' influence rises, they will:

[12]John B. Miner, "The OD-Management Development Conflict," *Business Horizons,* December 1973, p. 32.

> . . . lead corporations toward more openness and disclosure; more debate before making decisions; more emphasis on selecting, training and rewarding people, including more women and non-whites; and greater flexibility in life-styles.[13]

Butler notes that workers in our large rationalized industries and businesses are seeking more control over and involvement in the forces affecting their work lives.[14] In part because of the rising levels of education, changing aspirations, and shifts in values, especially among young people, we are witnessing a quiet revolution in what people expect from work, an expectation that goes beyond the economic and job-security issues that led to labor unrest in an earlier day.

It is difficult to judge how these changes will affect organizational productivity, efficiency, innovation, and other U.S. business organization concerns. But it is safe to assume that these young managers will bring about an uncomfortable ambience for many older managers and professionals.

If the foregoing characteristics of future young managers are valid, then we can expect these individuals to be less oriented to a vertical model of management and more concerned with playing a facilitating role in an organizational system. Indeed, these managers may well be the harbinger of the future, a manager-philosopher who, while performing the analytical role required in making management decisions, also truly sees the larger role of the organization in a greater system and is able to guide the organization to participate and contribute in its overall societal context.

Broad Cultural Changes

Concurrent with changes in management roles and in personal values and expectations have come broad cultural changes which have helped to break the vertical organizational culture. Some of these have been described by Delbecq as:

1 The western tradition of fragmented power, which has diluted the legitimacy of the hierarchical positions.

2 An emphasis on expertise and achieved status, as opposed to charisma and positional status.

3 Rise of the "scientific ethic," with its propensity toward evolution and change, including modification of conservative social institutions.

4 The knowledge explosion, making omniscience anachronistic and creating an imbalance between hierarchical authority and instrumental competence.

[13]"The '60's Kids as Managers," *Time,* Mar. 6, 1978.

[14]Arthur G. Butler, Jr.; "Behavioral Implications for Professional Employees of Structural Conflict Associated with Project Management in Functional Organizations," University of Florida 1969, *Business Administration,* p. 1938-A. For recent confirmation based on survey data over a period of 25 years, see M. R. Cooper, B. S. Morgan, P. M. Foley, and L. B. Kaplan, "Changing Employee Values: Depending Discontent," *Harvard Business Review,* January–February 1979, p. 117.

5 The impact of educated and mobile professional and technical personnel on organizational systems, including expertise and collaboration in decisions affecting working relationships.

6 The changed nature of transactional behavior and organizational goals, resulting in more complex technologies, whether programmed or unprogrammed.[15]

More autonomy for the individual, greater participation in determining both personal and organizational goals, and a greater appreciation for the ability of the nonmanager to effect organizational change are all elements which promise more experimentation with unique, perhaps revolutionary, management styles and organizational arrangements.

Changing Concepts of Leadership

Drucker has characterized three different kinds of managerial "work" within an organization:

> (1) the operating task, which is responsible for producing the results of today's business; (2) the innovative task, which creates the company's tomorrow, and (3) the top-management task, which directs, gives vision, and sets the course for the business of both today and tomorrow.[16]

Thus, he sees the role of top management as a leadership one, which is related to *both* the operating and innovative aspects of the organization.

Traditionally, leadership has been thought of more in one-dimensional terms, i.e., in terms of the *operating* task as it relates to the guidance of *subordinates*. In the past, leadership emphasized solitary decisions and depended on the followers to implement the decisions. Leaders played several roles: authorizer, delegator, coordinator, and evaluator. "As authorizer, he issued commands. As delegator and coordinator, he allocated credit or blame."[17] Rarely did the leader actively participate in carrying out the organizational assignment except in a *supervisory* capacity, to become personally involved only when control was indicated.

In today's organizations, people react more to ideas than to orders. More than ever, success depends on the leader's quality of information rather than experience. Lombard notes:

> The general manager's capacity for problem solving will be based more on his skills of *processing* information about a situation accurately than on his ability to contribute knowledge as an expert from a field such as engineering, law, or accounting. He will

[15]Andre L. Delbecq, *Matrix Organization—An Evolution beyond Bureaucracy*, The University of Wisconsin Press, Madison, p. 120.

[16]Peter F. Drucker, "New Templates for Today's Organizations," *Harvard Business Review*, January–February 1974, p. 51.

[17]Mack Hanan, "Make Way for the New Organization Man," *Harvard Business Review*, uly–August 1971, pp. 128–138.

work in many roles—conceptualizer, negotiator, counselor, arbiter, teacher—to improve the capacity of individuals to affect their environment.[18]

The authority of the leader that is respected is that which has been earned by performance and not merely granted by the legitimacy of an organizational position. Collaborative leadership is emphasized, so that the concept of a leader is broadened. Now *the effective leader is recognized to be one who carries influence with superiors, subordinates, and peers,* and also is able to effectively interface his or her organization with complementary organizations in the greater environmental system.

A leader can be described as having these major functions:

1 The effective supervision of resource application within a given organization

2 The perception of and reaction to the social, political, economic, legal, competitive, and technological forces in the broader environmental system

3 The facilitation of an organizational environment wherein the members of the organization can attain economic, psychological, and social satisfaction

4 The contrivance of survival and/or growth strategies for the organization

Each of these roles requires a different blend of conceptual, behavioral and analytical skills. Effectively applying resources within the parent organization requires human and conceptual skills; the contrivance of strategies for growth and survival in the broader competitive and environmental system demands analytical and conceptual skills. Providing an environment that is simultaneously conducive to innovation and satisfaction requires a wide range of skills.

Mashburn and Vaught suggest that dual leadership is needed. Drawing on the Edwin Fleishman work in the Ohio State studies which identified two types of leadership, namely, "task" and "social," they suggest that *two leaders are better than one.* The need for a contemporary manager to assess group needs, task structure, goals, and other variables exceeds the capacity of any individual except a "super leader." Put another way, "the manager who is both people-oriented and goal-oriented is a rarity." They further note:

> . . . we advocate just one goal-oriented manager for each work group in the organization, as is typical today. However, we also advocate the appointment of a relationship-oriented manager to act as a co-supervisor in the management of each group. Although their psychological approaches would differ, their efforts would not be at cross purposes.
>
> Thus two compatible supervisors (one task oriented, the other relationship oriented) can more effectively achieve both high productivity and employee job satisfaction than one supervisor working alone.[19]

What such an approach would do to management costs is open to speculation. But the notion has appeal.

[18]George F. Lombard, "Relativism in Organizations," *Harvard Business Review,* March–April 1972, p. 64.

[19]James I. Mashburn, and Bobby C. Vaught, "Two Heads Are Better Than One," *Management Review,* December 1980, pp. 53–54.

Organizational Implications

It is not at all clear that these changes in culture, managerial roles, personal values, and concepts of leadership should necessarily have led to anything more than increased dissatisfaction on the part of organizational participants. Traditionally, organization structures reflected a historical way of doing things, with the organizational design focusing on some pattern of functions, products, or geography. A pattern—of function, for example—once established, tended to perpetuate itself through amalgamating resources and encouraging emotional attachments which resulted in parochial attitudes and continuance of the established design.

Yet, in the recent past, managers and organization theorists have come to recognize the inadequacy of these traditional forms. For instance, in talking about his three kinds of managerial work—operating, innovative, and top-management—Drucker concludes that "no one organization design is adequate to all three kinds of work; every business will need to use several design principles side-by-side."[20] These recognitions, based on the perceptions of change, have led to a breaking down of traditional organizational barriers and to the development of project organizations and other horizontal forms.

Of course, project management is only one of many possible implications of how all of these forces will impact on the organization. The predictions made by *Business Week* and many others in the early 1970s have indeed come true:

> Project teams and task forces will become more common in tackling complexity . . . there will be more of what some people call temporary management systems as project management systems where the men who are needed to contribute to the solution meet, make their contributions, and perhaps never become a permanent member of any fixed and permanent management group.[21]

SOME HORIZONTAL ORGANIZATIONS

Having traced the broad elements which have set the stage for organizational change, let us now examine some of the specifics of some of those organizations which have been established to cope with the horizontal culture.

Weapons System Management

The Department of Defense is the largest single customer of the American industrial complex; the government depends almost exclusively on private industry for the development and acquisition of weaponry.[22] Unlike the situation

[20] Drucker, "New Templates for Today's Organizations," p. 51.

[21] "Business Says It Can Handle Bigness," *Business Week,* Oct. 17, 1970, p. 115.

[22] "Weaponry" is a general term connoting the varied instruments designed to inflict damage on an enemy through the destruction of physical or mental capabilities. The term "weapons system" refers to a highly sophisticated weapon, composed of a combination of equipment, skills, and managerial know-how, which as an integrated entity is capable of effectively destroying an enemy.

in the civilian consumer market, the magnitude and scale of the military market are determined by advance planning in the government. The federal budget plays a primary role in determining the direction for defense buying. Military expenditures are authorized as the result of continual interaction of many demands and requirements—not only of the defense sector of the economy, but of nondefense programs and policies as well.

Recent experience in military preparedness suggests that a requirement will continue to exist for a high level of military technology. The Office of the Secretary of Defense has received greater authority in the management of the military, economic, and social aspects of national defense. Traditional roles and missions of the military departments have changed drastically. Some functions of the Armed Forces have been merged and unified, and a single national system of defense may evolve. Technical breakthroughs, protracted weapon-development cycles, and increasing attention to the cost-effectiveness relationship of weaponry have stimulated a variety of innovations in management.

It is not surprising that one of these innovations is project management, because of the very nature of a weapons system—an interrelated system composed of hardware, software, and personnel. A missile system would include, for example, the missile subsystem itself, an integrated logistic support subsystem, and a personnel subsystem.

The development and production of a missile system requires the involvement of many industrial and government organizations. Many different people from different disciplinary backgrounds must be integrated to pursue a common objective. Even after the system is operational, many of these same complexities exist. In the Manhattan Project's development and production of the first atomic bomb, the Department of Defense found that a project type of organization was necessary in order to provide a unity of purpose and to establish a focal point for pulling together the cooperative efforts of literally dozens of relatively autonomous organizations. Then in the 1950s, as the United States began the development of the ballistic missile system, the Department of Defense again turned to this form of organization and required contractors in the aerospace industry to do likewise. Since then, project management has become a way of life in the NASA, defense, and aerospace contractor environment.

Project Management in the U.S. Department of Agriculture

In 1973, three forest insects were causing major damage to the forest resources in the United States. In the West, the Douglas-fir tussock moth was epidemic on over 800,000 acres. Across the South, from Texas to Virginia, the southern pine beetle was damaging or threatening over 74,000 square miles of susceptible pine-forest types. In the Northeast, the gypsy moth had defoliated over 1.7 million acres of hardwood forest in rural and urban areas. Damage from these three insects alone was estimated at over $100 million. Methods for suppressing these outbreaks were either unavailable or only partially effective and of short-term value.

A Combined Forest Pest R&D Program was initiated within the Department of Agriculture to pull together resources and develop a coordinated program to provide means for suppressing damage from these three pests within a short time frame. This R&D program was divided into three separate subprograms: the Gypsy Moth R&D Program, the Douglas Fir Tussock Moth R&D Program, and the Southern Pine Beetle R&D Program. Management of these programs was accomplished within the Office of the Secretary, Department of Agriculture.

Each of these three programs was directed by a program manager. These program managers had full authority to manage the programs across a wide spectrum of organizations within the Department of Agriculture, the Environmental Protection Agency, the forestry industry, and the university research community. Each program manager was responsible for selecting and funding R&D activities which met the requirements of the annual program Plan of Work and Budget. In addition to the program managers managing activities across several federal agencies, the R&D activities were carried out in 30 universities, colleges, and state agricultural-experiment stations. In addition, 9 state forestry organizations and a number of private companies were involved.

Project Management in the Construction Industry

Fluor Utah, Inc., a large construction company, operates under a strong task force[23] organization headed by a project manager. When the company was small, the owner acted as project manager over the construction projects performed at the company's beginning. In these early days, before the term "task force" was conceived, it was apparent that there was a real advantage to the owner and to the project by having a single point for management by an individual who had total responsibility for the project.

Within Fluor Utah, Inc., a task force is created under the direction of a project manager who is primarily responsible for the company's performance of the construction contract. The task force is composed of engineering, procurement, construction, project-control, and administrative personnel functioning under the direction of the project managers. These individuals are assigned to each task force by the various functional heads, who retain certain authority and responsibility for the quality of the work in accordance with established company procedures. A typical task-force organization chart at Fluor Utah, Inc., is reflected in Figure 8-1.

[23]Described in Robert K. Duke, "Project Management at Fluor Utah, Inc.," *Project Management Quarterly,* vol. 8, no. 3, September 1977. You should note that the task force is quite similar to the project team—a group of professionals are pulled together to accomplish some specific action or objective, with membership drawn mainly from the functional supporting organizations and supplemented as needed with people from other organizations. Once the project or task-force team objectives have been reached, the members are returned to their original organizational assignments. The task force, like the project team, enjoys a temporary existence within the organization.

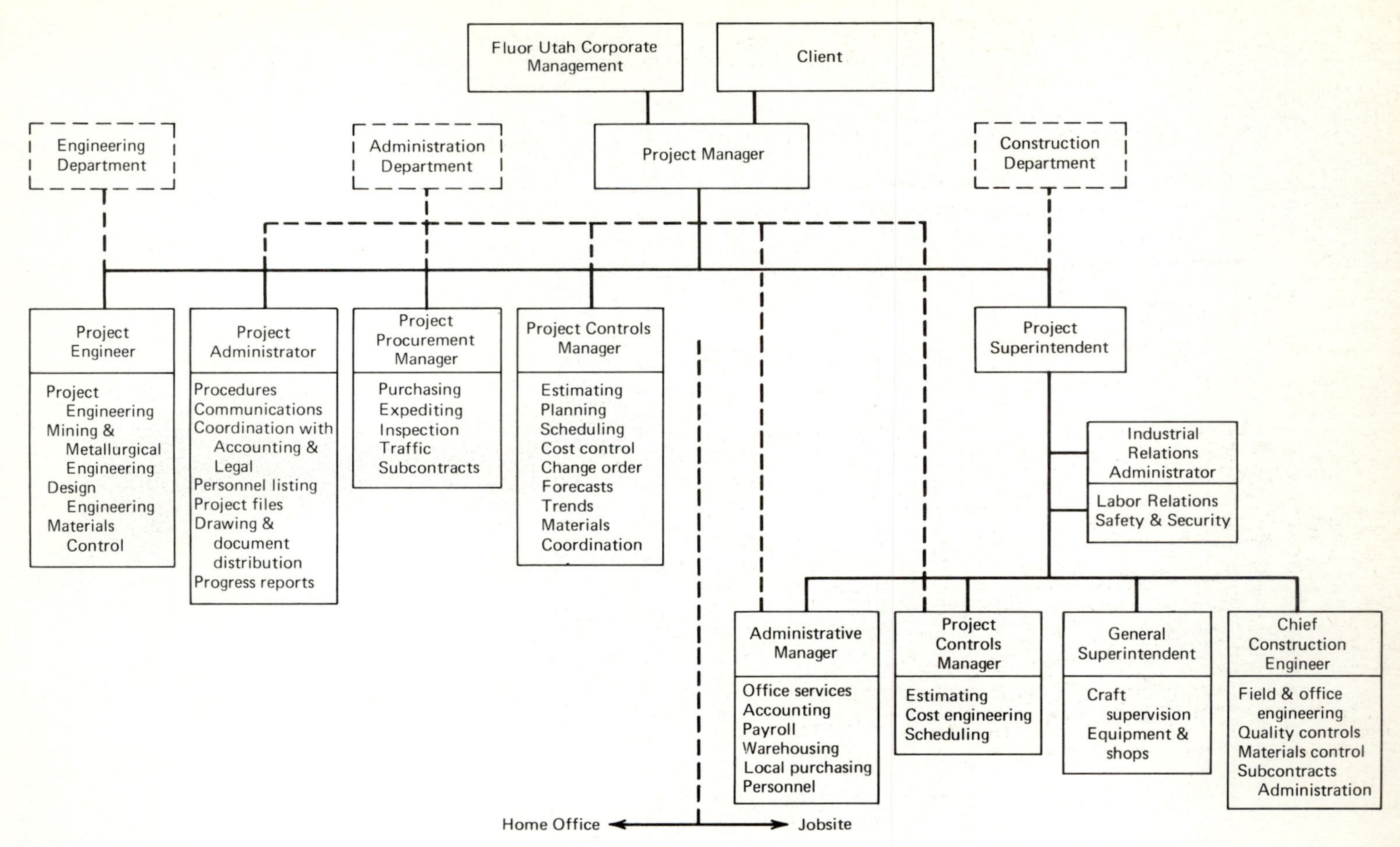

Fluor Utah Corporate Management
Client
Engineering Department
Administration Department
Project Manager
Construction Department
Project Engineer
Project Engineering
Mining & Metallurgical Engineering
Design Engineering
Materials Control
Project Administrator
Procedures
Communications
Coordination with Accounting & Legal
Personnel listing
Project files
Drawing & document distribution
Progress reports
Project Procurement Manager
Purchasing
Expediting
Inspection
Traffic
Subcontracts
Project Controls Manager
Estimating
Planning
Scheduling
Cost control
Change order
Forecasts
Trends
Materials
Coordination
Project Superintendent
Industrial Relations Administrator
Labor Relations
Safety & Security
Administrative Manager
Office services
Accounting
Payroll
Warehousing
Local purchasing
Personnel
Project Controls Manager
Estimating
Cost engineering
Scheduling
General Superintendent
Craft supervision
Equipment & shops
Chief Construction Engineer
Field & office engineering
Quality controls
Materials control
Subcontracts
Administration
Home Office
Jobsite

Intracompany Project Management

In one large U.S. corporation, the research and development laboratory is organized along project-management lines. Each R&D project has a project manager; in some cases, functional or department managers may also function as project manager on some development project. These R&D laboratory project managers have a project team in the laboratory that provides specialized support to the project. This corporation is organized on a profit-and-loss-center basis; each profit center "buys" research and development support from the corporate R&D laboratory. Each profit center will have a profit-center project manager having complete responsibility for the project. Specialized resources of the R&D laboratory are committed to the profit-center project manager by the R&D laboratory project manager. The R&D laboratory manager commits to the scope and objectives of the project, as specified by the profit-center management. The profit and loss management is kept informed of the project's progress to the level of detail desired through periodic design and status reviews as the project is carried out.

This type of intracorporate organizational arrangement is used for evaluating emerging strategic projects. Emerging strategic projects are those that are so important to a profit and loss center and so complex to manage that the project managers draw on the R&D laboratory resources to support the project. An intracorporate project-management system evolved to serve this need; Figure 8-2 illustrates this arrangement. The profit and loss center manager appoints a project manager to manage all aspects of the project within the P&L center as well as working with a project manager in the research and development laboratory who manages a project team in the laboratory. Technical guidance is provided to the P&L center project manager by the research and development technical and management personnel; all overall decisions on the project are made by the profit and loss management team. Engineers, technicians, draftspersons, and others are assigned to work on the research and development project, with the cost being borne by the customer—the profit and loss manager. The profit and loss project manager is furnished technical reports along with cost and schedule information in order that project visibility is maintained by the profit and loss manager.

Two important advantages result from this organizational arrangement:

1 Corporate research and development resources are brought into focus to support the profit center need.

2 Profit-center project managers can facilitate the introduction of technology

---

FIGURE 8-1
Typical Project Organization Chart for Fluor Utah, Inc. This organization chart shows the chain of command (dotted lines) between key personnel and their respective department or section managers. The latter has responsibility for furnishing personnel for the task force and giving necessary technical direction to ensure the quality of work. (Robert K. Duke, H. Frederick Wohlsen, and Douglas R. Mitchell, "Project Management at Fluor Utah, Inc.," *Project Management Quarterly*, vol. 8, no. 3, September 1977, p. 34.)

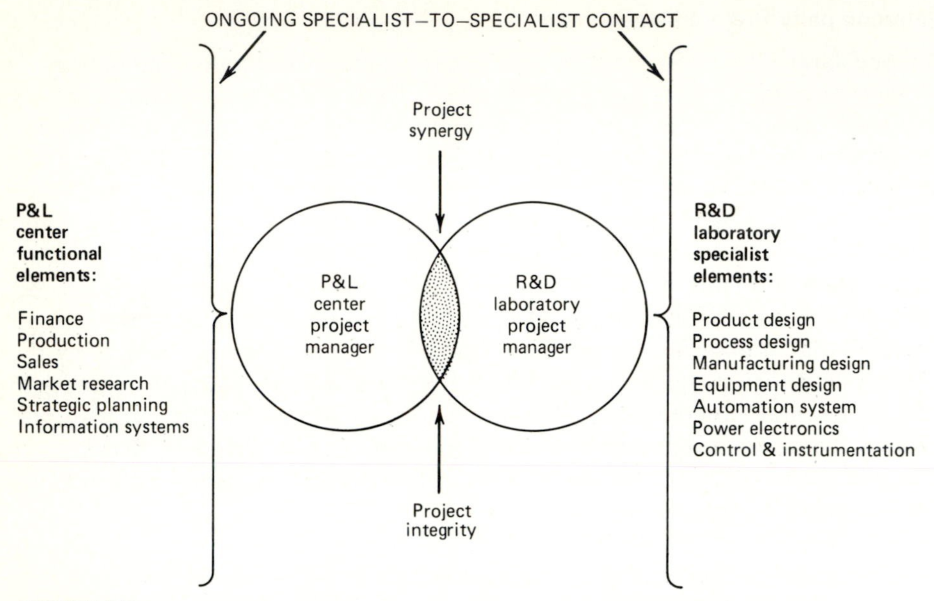

FIGURE 8-2
Relationship of Intracorporate Project Managers. Each project manager functions as an integrator-generalist responsible for ensuring project synergy. Each project manager is dependent on the other as they manage their respective project activities.

into strategic projects by virtue of their image as "local sponsors," i.e., members of the profit-center staff.

Project Management in Engineering

The General Motors Corporation, certainly one of the more conservative organizations in the U.S. industrial complex, adopted project management in its engineering function in 1974. Within General Motors, this adoption represents one of the most significant changes in organizational approaches since the profit-center decentralization concept in the 1920s and 1930s. Within General Motors Corporation, the use of the project-center concept is one of the most important managerial innovations in years. Engineering management was devised to coordinate the efforts of the five automobile divisions of General Motors. A project center consists of engineers temporarily assigned from the profit-center division to work on a new engineering design. If a major new effort is planned, e.g., a body changeover, a project center is formed across the automobile divisions which operates for the duration of the change. Project centers work on engineering problems which are common to all divisions, such as frames, steering gear, and electrical systems. The project center complements but does not replace the "lead division" concept where one division has primarily responsibility for taking innovation into production.[24]

In a similar vein, in 1971, the Microwave Cooking Division of Litton Industries, Inc., reorganized its engineering department from a functional structure to a task-team structure. For each engineering project, a task team was assigned, headed by a project manager. Members of the team included a design engineer, draftspersons, a stylist, technicians, quality and manufacturing engineers, a marketing manager, buyers, and a home economist. Each team was assigned specific responsibilities. As a project progressed through the development cycle, other personnel would be added, such as engineers, quality-control personnel, cost accountants, and production managers. The engineering project manager assigned at the beginning of a development cycle was responsible for managing the team until the product went into production and a steady level of production was reached. At that time, the task team as such was released and the team members went on to work on other tasks. In some cases, however, the project team might stay together for additional time in order to work on product improvement, cost reduction, and such ad hoc activities.

The task team becomes the central organizational mechanism by which the development activity is reviewed. In this review, members of top management and key department heads meet with the task team to assess progress against project goals, to discuss the problems encountered, and, in general, to get a good assessment of how things are going on the project. At these review meetings, decisions can be made as to whether the project should proceed and what changes in direction or objectives on the project should be taken. In the Microwave Cooking Division of Litton Industries today, all engineering projects are managed on this basis. A 90-person engineering department is organized into 14 task teams and can handle approximately 40 projects at any one time.[25]

"Special Projects"

Often, project management ideas are applied on a "special projects" basis, such as was discussed in Chapter 7, as one way of conducting systems analyses. When this is done, the implementation-oriented project-management approach is actually being applied to the planning phase.

For instance, sometimes project management is used to develop new business opportunities. Osgood and Wetzel have developed the concept of using project teams to identify a new business opportunity and turn it into a viable business. According to them:

> The team would have direct decision-making responsibility for taking a new venture

[24]For further information on the General Motors profit-center/project-center concept, see Charles G. Burck, "How GM Turned Itself Around," *Fortune,* Jan. 16, 1978, pp. 87–100.

[25]William W. George, "Task Teams for Rapid Growth," *Harvard Business Review,* March–April 1977.

> through the initial start-up stages, then gradually transferring this responsibility to administrative managers as the business approaches operating stability.[26]

In a similar vein, Allied Chemical Corporation established an acquisition task force in early 1980 to study potential acquisitions for the company. Within two months the company had agreed to acquire Eltra Corporation, a maker of typesetting equipment, batteries, and other electric and electronic equipment. The task force continued to look for other technology-based acquisitions in the waste-disposal, environmental-controls, and energy-controls fields.[27]

Project Management in School Administration

The opportunities for project-management techniques in school administration are many.

An educational organization, e.g., a public school system, might be viewed as a collection of teams composed of teachers, administrators, students, parents, and so forth, acting in a coordinated way. Each team has a leader and followers—each team has specialized activities to perform. Each individual must relate and coordinate efforts with members of the team and with other individuals on other teams. Administrators have to relate individual effort and team effort. The result is a "web of relationships" existing among the participating individuals and teams in the total organizational system. A matrix organizational approach is required.

The matrix view of school "organization" naturally results in many crossings of "chains of command" to manage team effort on projects which cut across many different subsystems. A useful way of describing a two-dimensional model of a matrix organization at a central administration level (such as a district superintendent's office) is by using a model like Figure 8-3. This model represents both functional departmentation (organizing along departmental lines such as business services, pupil services, and community services) and "projectization."

Each of the projects depicted in the left portion of Figure 8-3 is comprised of a project manager and work groups from functional departments. The managerial emphasis in these functional departments, because of the nature of their activity, would be primarily on improving current operating efficiency.

The level of participation by various functional entities in each project is highly variable, as each project progresses toward completion. In the early stages of a project, emphasis may be on planning and choice of strategies and alternatives. As the project is being implemented (when the end result of the project is being realized), emphasis may be placed on greater operating efficiency. This is the case because of the natural "life cycle" of a project. The

[26]W. R. Osgood and W. E. Wetzel, Jr., "A Systems Approach to Venture Initiation," *Business Horizons,* October 1977.

[27]Reported in *The Wall Street Journal,* Apr. 23, 1980.

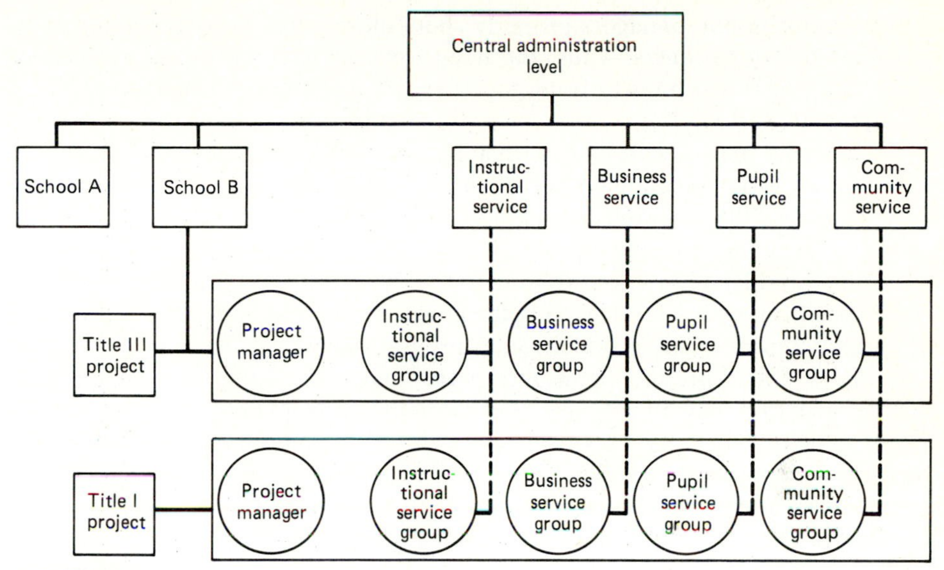

FIGURE 8-3
Project Management in a School District.

combination of people and resources that can best jointly pursue project goals changes from time to time in each of the projects. Thus, "organizational" changes will result.

Some examples of the use of project-management techniques in an educational system include:

1 Conception, design, and construction of a new school facility
2 Projects for improved reading and vocational education programs
3 Programs to advance creativity in education, such as experimental model schools and individual prescribed instruction
4 Regional planning and evaluation agencies to advance creativity in education
5 Development of new teaching equipment, performing community projects, testing out ideas generated by teachers, students, administrators, etc.[28]

Product Management

In the marketing function of consumer-product firms, horizontal management is widely practiced under the title *product management.* The creation of product-manager positions resulted from the recognition that top management could not be expected to know all the details about each of the organization's products.

[28]For further information on project management in school administration, see Desmond L. Cook, *Educational Project Management,* Charles E. Merrill Books, Inc., Columbus, Ohio, 1971.

Similarly, functional managers properly show more concern for their function than they do for products. Thus, the need for a manager who can cut across traditional functional lines to bring together the resources required to achieve product goals is clear.

The product manager is able to organize the diverse objectives and motivations of the various functional units of an organization so that the total effort is directed toward the accomplishment of overall product goals. Thus, controllable aspects of the enterprise can be viewed by one person (or group) whose concern is with the product and its contribution to the organization rather than with one of the specific methods for achieving the goals. In this way, effective implementation and control of the product's sales program may be enhanced.[29]

In most firms, product managers work much the same way as project managers, in that they do not have large numbers of people reporting to them in a line capacity. They must coordinate and integrate the efforts of others who themselves depend on functional managers as their "bosses" in the traditional sense. For instance, the product manager must normally integrate such supporting activities as pricing (done by financial people), promotion (done by advertising and sales), sales commission structures (sales), and sales and service training (personnel).

Production Task Forces

At the production level, the dissatisfaction of workers with doing highly routine, repetitive tasks is well known. This boredom has resulted in decline in productivity, quality problems, absenteeism, and labor increased interruptions and sabotage. Companies have tried to overcome this boredom by forming teams of workers who do their own work planning and control. A supervisor, in the usual sense of the word, does not exist. Rather, the use of a "facilitator" is provided to assist management in providing an environment for the production teams to work out the details of assuming responsibility for the manufacture of the entire product.

Huse and Beer describe the use of task forces in a plant engaged in manufacturing a wide variety of electrical and electronic instruments for medical and laboratory use.[30] The task forces were organized as a means of creating cohesive work groups around interrelated and interdependent activities. The motivation for the use of task forces stemmed also from the desire to create a greater meaning in work through identification with a particular group or product. The task forces brought together a close working relationship among marketing, product development, plant engineering, manufacturing, and other

[29]For discussion and evaluation of product management, see William R. King, *Quantitative Analysis for Marketing Management,* McGraw-Hill Book Company, New York, 1967, or Robert M. Fulmer, "Product Management: Panacea or Pandora's Box," *California Management Review,* Summer 1965, pp. 63–74.

[30]Edgar F. Huse and Michael Beer, "Electric Approach to Organizational Development," *Harvard Business Review,* September–October 1971, pp. 103–112.

functional elements. Eventually, the use of task forces led to plantwide changes in organization as well as changes within particular departments of the plant.

In the task-force approach used within this plant, as products moved through marketing, product development, and manufacturing, the teams were brought together to integrate the separate units and departments. A task force would typically consist of R&D, marketing, materials control, and production.

The production task-force team is used in many different contexts in industry. For example, General Electric uses worker teams of some 5 to 15 people to handle a particular responsibility of welding in a fabricating plant. The welders in the team were given responsibility for scheduling and planning their workload.

The General Foods Corporation at its Topeka pet food plant assigns production tasks to teams of 7 to 17 members. Each worker learns every job performed by a team, and pay is based on a single job rate for the team rate. There are no conventional departments in General Foods—no time clocks, no supervisors, just team leaders who work on equal terms with other team members.

TRW Systems Corporation created in one of its manufacturing plants a group of teams which it terms "semi-autonomous work teams." The workers are given the responsibility for assembling a product of the team rather than separately performing assembly line tasks. Once they were given the new assignment, the teams were allowed to schedule their own time as long as they did the job.[31]

Volvo auto has one auto-assembly plant at Kalmar, where workers on a mile-long assembly line have the option of organizing into groups of six or more persons working as teams. For example, one team of workers injects a sealing compound in all seams and installs sound-dampening insulation. The team rotates its 15 functions so that no one gets fatigued or too bored performing the same job. Pehr Gustaf Gyllenhammer, the chief executive of Volvo, argues:

> We have to change the organization so the job itself provides more for the individual. We will never build another production line as long as I am in command at Volvo.[32]

The plant where worker teams are used costs 10 percent more than one with a conventional assembly line; however, workers' morale improved, and there is less turnover and absenteeism than at other plants.[33]

At GM's Fisher Body Plant No. 2 in Grand Rapids, MI, the 2,000 employees have been organized into six "business teams," each team being essentially a business unto itself with its own maintenance, scheduling, and engineering personnel. Currently, salaried employees at all levels participate in deciding how to meet their team's objectives, but hourly-wage team members will be included in the decision-making process over the next two years.

[31]These examples are described in "Management Itself Holds the Key," *Business Week,* Sept. 9, 1972.

[32]David B. Tinnin, "Why Volvo Is Staking Its Future on Norway's Oil," *Fortune,* Feb. 12, 1979, p. 112.

[33]Ibid.

> "The plant is much more effective now than . . . when the process began," says a GM spokesman in Detroit. But he adds: "The point isn't to improve productivity. It's to improve the quality of work."[34]

In the Microwave Cooking Division of Litton Industries, Inc., the manufacturing function itself is organized into operating units. According to William George:

> Each unit is a self-contained task team responsible for all aspects of manufacturing a product series. The unit is led by a unit manager and includes two to four production supervisors, a product engineer, production engineers, quality inspectors, material handlers, schedulers or dispatchers, and 75 to 200 production workers.[35]

The task teams in Litton Industries operate as a unit and

> . . . make the day-to-day decisions in the manufacturing process. They are responsible for product quality, direct and indirect costs, production schedules, scrap, etc. They are assisted on an "as required" basis by specialists in industrial engineering, quality control, industrial relations, cost accounting, and design engineering. These specialists from the various functional departments lend expertise when required by the operating units, provide consistency in methods and processes, and insure the uniformity of standards.[36]

Since Litton formed task teams in manufacturing, production was increased fourfold in the space of 15 months, and product quality has increased. With the addition of 1,000 new production workers to a base of 400 people, unit production costs have declined 10–15 percent.[37]

Quality Circles

"Quality circles" and "quality of work life" programs include many forms of new work organization which generally mean involving workers (and sometimes professionals) in decisions through problem-solving committees. These new work organizations have come to mean a form of "participative management" that deals with a person's feeling about work-related economic rewards and benefits, security, working responsibility and conditions, organizational relationships, and interpersonal relationships in the employee's life. One of the first such successful programs was initiated at the General Motors plant at Tarrytown, New York, in 1970.[38] Today, the growing popularity of such programs is not in doubt. After a decade of experimentation and refinement, the basic principles and process have progressed from theory into practice in many U.S. companies.

[34] *The Wall Street Journal,* May 9, 1978.
[35] George, op. cit.
[36] Ibid.
[37] Ibid.
[38] See Robert H. Guest, "Quality of Work Life—Learning from Tarrytown," *Harvard Business Review,* July–August 1979.

The quality circle (QC) is a productivity- and quality-improvement technique that was implemented in Japan in the early 1960s. QCs are composed of small groups of employees who participate in improving their work and work environment. The formation of QCs in Japan grew out of the need to change the inferior image of the Japanese product. The Japanese introduced the concept of having the responsibility for quality reside with each member of the organization. Training in quality control was therefore extended to both supervisory and worker levels in the organization. A quality circle is a group of from 4 to 10 people with a common interest who meet regularly to participate in the solution of job-related problems and to identify opportunities for improvement. It is an ongoing group, operating in the work environment, that performs "surveillance" for the organization: searching for opportunities, defining problems by applying formal data collection and analysis, and arriving at solutions that are presented for acceptance and implementation by management.

The quality-circle working techniques include problem selection, brainstorming, cause and effect diagrams, data collection, histograms, check sheets, and graphs.[39] More advanced QC techniques include sampling, data collection and arrangement, control charts, scatter diagrams, and other statistical techniques. The services of professionals in the organization such as statisticians, industrial engineers, system engineers, and other staff services are made available to the QC groups. The use of statistical methods was originally highlighted in QCs. Today, the use of QCs has extended to participation in a broad range of work improvement. Westinghouse board chairman Robert Kirby emphasized the use of QCs in his corporation in the following manner:

> Throughout our corporation . . . in our offices as well as in our factories, employees are participating enthusiastically in programs such as Quality Circles. Here, employees tell their managers how they think quality, productivity and the work environment in general can be made better. And managers are listening, because nobody knows a job better than the person who is doing it.[40]

Although quality circles have been used primarily in the manufacturing environment to involve hourly production workers, there are opportunities for QCs in an organization composed primarily of professional employees. For example, a research report by D. N. Kitch, a University of Pittsburgh graduate student, notes that in the Nuclear Technology Division of the Westinghouse Electric Corporation, QCs will be working to:

Optimize use of computer time.
Improve efficiency of laboratory facilities.
Optimize mix of personnel skill levels.
Expand use of word processing equipment.
Reduce employee absenteeism.

[39]Portions of this section are adapted from David I. Cleland, "Matrix Management (Part II): A Kaleidoscope of Organizational Systems," *Management Review,* December 1981.
[40]Westinghouse Electric Corporation, Report to the Stockholders, 1980.

Reduce cost of utilities.
Reduce department supply costs.
Reduce quantity of paper flow.[41]

Decisions on matters in the quality-circle domain are shared, and results are usually shared either directly through incentive awards or indirectly through improved productivity. Multiple authority-responsibility-accountability influences are characteristic of the participative management ambience of quality circles.

"Participative management," as it is practiced at many U.S. companies, is a concept that had been in use long before quality circles and quality-of-life programs were developed. For example, at Motorola, Inc., a Participative Management Program is used to bring people together in work teams that regularly, openly, and effectively communicate ideas and solutions that help improve quality and productivity. In the process, many employees of the company claim that the program enhances their job satisfaction. The cultural ambience resulting in part from participative management within Motorola is described as:

- A structured, yet flexible way of operating our organizations to achieve our goals.
- Management encouragement and support—respect for the individual.
- Team work, trust.
- Idea sharing—releasing individual creativity—with suggestions being appreciated.
- Active participation—a return to the small company atmosphere.
- Increased communications—feedback—between management and workers.
- Common goals—in support of group goals.
- Increased skills through training.
- Employee responsibility—where direction, discipline, and control are self-directed and not externally imposed.
- Increased job satisfaction.
- Knowledge of the organization's business condition.
- A sharing (with the company) of the improvements in costs in the form of a cash bonus—with the bonus an addition to a fair, competitive base wage.
- A sharing (with the company) of the delta margin percent improvement in dollars in the form of a cash bonus—with the bonus in addition to a fair competitive base wage.
- A deeper investment of ourselves in something important to us—our jobs.[42]

The Motorola Participative Management Program was formulated in response to the need for increased productivity.

There is *no* question that participation management—by whatever name a company's program is called—deals with techniques for emphasizing teamwork. These techniques, properly applied, can improve organizational performance and individual satisfaction. Teamwork has always been an important value in

[41]David I. Kitch, "Quality Circles," graduate student research report, School of Engineering, University of Pittsburgh, Pittsburgh, Pa., April 1981.

[42]Pamphlet, "The Motorola Participative Management Program," Motorola, Inc., Schaumburg, Ill., undated.

U.S. organizational life. Participatory management is a cultural characteristic —and has worked in the United States. Project teams are by nature participative, since the team members share a common goal: the delivery of the project results on time and within budget. The successful project team identifies who is accountable for individual contributions; the realities of effective project integration and synergy dictate a team sharing of accountability, responsibility, and authority. A skillful project manager recognizes that each major decision involving the use of project resources must be "worked through" the members of the project team-indeed, a direct form of "participative management."

University Horizontal Management

The academic units of universities have traditionally been organized vertically, with deans and department heads in the academic "chain of command." Many schools have departed from this tradition in order to facilitate interdisciplinary research and teaching by abolishing the departmental structure in which professors with similar specialties are organized into "functional" departments, through which they obtain all rewards. The Southern Methodist University has organized the management of its graduate and research programs in engineering in a project matrix structure. Figure 8-4 shows the structure of Southern Methodist University organization of graduate and research programs.

The Graduate School of Business at the University of Pittsburgh has operated without departments for two decades. All 60 faculty members report directly to the Dean. Communication is facilitated through a representative committee of faculty who meet with the Dean and his senior staff to discuss strategic issues. Interdisciplinary study, research, and teaching benefit from the arrangement, since faculty members interact with a widely diverse group of other faculty members. "Interest groups"—ad hoc groups formed around a single teaching area or research project—form to focus attention on issues or areas and then disband when they are no longer useful. For instance, one interest group of faculty members from backgrounds as diverse as economics, psychology, operations research, computer science, and management formed to develop a new curriculum area in "management information systems." After the curriculum area had been introduced, a larger group of about 20 members formed to develop an organized research activity in the area. From that point, it is conceivable that a new interdisciplinary university unit—a "center"—might evolve to include faculty members from other schools.

Project Management in Mature Organizations

Project management has been described as one effective way of dealing with the growing size and complexity of organizations. In some situations, organizations have set up teams of workers to improve morale, reduce turnover and absenteeism, and improve productivity. A project team is a smaller entity than an entire organization. Within such a small entity, peer pressures doubtless

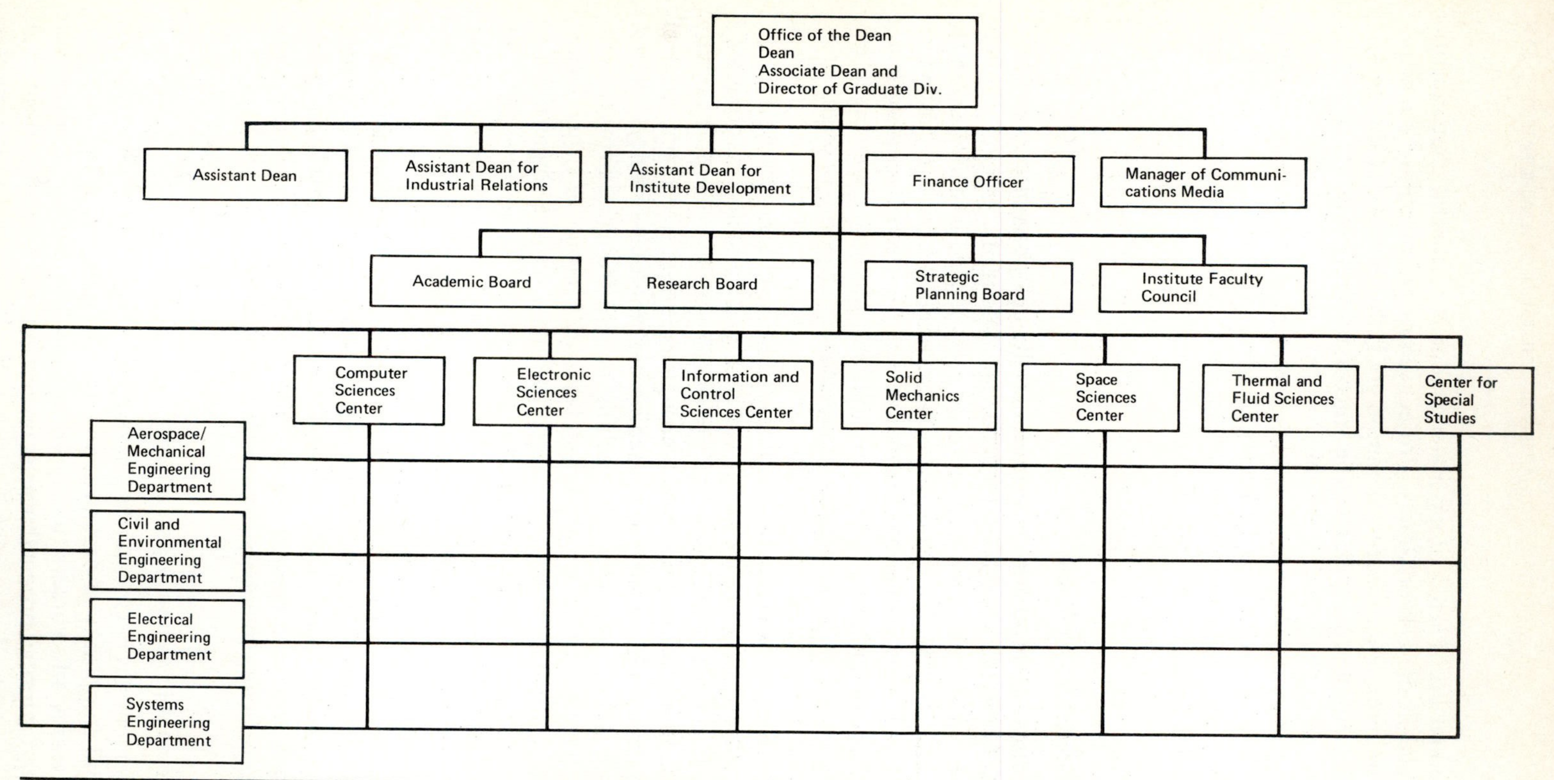

FIGURE 8-4
Southern Methodist University's Organization of Graduate and Research Programs in Engineering. (Robert E. Shannon, "Matrix Management Structures," *Industrial Engineering*, vol. 4, March 1972, p. 29.)

serve to motivate people and provide "checks and balances" on h[illegible] organizational resources are used. Also, a project team is an entity that can be dissolved when its purposes have been served. These advantages and many others provide the basis for using project teams to accomplish specific objectives and goals.

The fascinating aspect of a small project team is that it may herald the birth of a larger organization. An entrepreneur starts off with an idea and shares this idea with others, who become associated with it as disciples. As the idea grows in value, more people join the organization. We shall address this "life cycle" notion in greater depth in Chapter 9.

THE MANAGEMENT ENVIRONMENT[43]

One of the important contexts within which project-management ideas have been generated is that of traditional management thought. Hence, we shall seek to position project-management ideas within this framework of other modern management ideas which have resulted from the changes induced by the horizontal culture.

Traditional Management

To make a meaningful comparison of project management and traditional management, we must begin with the ideas behind traditional management theory. It should be recalled, however, that these ideas were developed for organizations that were smaller and environments that were simpler than those of today.

The Mainstream of Traditional Thought

The traditional theory of management evolved slowly over a period of time from the charismatic leader-follower arrangement. Some of the basic assumptions of traditional theory are those of bygone times.[44]

Pyramidal Structure The organization is viewed as a vertical pyramidal structure functioning as an integrated entity on a scalar basis. Implicit in this thought is the gradation of value placed on the different levels in the organization. The vertical levels approximate the gradations of competency. Therefore, the decisive and salient business is conducted up and down the hierarchy. Goals are established by assigning them as the responsibility of an official in the

[43]Some of the thoughts in this section have been adapted with permission from William H. Read, "The Decline of Hierarchy in Industrial Organizations," *Business Horizons,* Fall 1965.

[44]There have been many attacks on the traditional school, whose origin can be traced to Frederick W. Taylor, Henri Fayol, and Max Weber. For example, see Waino W. Suojanen, *The Dynamics of Management,* Holt, Rinehart and Winston, Inc., New York, 1966. We do not propose to continue this attack. What is desired is a comparison of the *traditional* and the *project* approaches.

hierarchy; this official exercises specific authority derived from the level of organizational position. The more crucial and important decisions are made at higher levels in the organization. Strategic decisions are combined with strategic policies and planning; routine decision making is delegated downward to a lower echelon in the hierarchy. Authority to execute decisions is passed down the hierarchy; information and responsibility are exacted upward through the intervening layers of executives.

Superior-Subordinate Relationships Since the traditional enterprise functions vertically, it relies almost entirely on superior-subordinate relationships.[45] Therefore, a strong superior-subordinate relationship is required to preserve unity of command and ensure unanimity of objective. If healthy relationships exist in the recurring chain of superiors and subordinates in the organization, the objective will be attained and the participants in the organization will gain economic, psychological, and sociological satisfaction in their jobs. The superior located higher in the structure presumably has more authority. Peer, associate, and informal relationships are present, but do not interfere with the legal distribution of power and influence in the organization.

Departmentation The alignment of a traditional organization is based on some technique of departmentation, such as functional homogeneity, similarity of product, territorial location, etc. In organization by functional homogeneity, certain functions are organic; that is, they are basic functions whose performance is vital to the perpetuation of the activity.[46] Separation of the business into organic functions encourages parochialism; each manager will be more concerned with individual area of effort than with the overall coordinated effort. However, the functional manager will be able to maintain integrated staff action through lateral staff coordination.

The Bureaucratic Syndrome Historically, organizational values have been built around the vertical structure. Principles of organization were drawn from military or church models and characterized by a form of bureaucracy. Division of labor, specialization, a visible chain of command, an objective system of policies and procedures—all these vestiges of the bureaucratic model are found in contemporary organizations and have as their purpose the assignment of subtasks to units of people who are expert and the creation of "departments" where people of similar skill, training, and attitudes are brought together. Under such an organizational arrangement, problems of *efficiency* and *control* in the *current operations* can be dealt with decisively.

[45]Frank J. Jasinski, "Adapting Organization to New Technology," *Harvard Business Review,* January–February 1959, p. 80.

[46]See Ralph C. Davis, *The Fundamentals of Top Management,* Harper & Row, Publishers, Incorporated, New York, 1951, for a full discussion of the concept of organic business functions.

Line and Staff Organizational groups in the traditional model have a basic dichotomy, i.e., the *line* and the *staff*. Line makes the salient decisions by exercising command prerogatives. Staff does the thinking and planning. Staff officials advise and counsel; their authority to command is limited by their ability (based primarily on technical competence) to influence the line official. Staff members do not issue orders to a line official, but facilitate and prescribe methods and procedures. Those in line positions plan and decide. Specialized activities apart from the line organization are staff functions. The nature of the line-staff relationships depends on a command-counsel parity. The staff is expected to influence the judgment of the line official; therefore, the staff exercises its authority by providing counsel to the line official.

The Scalar Chain The authority patterns in the traditional model of management follow the scalar chain of command. Authority flows from the highest to the lowest level, following every link in the chain. Subordinates receive orders from one superior only. Work is accomplished by relatively autonomous functional units of the organization. Individual authority is more or less constrained by the boundaries of the unit and by explicit delegation extended in the documentation. Everyday activities are set by the alignment reflected in the organizational chart. Horizontal relationships exist through informal organizations, committees, staff meetings, or formal coordination processes.

Goals Under the traditional viewpoint, goals are established only by making them the responsibility of some official and some office. When a new task is evolved, it must be assigned to an appropriate unit in the hierarchical chain. Higher-level individuals exercise authority and do most, if not all, of the directing and guiding of important matters. Authority patterns between managers and technicians in different independent organizations are ignored.

Decision Making Management theory has, in the main, approached decision making from the basis of several distinct phases:

Defining the problem
Analyzing the problem
Developing alternative solutions
Selecting the best solution
Converting the solution into effective action

Each of these phases has several steps and provides a framework for analyzing decision making in management. In the traditional approach, heavy reliance is placed on developing executive judgment for decision making through careful selection, education, and training of the individual. The role played by experience in the decision process has been stressed; i.e., experience in various management positions sharpens one's ability to select the most favorable alternative from among the choices available.

Committee Action Meetings, outside organizations, and committees are the means for achievement of the horizontal and external relationships needed to maintain the integrity of the traditional organization. Here, the time-consuming formal channels of communication are bypassed, and the organizational activities, both within the parent unit and between the parent unit and other organizations, are coordinated in the total managerial environment.

Organizational Position Each position within the traditional organization has a fixed and official area of jurisdiction, and this area is delineated in authority patterns and evidenced in job descriptions, policy manuals, etc. The specific description of a position within the hierarchy includes the facilities necessary to perform a task or group of tasks except for an element of supervision held in the superior position. To each position, except the lowest, a degree of authority is given, and a reciprocal degree of responsibility is exacted. A clear distinction must be drawn between the duty assignments for different jobs. Responsibilities for each job should be clearly defined, thereby encouraging everyone to conform to the job specification. No one should be responsible for many different activities except as those activities relate to the achievement of a common objective.

Rewards and Punishment The system of rewards and punishment in such organizations is copied from that of the church, the state, and the military establishments. This system relies on the assumptions stated by Shull:

> . . . motives and attitudes of people are the same without regard to the collectivity in which they perform or the nature of the external environment in which the organization exists;
>
> . . . leadership, communication, and participation needs of organizations are alike without regard to the nature of, and emphasis upon, specific survival and growth needs; and thus . . . the nature and type of the coordination and inducement systems in different organizations should be more similar than dissimilar.[47]

Management Principles Proponents of the traditional form of organization tend to explain and justify their organizational forms and their modus operandi in terms of principles of the organization.[48] The principles relating to the management function apply to the management of any kind of enterprise. They

[47]Fremont A. Shull, *Matrix Structure and Project Authority for Optimizing Organizational Capacity,* Southern Illinois University, Business Research Bureau, Carbondale, Ill., October 1965, p. 100.

[48]For example, see the principles of organization listed by James D. Mooney in his essay in L. H. Gulick and L. Urwick (eds.), *Papers on the Science of Administration,* New York Institute of Public Administration, New York, 1937. For a current discussion of management principles, see Harold D. Koontz and Cyril J. O'Donnell, *Principles of Management,* 3d ed., McGraw-Hill Book Company, New York, 1964.

provide the conceptual framework for the theory and are used as fundamental truths, applicable to any given environment and valuable in predicting results. The body of related principles is referred to as the theory. A bureaucracy is guided by general goals or objectives in a framework of a management theory and by a set of rules and principles which determine all conduct. The individual and the bureaucratic position are separated, with an impersonal, routine, rational result.

Span of Control Since areas of responsibility in the traditional organization are limited and fixed, and since each area of responsibility has its limits of authority, the number of subordinates that a supervisor controls must be correspondingly limited. Each organizational position (except, of course, the last one in the chain of echelons) has responsibilities that cannot be delegated; thus, the responsibilities of a position increase with the number of subordinate units it controls. In traditional theory, this span of control has received much attention, directed toward showing how restricting the span of control can improve executive effectiveness.[49] Several writers have attacked the validity of the span of control in terms of its theoretical soundness. Herbert Simon finds fault with the principle in that it produces excessive red tape.[50] Each contact between organizational members must be carried upward to a common superior—a needless waste of time and effort. Davis discusses the problem of "layering" in large organizations,[51] due to the inactive role of intervening layers of executives in the hierarchy. The search for an optimum span of control has, in our opinion, created intervening levels and increased the distance between the top-level executives and the front-line supervisors directly concerned with organizational objectives. Recent literature on the span of management (span of control) reflects a growing disenchantment with the concept and recognizes that many variables in the management environment affect the number of subordinates one can supervise effectively.

Component View Traditional theory tended to emphasize the functional components (finance, marketing, production) of the organization and neglected an analysis of the interfaces and the systems nature of the business organization. Management theory was taught in much the same way, with the role of integrating finance, marketing, etc., left to the student. Problems in industrial organizations were dealt with from a component basis, without an explicit evaluation of the total systems effect.

Industrial Parochialism Most traditional management theory developed in the industrial setting; hence, a form of parochialism developed. Although

[49]See, for example, A. V. Graicunas, "Relationship in Organization," in Gulick and Urwick, op. cit.

[50]Herbert A. Simon, *Administrative Behavior,* The Macmillan Company, New York, 1947, pp. 26–28.

[51]Davis, op. cit.

church and military models provided a reference point, the theory reflected the industrial milieu, and, unfortunately, many of the developing principles did not find early acceptance in educational and ecclesiastical organizations. Even today, many management books are written using industrial systems as the primary focus.

Horizontal Dimensions Traditional theory recognized the existence of horizontal relationships through the informal organization, in the operation of the manager's coordinative responsibilities, and in the doctrine of "completed staff action."[52] Formal matrix organizations did not exist, although there was the forerunner of project techniques found in military operations, e.g., in the naval task force organization.

Unilateral Objective Traditional management, in portraying the industrial complex, tended to emphasize the satisfaction of a single objective—the stockholder's claim—profitability. This claim was based on the belief that stockholders were the residual owners and thereby had a more direct claim on the management of the business than did other groups. Today, much of this belief still exists, but the business firm today must recognize that its success depends not only on stockholder and customer satisfaction, but also on discharging responsibilities to its employees, communities, and society in general. The objectives of today's business firm extend to a range of "clientele," each having its parochial objective, yet overall organizational effectiveness must be maintained.

Neglect of Long-Range Planning Long-range planning, as an activity arising from the organic management function of planning, had its early development in the military establishment. Although concepts of long-range planning in business circles existed long before businesspersons began writing about it, a conceptual framework for long-range planning was not developed in management literature until the 1950s. Traditional management theory neglected this dimension of organizational direction and integration. What long-range planning existed was responsive in nature; today's long-range planning techniques and philosophies tend to be contrived, i.e., a deliberate process of developing a sense of long-range direction and purpose for the organization.

Bureaucracy

A primary element of traditional management theory is the *bureaucracy*. The bureaucratic organization is an easy object of ridicule, but this is somewhat ironic, since bureaucratic organizations provide a significant proportion of today's employment. Bureaucratic organizations run our government and

[52]See, for example, "Completed Staff Action," *Army-Navy Journal,* Jan. 24, 1942.

manage our military forces. Some religious organizations contain vestiges of a bureaucracy, and heads of bureaucratic organizations shape our economic, social, and industrial worlds.

Characteristics An organization can be considered bureaucratic when it exhibits characteristics such as the following:

1 It is so large that the individuals cannot know all the other members.

2 Its members pursue a career in the organization and depend on it for most of their income. Individuals have a serious commitment to the organization and its provincial viewpoint. They feel restricted in voicing personal views, particularly if these views run counter to the prevailing modes of thought.

3 It includes many levels of management in the hierarchy, and promotions are based on how well the individual performs the organizational role. The individual's personal objectives are subordinated to the organizational goal. Efficiency, integrity, loyalty, and individual motivation are expected of the employee. Within a given bureau, however, are many types of officials—ranging from those who are motivated by self-interest ("What's in it for me?") to those motivated by loyalty and self-sacrifice.

4 It tends to perpetuate itself, to expand, regardless of whether or not there is any real need for its services. This phenomenon is aptly described in C. Northcote Parkinson's famous first law: "Work expands so as to fill the time available for its completion."[53] An organization's propensity to expand is in direct proportion to its ability to attract and retain capable personnel. An expanding organization normally provides its leaders with increased power, income, and prestige, so that leaders encourage growth. The growth of an organization is also a deterrent to internal conflict, since it enables the new members to improve their status without lowering that of the old.

Organization Size Increasing the size of an organization may very well improve the quality of its performance and its chances for survival. Therefore, the organizational leaders may seek expansion to reduce internal dissension and improve the morale of the organization. The advantages of being a large organization, in terms of continued life, are pointed out by William H. Starbuck in his analysis of organizational growth:[54]

> Large organizations have a better chance to survive than small ones. Large organizations are harder to destroy and harder to change than small ones (because they embody greater sunk costs); so they tend to be more resistant to external pressures. They also spend more on research and development (both in total and per employee), hence they can better develop new techniques useful in augmenting their power.

[53]C. Northcote Parkinson, *Parkinson's Law and Other Studies in Administration,* Houghton Mifflin Company, Boston, 1962, p. 2.

[54]William H. Starbuck, "Organizational Growth and Development," in J. G. March (ed.), *Handbook of Organizations,* Rand McNally & Company, Chicago, 1964.

> Very large organizations can impose a certain degree of stability upon their external environment, whereas smaller ones cannot. Increased environmental stability reduces uncertainty and anxiety and solidifies the control of high ranking officials.

Questioning the Traditional Model

The organizational characteristics of bureaucracy trace back to the church and military organizational models. Central to the bureaucratic form is the pyramidal organizational structure and the idea that authority is delegated downward. Many of the principles rest on speculation rather than empirical research and fall considerably short of being like the laws of the physical sciences.

In recent years there have been many attacks on the bureaucratic model of management. One of the earliest attacks in the popular press was waged by Toffler, who noted:

> . . . We are witnessing not the triumph, but the breakdown of bureaucracy. We are, in fact, witnessing the arrival of a new organizational system that will increasingly challenge, and ultimately supplant bureaucracy. This is the organization of the future. I call it "adhocracy."[55]

The lack of relevance of bureaucracy to modern society can be noted by looking at the experiences of an emerging nation:

> . . . China watchers believe that many of the management problems can be traced to the rigidity of the country's bureaucracy. The shifting winds of Communist ideology between pragmatists and ideologues under Mao Zedong, Liu Shaoqi, the Gang of Four, and now Deng Xiaoping and the reformists seem to have petrified official channels entirely.[56]

Management behavior appropriate in a tradition hierarchy is inappropriate in modern organizations. Changes in culture and in technical, economic, political, and legal systems have made the traditional model inappropriate. Traditional theory, when faced with increasing systems complexity and interdependency, counsels more centralization structure, decision making, and strengthening of staff. All of this places more limitations on a manager who must deal with local problems. The design of an organization encompasses structure, behavior, expectations, management processes, and an assessment of the work to be done. The idea that "structure follows strategy" sets the tone for the belief that success in organizational design is a convergence of systems, cultures, behavior, and management processes.

Much of the questioning of the bureaucratic or traditional theory of management centers around the notion of "crises centered" versus "knowledge centered" organizations. Unity of command may be essential on the battlefield, but it becomes necessary in many of our contemporary organizations when

[55]Alvin Toffler, *Future Shock,* p. 113.

[56]"International Business," *Business Week,* Sept. 8, 1980, p. 53. Reprinted by permission.

the environment is not so crises-centered, but rather resembles a society of coequals working in knowledge-centered organizations.

Crises-Centered Organizations[57] An executive in a crises-centered organization (for example, a military unit in combat conditions) uses direct command authority to effect strategy and management control. There may be little opportunity for judgment on the part of the subordinates except within the narrow definition of the objective. While the subordinates are expected to display initiative, there is less opportunity for the free discussions necessary for participative management. Crises-centered management is usually related to the military organization; yet only a very small proportion of the military forces engage in combat with the enemy. In recent years, the impact of technology has been felt in military weapons. The result has been the emergence of a military leader who is a manager-technologist and whose management techniques are very similar to those found in industrial and educational organizations. To an increasing extent, the military organization is faced with a management situation where the problems are the same as those found in knowledge-centered organizations. This coexistence of a crises-centered and a knowledge-centered organization in the military establishment has resulted in a modification of the image of military officers; rather than decisive, intuitive people who must cope with their environment under adverse conditions, they are now conceived as individuals who must also use deliberate analysis to deal with the environment.

The present-day military organization is necessarily composed of varied forms, ranging from pure military complements to those made up of military and civilian personnel in varying proportions. In these mixed organizations, and in many pure military organizations as well, the management techniques must be designed to reflect a knowledge-centered environment.

Knowledge-Centered Organizations In the knowledge-centered organization, such as a project team whose goal is to develop and produce a weapons system, there are closely coordinated, integrated teams that circumvent chains of command and depend on a high degree of reciprocity between the participants. In this type of organization, the traditional functional theory has some application, but if followed slavishly, it results in an authoritative environment which can offend and stifle the creative bent of the members. The knowledge-centered activity is more participative—there is greater reliance on peers and associates within a complex of organizations, and there is more colleague authority and responsibility.

Going beyond Traditional Theory For many years, the traditional theory of management was the model taught in management courses. Today, however,

[57]This discussion of crises-centered organizations and knowledge-centered organizations is based on Suojanen, op. cit., pp. 108–111.

traditional theory serves merely as the point of departure in the development of a philosophy of management. The textbook principles of organization—hierarchical structure, authority, unity of command, task specialization, span of control, line-staff division, parity of responsibility and authority, the sanctity of the superior-subordinate relationship, etc.—comprise a complex of assumptions which have had a profound influence on management thought over several generations. Many of the principles were derived from the military and church establishments, which differ from modern business organizations in many respects, particularly in that they may ignore some of the social and economic realities of this changing environment.

An example is the traditional principle of operating by the organizational chart, which is, at best, a picture of how the organizational groups relate to one another at a given moment in time. In spite of the organizational chart, managers at all levels find their behavior and their modus operandi controlled by many others in the environment besides their superior(s).

The traditional theory assumes some factors about human behavior and organizational interdependency that are not as true today as they were in the simpler organizations of the past, when the system of rewards and punishments provided for more negative than positive motivation. There has been insufficient attention given to the interdependencies that exist between organizations and individuals in their environment. More and more attention must be paid to the management problem of providing an environment in which the contributors can support one another in a reciprocal arrangement. The principles of the feudal hierarchy, the military authority, and the patriarchy of the church were transferred to Max Weber's bureaucracy, where public and private business is carried out "according to calculable rules and without regard for persons." The realities of modern industrial competition cause serious doubt about the universality of the traditional ideas.

For some time, it has been realized that the flow of work and the use of authority have significant lateral and horizontal relations. The role of the superior has changed from that of a powerful executive who controls people to that of a manager who provides an environment in which people can work with the many different groups in the total environment.

Bureaucratic theory considers that the main problems of management exist only within the boundaries of the parent organization. Little attention has been given to the manager's effect on contacts and negotiations outside the company. These contacts (with bankers, suppliers, customers, and civic organizations, to name just a few) can be time-consuming and limit the time available for internal company affairs. The success of a company may well depend on its executives' accepting their role in the environment in which the company competes and on their relating that role to the company's needs.

Theory X and Theory Y The traditional view of management was seriously challenged by Douglas McGregor in his discussion of two managers, one of whom operates under *Theory X* and the other under *Theory Y*. According to McGregor, behind every managerial decision or action are assumptions about human nature and human behavior; for Theory X these assumptions are:

> The average human being has an inherent dislike of work and will avoid it if he can.
> Because of this human characteristic of dislike of work, most people must be coerced, controlled, directed, threatened with punishment to get them to put forth adequate effort toward the achievement of organizational objectives.
> The average human being prefers to be directed, wishes to avoid responsibility, has relatively little ambition, wants security above all.[58]

Theory Y assumptions about human behavior are in dramatic contrast to these:

> The expenditure of physical and mental effort in work is as natural as play or rest. The average human being does not inherently dislike work. Depending upon controllable conditions, work may be a source of satisfaction (and will be voluntarily performed) or a source of punishment (and will be avoided if possible).
> External control and the threat of punishment are not the only means for bringing about effort toward organizational objectives. Man will exercise self-direction and self-control in the service of objectives to which he is committed.
> Commitment to objectives is a function of the rewards associated with their achievement. The most significant of such rewards, e.g., the satisfaction of ego and self-actualization needs, can be direct products of effort directed toward organizational objectives.
> The average human being learns, under proper conditions, not only to accept but to seek responsibility. Avoidance of responsibility, lack of ambition, and emphasis on security are generally consequences of experiences, not inherent human characteristics.
> The capacity to exercise a relatively high degree of imagination, ingenuity, and creativity in the solution of organizational problems is widely, not narrowly, distributed in the population.
> Under the conditions of modern industrial life, the intellectual potentialities of the average human being are only partially utilized.[59]

McGregor's contribution to the dynamics of human behavior leads to serious doubt about the ability of a traditional manager to get the maximum support from people. It also points out vividly the inadequacies of a theory which limits human collaboration in the organization. The Theory Y approach implies that the failure to reach desired objectives lies in management's method of operation and motivation, rather than in an inadequate scalar organizational posture.

Serious questions can be raised about the ability of traditional theory to provide a suitable answer to the complex industrial problems of today. We have seen remarkable economic and social changes, and these changes are forcing the development of new ideas about how to manage organized activity.

[58]Douglas McGregor, *The Human Side of Enterprise,* McGraw-Hill Book Company, New York, 1960, pp. 33–34.
[59]Ibid., pp. 47–48.

PROJECT MANAGEMENT VERSUS TRADITIONAL MANAGEMENT

The form of a bureaucracy is almost universally hierarchical, reflecting the scalar principle referred to by Mooney and Reiley.[60] The management of project activities such as exist in a research and development organization, however, requires horizontal and diagonal relationships. In such an organization, managers and technicians deal horizontally with peers and associates at different levels in the same organization and with outside organizations. To always follow the chain of command would be unwieldy, time-consuming, and costly and would disrupt and delay the work. Horizontal and vertical contacts grow out of the necessity to get the job done; they are seldom charted, and yet they are necessary to a smooth flow of work in the organization. These relationships have been called the informal organization, but this is a misnomer. There may be little informality; the standards of performance may be just as stringent as those in the formal (hierarchical) structure. In many cases, these relationships have sufficient strength and permanency to become de facto the modus operandi of the organization.

The acceptance of horizontal-vertical relationships between members of an organization requires changes in the organizational form. The realignment of tasks, the restructuring of the formal hierarchical structure, and the de jure recognition of a hybrid organizational form have been accomplished in many of today's corporations. In weapons-system management, rigid hierarchical structuring has been abandoned in favor of closely integrated project groups.

To understand the concept of project management, we must first understand the framework of the project environment and the phenomena found in it. This framework points up the salient differences between the role of the project manager and that of the traditional functional manager. These differences are possibly more theoretical than actual, because the role of the traditional manager has naturally shifted. However, differences do exist, and they affect the manager's modus operandi and philosophy. The differences in the viewpoints of the *project* and the *functional* managers are outlined in Table 8-1. This comparison highlights a singular characteristic of project managers; i.e., they must manage activities that include extensive participation by organizations and people not under their direct (line) control.

SUMMARY

Project management is undoubtedly one of the most complex and demanding management concepts in existence. A large project has all the elements of an enterprise which has been conceived and built, reaches maturity, completes its mission, and phases out, perhaps all in a period of three to five years. The task of

[60]James D. Mooney and Alen C. Reiley, *Onward Industry,* Harper & Row, Publishers, Incorporated, New York, 1931.

TABLE 8-1
COMPARISON OF THE FUNCTIONAL AND THE PROJECT VIEWPOINTS

Phenomena	Project viewpoint	Functional viewpoint
Line-staff organizational dichotomy	Vestiges of the hierarchical model remain, but line functions are placed in a support position. A web of authority and responsibility relationships exists.	Line functions have direct responsibility for accomplishing the objectives; line commands, and staff advises.
Scalar principle	Elements of the vertical chain exist, but prime emphasis is placed on horizontal and diagonal work flow. Important business is conducted as the legtimacy of the task requires.	The chain of authority relationships is from superior to subordinate throughout the organization. Central, crucial, and important business is conducted up and down the vertical hierarchy.
Superior-subordinate relationship	Peer-to-peer, manager-to-technical-expert, associate-to-associate, etc., relationships are used to conduct much of the salient business.	This is the most important relationship; if kept healthy, success will follow. All important business is conducted through a pyramiding structure of superiors and subordinates.
Organizational objectives	Management of a project becomes a joint venture of many relatively independent organizations. Thus, the objective becomes multilateral.	Organizational objectives are sought by the parent unit (an assembly of suborganizations) working within its environment. The objective is unilateral.
Unity of direction	The project manager manages across functional and organizational lines to accomplish a common interorganizational objective.	The general manager acts as the one head for a group of activities having the same plan.
Parity of authority and responsibility	Considerable opportunity exists for the project manager's responsibility to exceed authority. Support people are often responsible to other managers (functional) for pay, performance reports, promotions, etc.	Consistent with functional management; the integrity of the superior-subordinate relationship is maintained through functional authority and advisory staff services.
Time duration	The project (and hence the organization) is finite in duration.	Tends to perpetuate itself to provide continuing facilitative support

Source: David I. Cleland, "Understanding Project Authority," *Business Horizons,* Spring 1966.

project managers is enormously complicated and diverse; they tie together the efforts of many organizations. They deal with technical and administrative disciplines in pulling together a project team to act as an entity rather than as a fragmented group of functional experts.

Project managers deal with the concepts of management in general. Many of the classical management principles apply; many project techniques may be used to relate these principles. Careful attention must be given to the division of tasks among the project participants. That division of work should be made which offers the fewest technical and contractual interfaces among the participants.

A bureaucratic organization generally does not provide the environment essential to project success. In the most flexible of traditional bureaucratic organizations, it is difficult to maintain a large number of people working in close harmony on creative abstract work. Creative people do not fit into a precise and orderly bureaucratic organization where all work is thoroughly organized and all assignments are rigidly controlled; where each individual has a definite area to cover, definite information to work with, and a definite schedule to meet; where superiors must be reported to and subordinates directed. Such an organization may soon have only the few creative ones who lead the others. Innovations are difficult to come by, since each one must be introduced and explained in detail at every one of the successive levels of the hierarchical chain.

Project management is an outgrowth of the need to develop and produce large projects in the shortest possible time. It has been developed from a need, with an evolving theoretical formulation. In the next chapter, we will present project management in the setting of the matrix organization.

DISCUSSION QUESTIONS

1 What is a project? Give some examples. How would you distinguish between a project and a program? Why is it important to make such a distinction?

2 Projects vary in size and scope. Are the management processes in a project different depending on the project's size and complexity? Would the qualifications of the project manager differ depending on the nature of the project? Discuss.

3 Project management typically evolves in an organization. What is the usual nature of its genesis and evolution?

4 A family camping trip might be considered a project. Would you agree? If so, what are some of the project-management concepts that can be applied to such a trip?

5 What are some of the more salient features that characterize the application of project management? Select an organization of your choosing and determine if any projects in that organization reflect these salient characteristics.

6 The people assigned to a project team usually come from different functions or disciplines in the organization. What challenge does that present for the project manager? For the professionals on the project team?

7 Trace the conceptual evolution of project management in contemporary management, thought, and theory. Why is an understanding of this evolution useful for an individual who is involved in project management today?

8 Project management seems to be testing the traditional concepts of bureaucracy. Is this really so? Can a project-management system exist successfully within a bureaucratic organization? If you believe this to be so, give some examples.

9 The General Electric Co. has organized its strategic management process in such a way that a horizontal culture has been created. What is meant by the notion of a horizontal culture? Does the existence of such a culture have any impact on the way a senior executive in General Electric would manage the organization?

10 In the last decade, managers have found themselves in somewhat of an identity crisis. What have been some of the changes in the management of organizations in the last ten years that draw upon different knowledge, skills, and attitudes of today's successful managers?

11 In the text of this chapter we have cited the notion of dual leadership in contemporary organizations. Do you think this makes any sense? Can you identify any contempo-contemporary organizations where dual leadership is appropriate?

12 Identify and describe some of the horizontal organizations discussed in this chapter. What are some of the common characteristics of these organizations? What are some of the differences in these organizations?

13 When production teams are used in the manufacturing environment, the role of the first-line supervisor is changed. What is the nature of this change? What new knowledge, skills, and attitudes must these new first line supervisors develop?

14 Identify and discuss some of the changes that have been introduced in contemporary organizations by the horizontal culture.

15 Douglas McGregor wrote about Theory X and Y concepts over twenty years ago. Are these concepts still applicable to today's organizations?

16 Bureaucratic approaches to the management of contemporary organizations have outlived their usefulness. Defend or refute this statement.

RECOMMENDED READINGS

Burt, David N.: "A Multimatrix Approach to Project Management," *Project Management Quarterly,* September 1980.

Butler, Arthur G., Jr.: "Behavioral Implications for Professional Employees of Structural Conflict Associated with Project Management in Functional Organizations," University of Florida, *Business Administration,* 1969.

Cathey, Paul: "Making Profit Centers Work Through Matrix Managing," *Iron Age,* October 15, 1979.

Cleland, David I.: "Understanding Project Authority," *Business Horizons,* Spring 1966.

Cook, Desmond L.: *Educational Project Management,* Charles E. Merrill Books, Inc., Columbus, Ohio, 1971.

Cooper, M. R.; B. S. Morgan; P. M. Foley; and L. B. Kaplan: "Changing Employee Values: Deepending Discontent," *Harvard Business Review,* January–February 1979.

Davis, Keith: "The Role of Project Management in Scientific Manufacturing," *IEE Transactions on Engineering Management,* vol. 9, no. 3, 1962.

Davis, Stanley M.: "Two Models of Organization: Unity of Command versus Balance of Power," *Sloan Management Review,* no. 1, Fall 1974.

——— and P. R. Lawrence: *Matrix,* Addison-Wesley, Reading, Mass. 1977.

Delbecq, Andre L.: *Matrix Organization—An Evolution beyond Bureaucracy,* The University of Wisconsin Press, Madison.

Drucker, Peter F.: "New Templates for Today's Organizations," *Harvard Business Review,* January–February 1974.

Duke, Robert K.: "Project Management at Fluor Utah, Inc.," *Project Management Quarterly,* vol. 8, no. 3, September 1977.

Fulmer, Robert M.: "Product Management: Panacea or Pandora's Box," *California Management Review,* Summer 1965.

George, William W.: "Task Teams for Rapid Growth," *Harvard Business Review,* March–April 1977.

Goggin, W. C.: "How the Multidimensional Structure Works at Dow Corning," *Harvard Business Review,* January–February 1974.

Graicunas, A. V.: "Relationship in Organization," in Gulick and Urwick (eds.), *Papers on the Science of Administration,* New York Institute of Public Administration, New York, 1937.

Gray, J. L.: "Matrix Organizational Design as a Vehicle for Effective Delivery of Public Health Care and Social Services," *Management International Review,* 1974.

Greiner, Larry E.: "Evolution and Revolution as Organizations Grow," *Harvard Business Review,* July–August 1972.

Hanan, Mack: "Make Way for the New Organization Man," *Harvard Business Review,* July–August 1971.

Hlavacek, J. D., and V. A. Thompson: "Bureaucracy and Venture Failures," *Academy of Management Review,* April 1978.

Huse, Edgar F., and Michael Beer: "Electric Approach to Organizational Development," *Harvard Business Review,* September–October 1971.

Janger, Allen R.: *Organization of International Joint Ventures,* Conference Board Report No. 787, The Conference Board, Inc., New York, 1980, p. 14.

Koontz, Harold D., and Cyril J. O'Donnell: *Principles of Management,* 3d ed., McGraw-Hill Book Company, New York, 1964.

Labovitz, George: "Organizing for Adaptation," *Business Horizons,* June 1971.

Lee, James A.: "Behavioral Theory versus Reality," *Harvard Business Review,* March–April 1971.

Lombard, George F.: "Relativism in Organizations," *Harvard Business Review,* March–April 1972.

Mashburn, James I., and Bobby C. Vaught: "Two Heads Are Better Than One," *Management Review,* December 1980.

McGregor, Douglas: *The Human Side of Enterprise,* McGraw-Hill Book Company, New York, 1960.

Mee, John F.: "Speculation about Human Organization in the 21st Century," *Business Horizons,* February 1971.

Miller, Barry M., and Charles D. Williams: "Management Action through Effective Project Controls: A Case Study of a Nuclear Power Plant Project," 1978 Proceedings of the Project Management Institute, Los Angeles, October 8–11, 1978.

Miner, John B.: "The OD-Management Development Conflict," *Business Horizons,* December 1973.

Mooney, James D., and Alen C. Reiley: *Onward Industry,* Harper & Row, Publishers, Incorporated, New York, 1931.

O'Brien, James J.: "Project Management: An Overview," *Project Management Quarterly,* vol. 3, no. 3, September 1977.

Osgood, William R., and William E. Wetzel, Jr.: "A Systems Approach to Venture Initiation," *Business Horizons,* October 1977, p. 42.

Parkinson, C. Northcote: *Parkinson's Law and Other Studies in Administration,* Houghton Mifflin Company, Boston, 1962.
Prahalad, C. K.: "Strategic Choices in Diversified MNCs," *Harvard Business Review* July–August 1976, p. 72.
Read, William H.: "The Decline of Hierarchy in Industrial Organizations," *Business Horizons,* Fall 1965.
Seneker, Harold: "Mr. Nice Guy He Wasn't," *Forbes,* March 31, 1980.
Shannon, Robert E.: "Matrix Management Structures," *Industrial Engineering,* vol. 4, March 1972.
Shull, Fremont A.: *Matrix Structure and Project Authority for Optimizing Organizational Capacity,* Southern Illinois University, Business Research Bureau, Carbondale, Ill., October 1965.
Simon, Herbert A.: *Administrative Behavior,* The Macmillan Company, New York, 1947.
Smith, Robert F.: *The Variations of Matrix Organization,* Special Study No. 73, The Presidents Association, the Conference Board, New York, p. 11.
Starbuck, William H.: "Organizational Growth and Development," in J. G. March (ed.), *Handbook of Organizations,* Rand McNally & Company, Chicago, 1964.
Suojanen, Waino W.: *The Dynamics of Management,* Holt, Rinehart and Winston, Inc., New York, 1966.
Tinnin, David B.: "Why Volvo Is Staking Its Future on Norway's Oil," *Fortune,* February 12, 1979.
Urwick, Lyndall F.: "The Manager's Span of Control," *Harvard Business Review,* May–June 1956.
Werner, H.: "Project Management in the Year 2000," *Journal of Systems Management,* vol. 33, October 1981.
Williams, Earle C.: "Matrix Management Offers Advantages for Professional Services Firms," *Professional Engineer,* February 1978.
"As Costs Climb, a Rising Demand for Expediters," *Business Week,* February 23, 1981.
"Completed Staff Action," *Army-Navy Journal,* January 24, 1942.
"Guideline for Use of Program/Project Management in Major Appliance Business Group," General Electric Corporation, March 14, 1977.
"How a New Chief is Turning Interbank Inside Out," *Business Week,* July 14, 1980, pp. 109, 111.
"How to Stop the Buck Short of the Top," *Business Week,* January 17, 1978, pp. 82–83.
"International business," *Business Week,* September 8, 1980.
"Kaiser Aluminum Flattens Its Layers of Brass," *Business Week,* February 24, 1973.
"Management: GE's Search for Synergy," *The New York Times,* April 16, 1978.
"Management Itself Holds the Key," *Business Week,* September 9, 1972.
"The '60's Kids as Managers," *Time,* March 6, 1978.

CASE 8-1: Product/Functional/Geographic Interface Model for a Corporation

The Generic International Corporation has recently undergone a realignment to emphasize its competitive posture in the international marketplace. Since this realignment, questions have come up concerning the manner in which certain formal authority and responsibility relationships should be established. There is a growing appreciation of the need for the documentation of this relationship. Particular concern has been raised about how and what the authority/responsibility relationships among profit-center managers and geographic and functional managers should be.

It is believed these relationships would be centered around the "work breakdown structure" of the product/functional/geographic interface created when the International realignment was set in motion.

Your Task Develop such a document for the Generic International Corporation. Some things you might wish to consider in developing this document include the relative authority/responsibility patterns involving:

Strategic planning
Product planning
Human resources
Project management
Facility management

You might wish to use the geographic manager's and the profit-center manager's roles as the focal positions around which to develop these formal authority and responsibility patterns.

Prepare a summary briefing of your ideas for presentation.

CASE 8-2: "Advantages/Disadvantages Profile" of Project Management

Project management is not a panacea—it's not for every organization. Even where it is used, it has advantages and disadvantages. An appreciation of both aspects is important for the manager/professional who is responsible for producing results within a cultural context.

Situation The Generic International Corporation's design and development of a project management system continues in the environment. Some of the cultural changes accompanying this management system appear to emerge as advantages/disadvantages, as perceived by the people involved. Your general manager would like to have a "workshop" to discuss these advantages/disadvantages in the near future. The manager is hopeful that such a discussion will lead to a better appreciation by all concerned of what project management can do and cannot do for the organization.

Your Task Develop a profile of the advantages/disadvantages of project management in the context of a general manager's environment. Some factors you might wish to consider in developing a profile include the impact of project management on:

Strategic decision making
Operational decision making
Resource control
Interpersonal relationships
Authority and responsibility patterns
Organizational effectiveness

Make whatever assumptions you feel are necessary to carry out this assignment.

Prepare a summary briefing of your ideas and be prepared to share your ideas with other members of the organization.

CHAPTER 9

THE DYNAMICS OF PROJECT MANAGEMENT

And one man in his time plays many parts, / His acts being seven ages.[1]

Chapter 8 relates some of the elements of systems thinking to the emergence of an environment that is amenable to the development of a horizontal dimension in modern organizations. This chapter examines some of the characteristics of project management that are dynamic in nature. This dynamism is manifest in a project, its life cycle, the organizational underpinnings that support the project, the fluid assignment of members on the project team, and the continuing interface of organizational elements.

The dynamics of the project environment, as discussed in Chapter 8, are amplified by the dynamic nature of systems and of the projects that are created to invent, develop, implement, or sell systems. These dynamics create the need for the management approach that has come to be called "project," or "matrix," management.

This chapter discusses those dynamics, how they affect organizations, and how they have been addressed through project management.

[1]Shakespeare, *As You Like It,* act 2, sc. 7, lines 142–143.

SYSTEM LIFE CYCLES

The primary reason for the complexity of the implementation function has to do with *the dynamic nature of systems.* Solutions to complex problems, once decided upon, are not immediately available. Transportation systems, for instance, go through long years of detailed design and development after the planning process is completed. Thus, the systems which are the proposed solutions to problems go from state to state as they evolve from idea to proposal to fruition.

The same dynamic evolution is descriptive of a wide variety of systems—whether they are thought of as problem solutions or not. For instance, new products, management information systems, and social programs share this same dynamic evolutionary characteristic. So, too, do organizations and societies exhibit traits of a cyclical existence. They are born, grow, and decline.[2]

Organizational Life Cycles

According to Greiner,[3] organizations exhibit life cycles as they move through five distinguishable phases of development, each phase containing a relatively calm period of growth which ends in a management crisis. The phases which he describes begin with "creativity," which occurs in the first stage of an organization, when emphasis is placed on creating both a product and a market. In the second phase, "direction," the business organization embarks on a period of sustained growth under an able and directive leadership. The third era of growth, "delegation," evolves from the successful application of a decentralized organization structure. Phase four, "coordination," is characterized by the use of formal systems for achieving greater coordination within the organization. Finally, in phase five, "collaboration," a management "crisis" develops in which there is a lot of "red tape." In this phase, a strong team collaboration is needed in order to overcome the red-tape crisis. In phase five, the organization functions around a management philosophy of flexibility and a behavioral approach. Some of the characteristics of this phase are (1) solving problems through team action; (2) organization of teams across functions; and (3) the frequent use of a matrix-type function to assemble the right teams for the appropriate problems.

The important part of Greiner's analysis is both the general life cycle notion and his observation that organizations start as a small team effort, grow in complexity, and eventually return to a small-project team approach in collaborating in those things that have to be done to sustain the organization in a dynamic environment. Thus, the organization has gone full cycle, from birth to maturity, with project teams dealing both with innovation at the beginning and with complexity and size at the end.

[2]For a thorough treatment of organization life cycles, see John R. Kimberley, et al., *The Organizational Life Cycle,* Jossey-Bass, Inc., Publishers, San Francisco, 1980. A similar approach applied to cultures is the classic work by Oswald Spengler, *The Decline of the West,* Knopf, 1980.

[3]Larry E. Greiner, "Evolution and Revolution as Organizations Grow," *Harvard Business Review,* July–August 1972.

Product Development Life Cycles

The design and development of an automobile goes through a similar life cycle. The beginning stages of this process are described as follows:

> In the first stage, which we call "concepting," experimental engineers, environmental scientists, forward planners, and marketing experts pool their thinking. Their objective: what the marketplace will require. This is the most important stage. Here we must determine not only what kind of car, but how many we might be able to build and sell years later. Economics, customer tastes, availability of various kinds of fuels must be compared with state-of-the-art technology—and what steps must be taken to advance that technology quickly yet surely.
>
> In the "concepting" stage, a new car is conceived. If the car is to be sold to customers three years later, construction of new plants must begin and basic tooling must be ordered.
>
> The second phase . . . takes 24 to 30 months. It encompasses development, design, structural analysis, handling analysis, emissions, noise and vibration, safety, reliability, serviceability and repairability, manufacturing, assembly, marketing, financing.
>
> Advanced product engineers and research scientists work with the one hundred fifty to two hundred people at the Project Center and thousands more in the staffs and divisions to transfer new science and technology to the new car. Components are hand-built and "cobbled" into existing models for road testing.
>
> Prototype cars are handbuilt at a cost of more than $250,000 each. These enable the Project Center team to determine how newly developed, pretested components operate as a unit. Then, pilot models will be built from production tooling and tested some more. Finally, the car is committed to production.[4]

Then, of course, the auto enters into its *sales life cycle.*

Sales Life Cycle

Every dynamic system has natural phases of development. Recognition of these phases permits the manager to properly control what is happening and to use characteristics of the various phases to advantage. A product, for example, moves through various phases of sales life cycle after it has been placed on the market. These life cycle phases have been referred to as *establishment, growth, maturation,* and *declining sales* phases.[5] Figure 9-1 shows these phases in terms of the sales revenue generated by the product during its period of slow establishment in the marketplace, followed by a period of rapid sales increase, a peaking, and a long, gradual decline. Virtually every product displays these dynamic characteristics, although some may have a sales life cycle which is so long or short that the various phases are not readily distinguishable. For

[4]"How GM 'Project Centers' Create Cars," Customer Information Sheet from General Motors to GM customers on cars and trucks. See Chapter 8 for a review of how the project-center concept works at General Motors Corporation.

[5]William R. King, *Quantitative Analysis for Marketing Management,* McGraw-Hill Book Company, New York, 1967, p. 113.

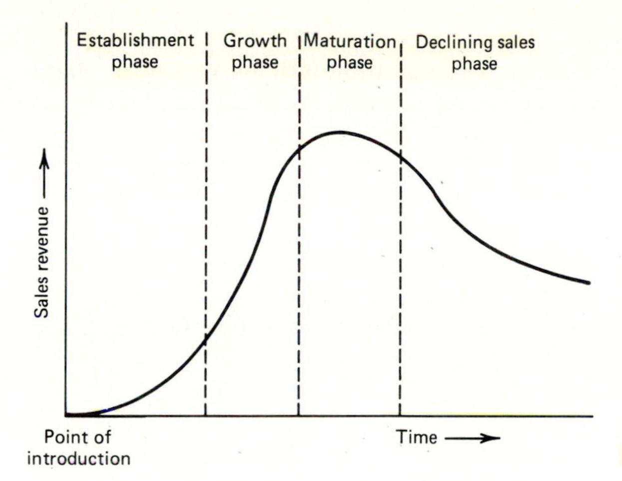

FIGURE 9-1
Product sales life cycle.

example, faddish products such as Hula-Hoops, "pet rocks," or a poster of E. T. will have a very high-peaked sales curve with a rapid decline. Many such products will have a long, slow decline after an initially rapid decline from the peak. With other products, the maturation phase is very long and the declining sales phase very gradual. But the general life cycle concept is virtually unavoidable for a successful product, for without product improvements its competition will eventually lure away customers because consumers' attitudes, habits, and needs will change as time passes.

Even some nonfaddish products have a short sales life cycle because of the fast pace of technical advances. This shortness affects the management style practiced in the organization. According to the chief executive of the Digital Equipment Corporation:

> We consciously designed DEC for ideas to flow from the bottom up. I think that's necessary because in minicomputer things move fast—average product life-cycle is two years—that if you try to innovate from the top of the company this big your decisions will probably be inappropriate by the time they're carried out.[6]

Of course, the sales portion of the life cycle of a product is really one aspect of its entire life. Indeed, only products which are marketing successes ever get to experience the sales life cycle of Figure 9-1. Most new consumer products have from the beginning of their sales period "an infinitely descending curve. The product not only doesn't get off the ground; it goes quickly under ground—six feet under."[7]

[6]Seneker, Harold, "If You Gotta Borrow Money, You Gotta," *Forbes,* Apr. 28, 1980, p. 120.

[7]Theodore Levitt, "Exploit the Product Life Cycle," *Harvard Business Review,* November–December 1965, p. 82. For instance, it has been estimated that 80 to 90 percent of newly introduced packaged grocery products are marketing failures. See Peter J. Hilton, *New Product Introduction for Small Business Owners,* Small Business Management Series no. 17, U.S. Government Printing Office, Washington, D.C.

Systems Development Life Cycle

All systems—be they weapons systems, transportation systems, or new products—begin as a gleam in the eye of someone and undergo many different phases of development before being deployed, made operational, or marketed. For instance, the U.S. Department of Defense (DOD) uses a system life cycle concept in the management of the development of weapons systems and other defense systems. In the U.S. Air Force version of this systems development life cycle, there are a variety of phases that are identified. Each phase has specific content and management approaches. Between the various phases are *decision points* at which an explicit decision is made concerning whether the next phase should be undertaken, its timing, etc.

These phases are:

1 *The conceptual phase.* During this phase, the technical, military, and economic bases are established, and the management approach is formulated.

2 *The validation phase.* During this phase, major program characteristics are validated and refined, and program risks and costs are assessed, resolved, or minimized. An affirmative decision concerning further work is sought when the success and cost realism become sufficient to warrant progression to the next phase.

3 *The full-scale development phase.* In the third phase, the design, fabrication, and testing are completed. Costs are assessed to ensure that the program is ready for the production phase.

4 *The production phase.* In this period, the system is produced and delivered as an effective, economical, and supportable weapons system. When this phase is entered into, it denotes that the weapons system has reached its operational ready state and is turned over to the using command. During this period, responsibility for program management is transferred to an Air Force logistics supporting capability within the Air Force.

5 *The deployment phase.* In this phase, the weapons system is actually deployed as an integral organizational combat unit somewhere within the Air Force.[8]

In the Pratt and Whitney Aircraft Group of the United Technologies Corporation, a life cycle approach is used in the design and development of its basic-product advanced-technology gas-turbine aircraft engines. Figure 9-2 illustrates this life cycle. Six fundamental phases are shown in the figure. If the scope of a program is less than the full development of an engine, only certain of the steps will be applicable.

The system development life-cycle concept recognizes a natural order of

[8]For those interested in reading more on the Air Force system of program management, see Air Force Systems Command Pamphlet 800-3, Apr. 9, 1976, Headquarters, Air Force Systems Command, Andrews A.F. Base, D.C. 20334.

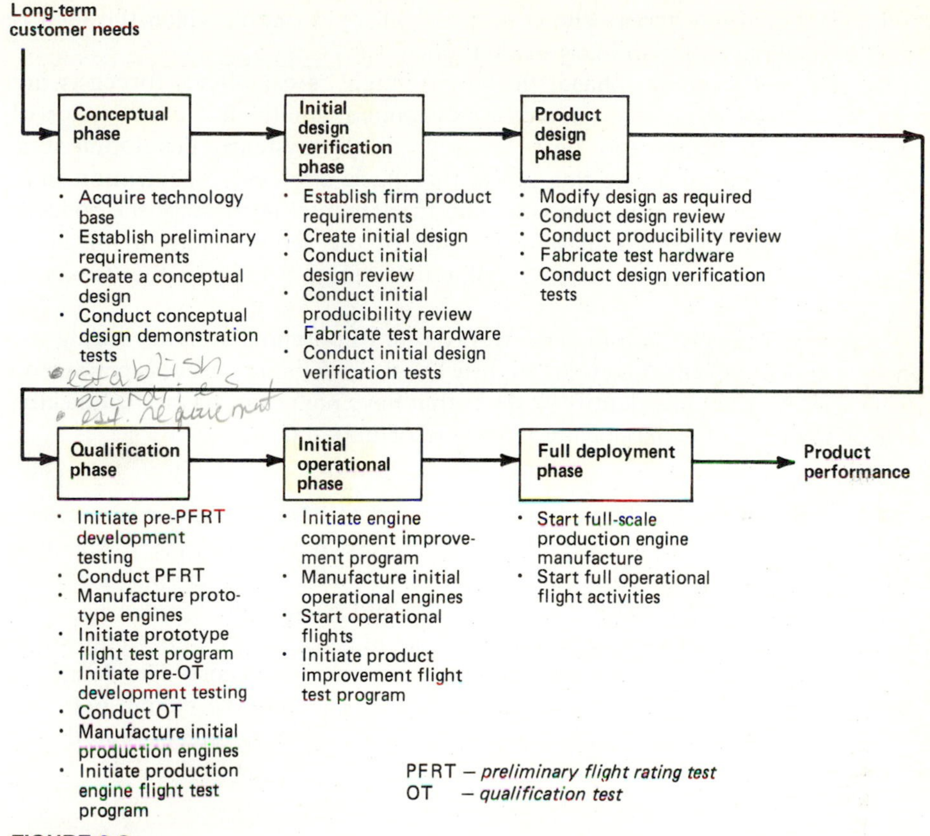

FIGURE 9-2
The flow of the design/development process. (Management Guide, Pratt & Whitney Aircraft Group Government Products Division, United Technologies, p. 39.) Used with permission.

thought and action which is pervasive in the development of many kinds of systems, whether commercial products, space-exploration systems, or management systems.

New products, services, or roles for the organization have their genesis in ideas evolving within the organization. Typically, such systems ideas go through a distinct life cycle, i.e., a natural and pervasive order of thought and action. In each phase of this cycle, different levels and varieties of specific thought and action are required within the organization to assess the efficacy of the system. The generic phases of this cycle serve to illustrate the life-cycle concept and its importance.

The Conceptual Phase The germ of the idea for a system may evolve from other research, from current organizational problems, or from the observation of organizational interfaces. The conceptual phase is one in which the idea is conceived and given preliminary evaluation.

During the conceptual phase, the environment is examined, forecasts are prepared, objectives and alternatives are evaluated, and the first examination of the performance, cost, and time respects of the system's development is performed. It is also during this phase that basic strategy, organization, and resource requirements are conceived. The fundamental purpose of the conceptual phase is to conduct a feasibility study of the requirements in order to provide a basis for further detailed evaluation. Table 9-1 shows the details of these efforts.

There will typically be a high mortality rate of potential systems during the conceptual phase of the life cycle. Rightly so, since the study process conducted during this phase should identify projects that have high risk and are technically, environmentally, or economically infeasible or impractical.

The Definition Phase The fundamental purpose of the definition phase is to determine, as soon as possible and as accurately as possible, cost, schedule, performance, and resource requirements and whether all elements, projects, and subsystems will fit together economically and technically.

The definition phase simply tells in more detail what it is we want to do, when we want to do it, how we will accomplish it, and what it will cost. The definition phase allows the organization to fully conceive and define the system before it starts to put the system physically into its environment. Simply stated, the

TABLE 9-1
CONCEPTUAL PHASE

1 Determine existing needs or potential deficiencies of existing systems
2 Establish system concepts which provide initial strategic guidance to overcome existing or potential deficiencies
3 Determine initial technical, environmental, and economic feasibility and practicability of the system
4 Examine alternative ways of accomplishing the system objectives
5 Provide initial answers to the questions:
   a What will the system cost?
   b When will the system be available?
   c What will the system do?
   d How will the system be integrated into existing systems?
6 Identify the human and nonhuman resources required to support the system
7 Select initial system designs which will satisfy the system objectives
8 Determine initial system interfaces
9 Establish a system organization

definition phase dictates that we stop and take time to look around to see if this is what we really want before the resources are committed to putting the system into operation and production. If the idea has survived the end of the conceptual phase, a conditional approval for further study and development is given. The definition phase provides the opportunity to review and confirm the decision to continue development, create a prototype system, and make a production or installation decision.

Decisions that are made during and at the end of the definition phase might very well be decisions to cancel further work on the system and redirect organizational resources elsewhere. The elements of this phase are described in Table 9-2.

Production or Acquisition Phase The purpose of the production or acquisition phase is to acquire and test the system elements and the total system itself using the standards developed during the preceding phases. The acquisition process involves such things as the actual setting up of the system, the fabrication of hardware, the allocation of authority and responsibility, the construction of facilities, and the finalization of supporting documentation. Table 9-3 details this phase.

The Operational Phase The fundamental role of the manager of a system during the operational phase is to provide the resource support required to accomplish system objectives. This phase indicates the system has been proved economical, feasible, and practicable and will be used to accomplish the desired ends of the system. In this phase the manager's functions change somewhat. The manager is less concerned with planning and organizing and more concerned with controlling the system's operation along the predetermined lines of

TABLE 9-2
DEFINITION PHASE

1 Firm identification of the human and nonhuman resources required
2 Preparation of final system performance requirements
3 Preparation of detailed plans required to support the system
4 Determination of realistic cost, schedule, and performance requirements
5 Identification of those areas of the system where high risk and uncertainty exist, and delineation of plans for further exploration of these areas
6 Definition of intersystem and intrasystem interfaces
7 Determination of necessary support subsystems
8 Identification and initial preparation of the documentation required to support the system, such as policies, procedures, job descriptions, budget and funding papers, letters, memoranda, etc.

TABLE 9-3
PRODUCTION PHASE

1 Updating of detailed plans conceived and defined during the preceding phases
2 Identification and management of the resources required to facilitate the production processes, such as inventory, supplies, labor, funds, etc.
3 Verification of system production specifications
4 Beginning of production, construction, and installation
5 Final preparation and dissemination of policy and procedural documents
6 Performance of final testing to determine adequacy of the system to do the things it is intended to do
7 Development of technical manuals and affiliated documentation describing how the system is intended to operate
8 Development of plans to support the system during its operational phase

performance. Responsibilities for planning and organization are not entirely neglected—there are always elements of these functions remaining—but the manager places more emphasis on motivating the human element of the system and controlling the utilization of resources of the total system. It is during this phase that the system may lose its identity per se and be assimilated into the "institutional" framework of the organization.

If the system in question is a product to be marketed, the operational stage begins the sales life-cycle portion of the overall life cycle, for it is in this phase that marketing of the product is conducted. Table 9-4 shows the important elements of this phase.

The Divestment Phase The divestment phase is the one in which the organization gets out of the business which it began with the conceptual phase. Every system—be it a product system, a weapons system, a management system, or whatever—has a finite lifetime. Too often, this goes unrecognized, with the result that outdated and unprofitable products are retained, inefficient

TABLE 9-4
OPERATIONAL PHASE

1 Use of the system results by the intended user or customer
2 Actual integration of the project's product or service into existing organizational systems
3 Evaluation of the technical, social, and economic sufficiency of the project to meet actual operating conditions
4 Provision of feedback to organizational planners concerned with developing new projects and systems
5 Evaluation of the adequacy of supporting systems

management systems are used, or inadequate equipment and facilities are "put up with." Only by the specific and continuous consideration of the divestment possibilities can the organization realistically hope to avoid these contingencies. Table 9-5 relates to the divestment phase.

Taken together, Tables 9-1 through 9-5 provide a detailed outline of the overall systems development life cycle. Of course, the terminology used in these tables is not applicable to every system which might be under development, since the terminology generally applied to the development of consumer product systems is often different from that applied to weapons systems. Both in turn are different from that used in the development of a financial system for a business firm. However, whatever the terminology used, the concepts are applicable to all such systems.

Table 9-6 indicates the project phases and stages suggested by the United Nations Industrial Development Organization, together with objectives for each phase that are typical of a development project life cycle.

PROJECT DYNAMICS AND THEIR MANAGEMENT IMPLICATIONS

The dynamic life cycles of systems place demands on the organization which are different from those traditionally felt by managers. These demands have resulted in the creation of ad hoc "teams" or "projects" as organizational devices for coping with these new phenomena. We shall subsequently deal in detail with these organizational approaches. For the moment, we shall deal only with the implications of the system dynamics to the organizational entities created to cope with them—the projects.

TABLE 9-5
DIVESTMENT PHASE

1. System phasedown
2. Development of plans transferring responsibility to supporting organizations
3. Divestment or transfer of resources to other systems
4. Development of "lessons learned from system" for inclusion in qualitative-quantitative data base to include:
   - a Assessment of image by the customer
   - b Major problems encountered and their solution
   - c Technological advances
   - d Advancements in knowledge relative to department strategic objectives
   - e New or improved management techniques
   - f Recommendations for future research and development
   - g Recommendations for the management of future programs, including interfaces with associate contractors
   - h Other major lessons learned during the course of the system

TABLE 9-6
PHASES, STAGES, AND OBJECTIVES OF INDUSTRIAL PROJECTS

Phase	Stage	Objective
Preparation or initiation	1 Identification of project idea (preliminary analysis)	Project(s) and programme goals identified and analysed Project objectives and preliminary global schedule and cost estimate determined
	2 Preliminary selection	Ideas for possible solutions developed into alternative concepts; desirable technical solutions identified and classified
	3 Feasibility (formulation)	Feasibility of the envisaged concepts or solutions and relevant alternatives assessed, evaluated and classified
	4 Evaluation (post-feasibility evaluation) and decision-to-invest	Decision on adoption of the most promising alternative solution; funding provided
Implementation (construction)	5 Initial project implementation, scheduling and detailed project design and engineering	All detailed drawings, specifications, bills of materials, schedules, plans, cost estimates and other relevant documentation checked and approved
	6 Contracting and purchase	Appropriate manpower, machinery, manufacturing and construction facilities, utilities, materials, documentation and all other relevant infrastucture components mobilized and available
	7 Facility construction and pre-operations (system implementation, start-up)	Completed, tested, "debugged" and accepted product, facility or system (optimum performance, time and cost)
Operation	8 Operations (not a project phase but listed for interface purposes and programme continuity)	Product, facility or system operational at all times and at optimum cost

Source: United Nations Industrial Development Organization, *The Initiation and Implementation of Industrial Projects in Developing Countries: A Systematic Approach,* United Nations, New York, 1975, p. 6.

Project Life Cycles

The natural life cycles of systems induce similar life cycles in project organizations. This is particularly apparent at the level of considering the "state variables" which may be used to characterize a project as it evolves.

Such measures vary widely. For instance, in developing a new product, we might characterize the various phases of the project life cycle in terms of the proportional composition of the work force assigned to the activity. In the beginning, research personnel predominate; subsequently, their role diminishes and engineers come to the forefront; finally, marketing and sales personnel become most important. Alternatively, the level of expenditures on the development of the product may well be an appropriate way to characterize various phases of development.

Basic life-cycle concepts hold for all projects and systems. Thus, an organizational system develops and matures according to a cycle which is much like that of a product. The measures used to define various phases of an organization's life cycle might focus on its product orientation, e.g., defense versus nondefense; its personnel composition, e.g., scientists versus nonscientists; its per-share earnings, etc. For a management information system, the life cycle might be characterized by the expenditure level during the developmental phase together with the performance characteristics of the system after it becomes operational.

A hardware system displays no sales performance after it is in use, but it does display definite phases of operation. For example, Figure 9-3 shows a typical failure rate curve for the components making up a complex system. As the system is first put into operation, the failure rate is rather high because of "burn in" failures of weak components. After this period is passed, a relatively constant failure rate is experienced for a long duration; then, as wear-outs begin to occur, the component failure rate rises dramatically.

Perhaps a comparison of Figures 9-1 and 9-3 best illustrates the pervasiveness of life-cycle concepts and the importance of assessing the life cycle properly. Figure 9-1 represents a sales life cycle for a product. The most appropriate measure to be applied to this product's sales life cycle is "sales rate." Figure 9-3 shows the operating life cycle of a hardware system—for instance, a military weapons system. The concept is the same as that of Figure 9-1, but the

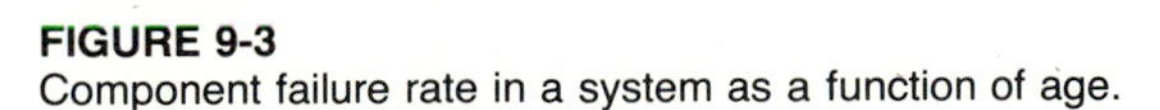
FIGURE 9-3
Component failure rate in a system as a function of age.

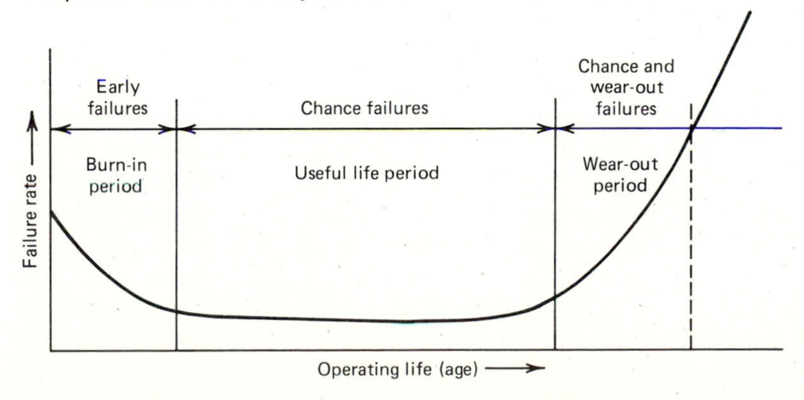

appropriate measurement is different. In Figure 9-3 the "failure rate" is deemed to be the most important assessable aspect of the life cycle for the purpose for which the measurement will be used.

Critical Project Dimensions

By viewing project life cycles in terms of the measures which may be applied at the various stages of evolution, one is naturally led to question whether there are not a number of universal measures which can be applied to projects regardless of their nature or specific context.

Three critical general dimensions which can be used for assessing the progress of most projects are *cost, time,* and *performance.* Cost refers to the resources being expended. One would want to assess cost sometimes in terms of an expenditure rate (e.g., dollars per month) and sometimes in terms of total cumulative expenditures (or both). Time refers to the timeliness of progress in terms of a schedule which has been set up. Answers to such questions as: "Is the project on schedule?" or "How many days must be made up?" reflect this dimension of progress.

The third dimension of project progress is performance; i.e., how is the project meeting its objectives or specifications? For example, in a product development project, performance would be assessed by the degree to which the product meets the specifications or goal set for it. Typically, products are developed by a series of improvements which successively approach a desired goal, e.g., soap powder with the same cleaning properties but less sudsiness. In the case of an airplane, certain requirements as to speed, range, altitude capability, etc., are set, and the degree to which a particular design in a series of successive refinements meets these requirements is an assessment of the performance dimension of the aircraft design project.

Managing the Life Cycle

As the project progresses through its life cycle, ever-changing levels of cost, time, and performance are exhibited. Correspondingly dynamic responses are required in the form of a changing mix of resources that must be assigned to the project as a whole and to its various elements.

Thus, budgets will vary substantially in total and in terms of the allocation to the various project elements. The need for personnel with various kinds of expertise will similarly vary, as will virtually everything else. This is portrayed in Figure 9-4, which shows changing levels of budget and changing levels of engineering and marketing personnel for various stages of the life cycle.

This constantly changing picture of "peaks and valleys" is an underlying structural rationale for project management. The traditional hierarchical organization is not designed to cope with managing such an always changing mix of resources. Rather, it is designed to control and monitor a much more static entity that, day to day, involves very similar levels of expenditures, numbers of persons, etc.

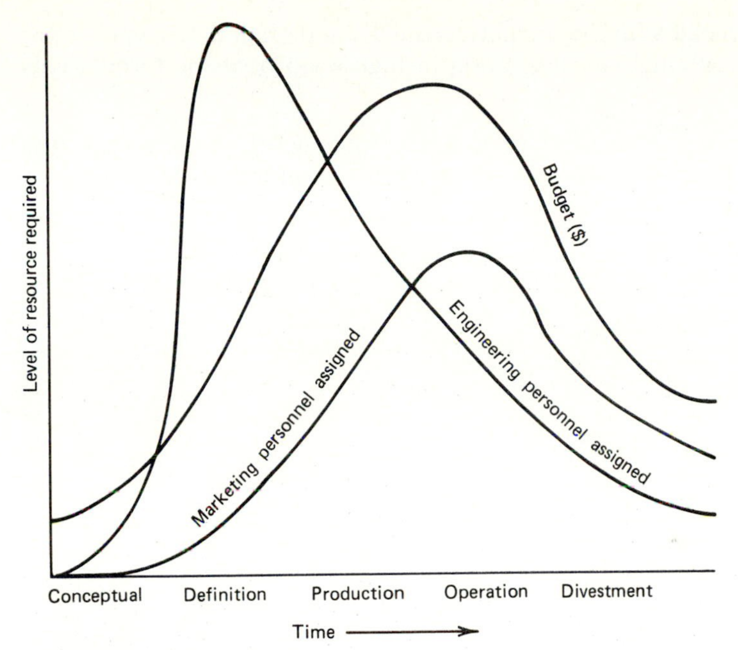

FIGURE 9-4
Changing resource requirements over the life cycle.

Project Life Cycles and Uncertainty

As the project life cycle progresses, the cost, time, and performance parameters must be "managed." This involves continuous replanning of the as yet undone phases in the light of emerging data on what has actually been accomplished. As the United Nations Industrial Development Organization prescribes:

> Each stage has its own particular characteristics and thus requires information and control procedures suited to it. On the basis of the maximum available knowledge about the entire project in a given project stage, the remaining stages must be planned and replanned so that at the end of each stage an acceptable, realistic final plan to control the succeeding stage is available, plus a revised version of the preliminary plans for all future project stages. As the project passes through the various stages, the planning for the last and often most critical stage (construction/preoperations) becomes more and more comprehensive and precise, and eventually permits improved control of the pertinent project activities. A clear identification of the project stages and pertinent objectives permits the responsible project initiator or other body to assign responsibilities systematically.[9]

Archibald[10] notes how this continuous rethinking reduces the overall uncer-

[9]United Nations Industrial Development Organization, op. cit., p. 5.
[10]Russell D. Archibald, *Managing High-Technology Programs and Projects,* John Wiley & Sons, Inc., New York, 1976, p. 23.

tainty that is associated with the ultimate project completion date (time), cost, and performance (although his illustration in Figure 9-5 treats performance as fixed).

> The uncertainty related to each factor is reduced with completion of each succeeding life-cycle phase. Figure [9-5] illustrates the reduction of uncertainty in the ultimate

FIGURE 9-5
Relative uncertainty of ultimate time and cost by life-cycle phase. (Russell D. Archibald, *Managing High-Technology Programs and Projects,* John Wiley & Sons, Inc., New York, 1976, p. 23.)

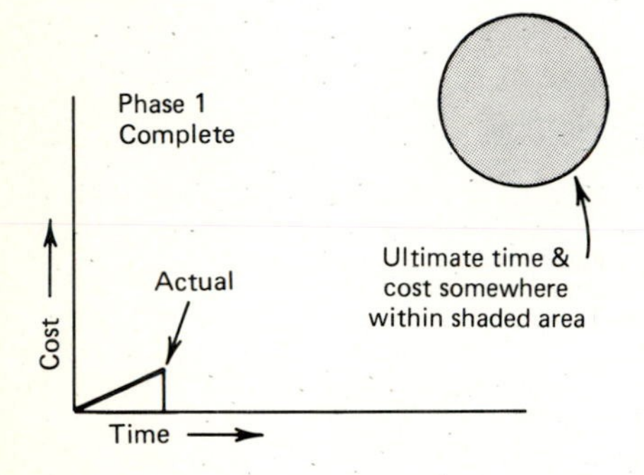

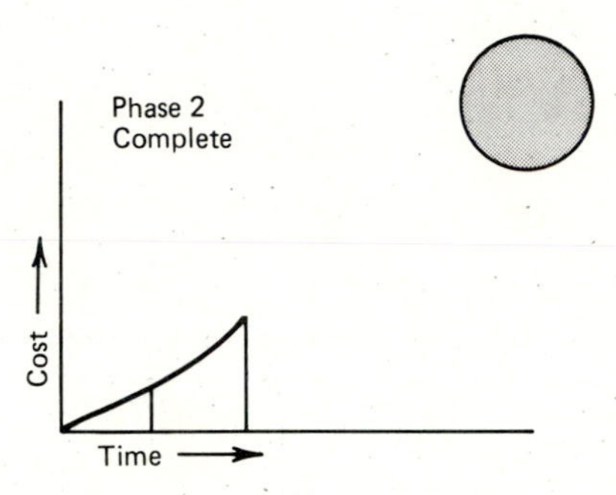

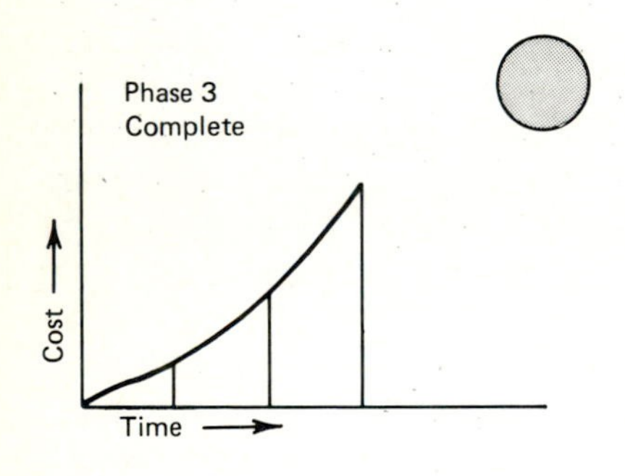

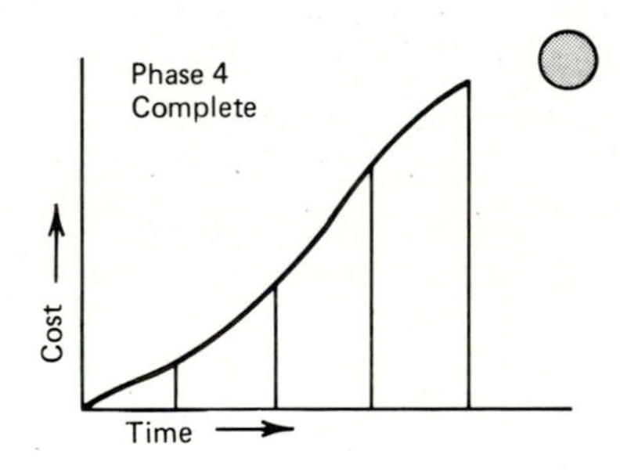

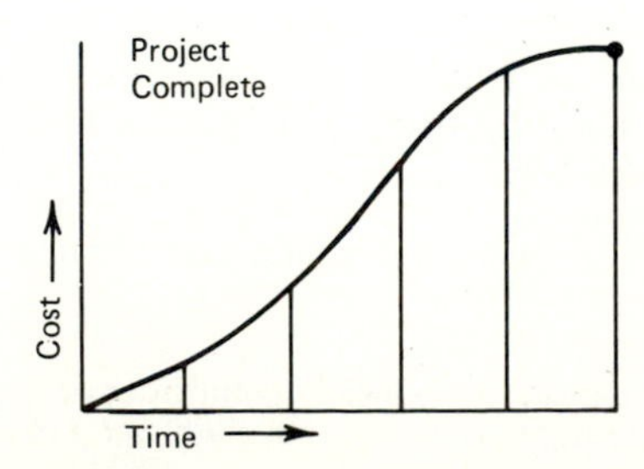

time and cost for a project at the end of each phase. In phase 1, the uncertainty is illustrated by the largest circle. The area of uncertainty is reduced with each succeeding phase until the actual point of completion is reached.

Life-Cycle Management: A Systems View

The monitoring of time, cost, and performance parameters is a very abstract way of describing life-cycle management. Table 9-7 shows a set of management strategies, prescribed by Fox[11], that are associated with a five-stage life cycle:

1 Precommercialization
2 Introduction
3 Growth
4 Maturity
5 Decline

The first stage of Fox's life cycle may be roughly thought of as the development life cycle, which has itself previously been treated in terms of a number of stages. The remaining four stages represent the sales life cycle.

Table 9-7 clearly indicates the extreme variability in management strategy and outlook necessitated by the dynamics of the life cycle. The prospect of such flexibility's being developed in the context of a traditional hierarchical organization, designed primarily to ensure efficiency and control, is remote. Therefore, the implications of life cycles to both the need for, and practice of, project management are straightforward.

Overall Organizational Management Implications

Many organizations can be characterized at any instant in a given time by a "stream of projects" which place demands on its resources. The combined effect of all the "projects" facing an organization at any given time determines the overall status of the organization at that time.

The projects facing a given organization at a given time typically are diverse in nature—some products are in various stages of their life cycles, other products are in various stages of development, management subsystems are undergoing development, organizational subsystems are in transition, major decision problems such as merger and plant location decisions have been "projectized" for study and solution, etc.

Moreover, at any given time, each of these projects will typically be in a different phase of its life cycle. For instance, one product may be in the conceptual phase undergoing feasibility study, another may be in the definition phase, some are being produced, and some are being phased out in favor of oncoming models.

[11]Harold W. Fox, "A Framework for Functional Coordination," *Atlanta Economic Review*, vol. 23, no. 6, pp. 10–11, 1973. Used with permission.

TABLE 9-7
CHANGING MANAGEMENT STRATEGIES OVER THE LIFE CYCLE

Stage of Product Life Cycle	Functional Focus	R&D	Production
Precommercialization	Coordination of R&D and other functions	Reliability tests Release blueprints	Production design Process planning Purchasing dept. lines up vendors & subcontractors
Introduction	Engineering: debugging in R&D production, and field	Technical corrections (Engineering changes)	Subcontracting Centralize pilot plants; test various processes; develop standards
Growth	Production	Start successor product	Centralize production Phase out subcontractors Expedite vendors output; long runs
Maturity	Marketing and logistics	Develop minor variants Reduce costs thru value analysis Originate major adaptations to start new cycle	Many short runs Decentralize Import parts, low-priced models Routinization Cost reduction
Decline	Finance	Withdraw all R&D from initial version	Revert to subcontracting; simplify production line Careful inventory control; buy foreign or competitive goods; stock spare parts

Adapted from Harold W. Fox, "A Framework for Functional Coordination," *Atlantic Economic Review*, vol. 23, no. 6, pp. 10-11, 1973. With permission.

Marketing	Physical Distribution	Personnel	Finance
Test marketing Detailed marketing plan	Plan shipping schedules, mixed carloads Rent warehouse space, trucks	Recruit for new activities Negotiate operational changes with unions	LC plan for cash flows, profits, investments, subsidiaries
Induce trial; fill pipelines; sales agents or commissioned sales reps; publicity	Plan a logistics system	Staff and train middle management Stock options for executives	Accounting deficit; high net cash outflow Authorize large production facilities
Channel commitment Brand emphasis Salaried sales force Reduce price if necessary	Expedite deliveries Shift to owned facilities	Add suitable personnel for plant Many grievances Heavy overtime	Very high profits, net cash outflow still rising Sell equities
Short-term promotions Salaried salesmen Cooperative advertising Forward integration Routine marketing research; panels, audits	Reduce costs and raise customer service level Control finished goods inventory	Transfers, adancements; incentives for efficiency, safety, and so on Suggestion system	Declining profit rate but increasing net cash flow
Revert to commission basis; withdraw most promotional support Raise price Selective distribution Careful phase-out considering entire channel	Reduce inventory and services	Find new slots Encourage early retirement	Administer system retrenchment Sell unneeded equipment Export the machinery

The typical situation with products which are in the sales portion of their overall life cycle is shown in Figure 9-6, as projected through 1995 for the sales levels of three products, A, B, and C. Product B is expected to begin sales in 1988 and to be entering the declining sales phase of its cycle after 1991. Product A is already in the midst of a long declining sales phase. Product C is in development and will not be marketed until 1990. At any moment in time, each is in a different state. In 1991, for example, A is in a continuing decline, B is beginning a rather rapid decline, and C is just expanding rapidly.

Whatever measure is chosen to represent the activity level or state of completion of each of the projects in the stream facing an organization—whether products, product-oriented projects, management systems development projects, or decision-oriented projects—the aggregate of all of the projects facing the organization represents a stream of projects which it must pursue. Although the same measure (e.g., revenues, resources employed, percent completed) will not normally be applicable to all projects, the idea of a stream of projects—each at a different phase of its life cycle—is applicable to assessing the state of any dynamic organization.

The overall management implications of the stream of projects are clear from Figure 9-6. Top managers must plan in terms of the project stream. Since overall results are the sum of the results produced by the various projects, this planning must be not only in terms of long-run goals, but also in terms of the various steps along the way.

Most organizations do not wish to have their overall results appear to be erratic, so they must be concerned with the "sum" of the project stream at various points in time. A number of projects, each of which is pursuing a future goal quite well, might, in sum, appear to be not performing well at some points before they have reached their respective goals. In each case, for instance, although individual progress may be adequate, each may still be in a phase of its life cycle where it is consuming more resources than it is producing results.

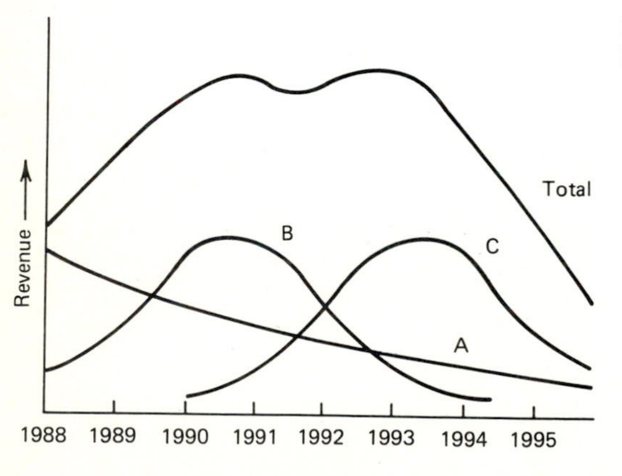

FIGURE 9-6
Life cycles for several products.

Thus, the problems associated with the overall management of an organization that is involved in a stream of projects are influenced by life cycles just as are the problems associated with managing individual projects.

There are even broader overall organizational implications than those that can be demonstrated with a single revenue measure for multiple products. For instance, Nolan[12] has specified six "stages of growth" for an organization's data processing (DP) function. Each of these life-cycle stages is characterized by different levels of sophistication in computer applications, different roles for the data processing organization, different levels of formality in the required degree of DP planning and control, and different roles for the user of the DP system.

The demanding requirements placed on overall organizational management by this life-cycle diversity are apparent. One of the approaches that has been suggested for dealing with this in the planning phase is that termed "strategic planning for management information systems."[13] In the implementation phase, some form of project management is essential to ensure that complex overall plans are, in fact, carried out.

CRITICAL PROJECT ELEMENTS: THE WORK BREAKDOWN STRUCTURE

Just as a project can be viewed as experiencing various phases, so, too, is a complex project made up of numerous elements. One of the fundamental bases of project management is the definition of these elements. The form of this project definition is the *work breakdown structure* (WBS).

The WBS divides the overall project into elements that represent assignable work units; these work units then become the common denominator around which the project management system is built.

Each project must be subdivided into tasks that are capable of being assigned and accomplished by some organizational unit or individual. These tasks are then performed by specialized functional organizational components. The overall "map" of the project represented as the collection of these units gives the project manager many organizational and subsystem interfaces to manage.

The underlying philosophy of the work breakdown structure is to break the project down into "work packages" which are assignable and for which accountability can be expected. Each work package is a performance-control element; it is negotiated and assigned to a specific organizational manager usually called a "work package manager." The work package manager is responsible for a specific objective (which should be measurable), detailed task descriptions, specifications, scheduled task milestones, and a time-phased

[12] R. L. Nolan, "Managing the Crises in Data Processing," *Harvard Business Review,* March–April 1979, pp. 115–126.

[13] W. R. King, "Strategic Planning for Management Information Systems," *MIS Quarterly,* March 1978.

budget in dollars and manpower. Each work package manager is held responsible for the completion of the work package in terms of objectives, schedules, and costs by both the project and the functional managers.

The process of developing the WBS is to establish a schema for dividing the project into major groups, then divide the major groups into tasks, subdivide the tasks into subtasks, and so forth.[14] The lowest level of the WBS becomes the unit around which project planning, organizing, and control can be carried out. The organization of the WBS should follow some orderly identification scheme; each WBS element is given a distinct identifier. Using an aircraft as an example, this can be done along the following basis:

1.0 Aircraft
  1.1 Final Assembly
  1.2 Fuselage
  1.3 Tail
  1.4 Wing
  1.5 Engines

A graphic representation of the WBS can facilitate its understanding. Using numerical listings with progressive indentation for successively lower levels can aid in communications and in developing understanding of the total project and its integral subsystems, sub-subsystems, etc. For instance:

1.0 . . . .
  1.1 . . . .
    1.1.1 . . . .
      1.1.1.1. . . . .
      1.1.1.2. . . . .
      1.1.1.3. . . . .
    1.1.2 . . . .
    Etc.

Figure 9-7 shows a hydroelectric electric plant with its component parts. Each part, e.g., intakes, power house, dam, and control panel, can be thought of as a "work package."

A work package grows out of having performed a work breakdown structure analysis on the project. When the work breakdown structure analysis is completed and the work packages are identified, a work breakdown structure comes into existence. A work breakdown structure can be represented by a pyramid similar to that used for describing the traditional organizational structure, such as depicted in Figure 9-7. Stated another way, the work breakdown structure presents the breakdown of the project objectives.

[14]Abramson and Kennedy have developed a WBS for a family camping trip and then portrayed this WBS in a flow chart. See Bertram N. Abramson and Robert D. Kennedy, *Managing Small Projects,* TRW Systems Group, 1969.

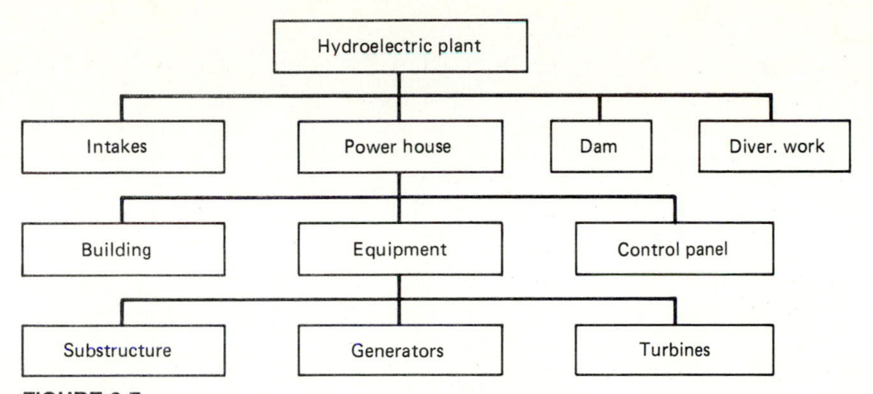

FIGURE 9-7
Hydroelectric plant, showing integral "work packages."

In the context of a project, the work breakdown structure and the resulting work packages provide a model of the products (hardware, software, services, and other elements) that completely define the project. Such a model enables project engineers, project managers, functional managers, and general managers to think of the totality of all products and services comprising the project, as well as its component subsystems. This model can be used as the focus around which the project is managed—reporting progress and status of engineering efforts, resource allocations, cost estimates, procurement actions, etc.

More particularly, the development of a work breakdown structure with accompanying work packages is necessary to accomplish the development of a *project management system* for the project. Such a system is meant to provide the means for:

1 Summarizing all products and services comprising the project, including support and other tasks
2 Displaying the interrelationships of the work packages to each other, to the total project, and to other engineering activities in the organization
3 Establishing the authority-responsibility matrix organization
4 Estimating project cost
5 Performing risk analysis
6 Scheduling work packages
7 Developing information for managing the project
8 Providing a basis for controlling the application of resources on the project
9 Providing reference points for getting people committed to support the project

The purpose of establishing a work breakdown structure is to provide a basis around which the project-management system can be built. The project-management system is a dynamic goal-seeking, purposeful system existing to provide a base for the project-management *process* to be carried out. This

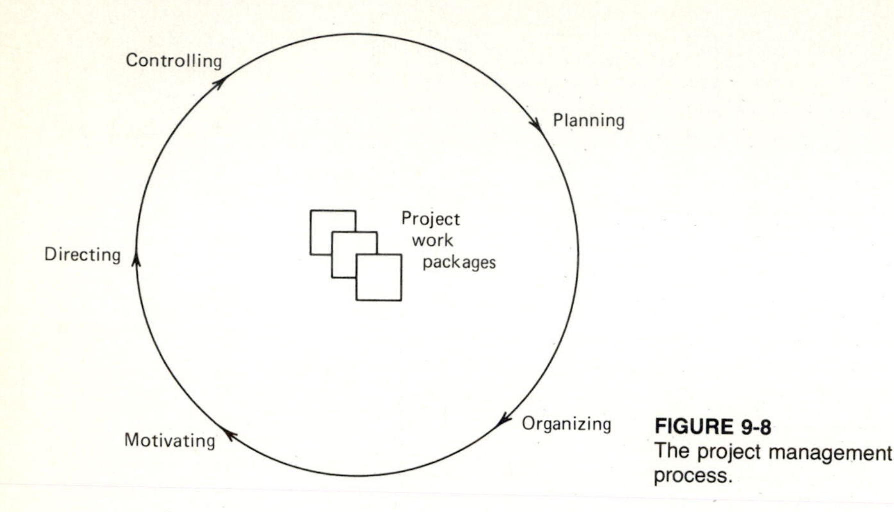

FIGURE 9-8
The project management process.

process can be portrayed through the use of a simple pictorial model; such a model is reflected in Figure 9-8.

In Figure 9-8, the model carries an important message: the functions of management, carried out by all managers and professionals, operate through the work packages of the work breakdown structure. Planning of a work package entails such matters as engineering design, materials acquisition, and procurement. Organizing of a work package means doing such things as assembling the required resources, and delegating authority and responsibility to an individual to manage the development and production of the work package. Motivating the people who are responsible for putting a package together is a challenging job for all the professionals and managers involved. Direction deals with the making of decisions with respect to the work package. Decisions regarding quality, reliability, specification compliance, etc., have to be made by someone. Finally, control of the work package in such matters as cost accrual, schedule compliance, and performance tolerances have to be carried out.

The work package is a unit of identification—a focal point around which the project is managed. If a project has not been broken down into its integral work packages, then it is doubtful if it can ever be managed effectively, or be accomplished on time, within budget, and meeting the performance specifications.

WHEN TO USE PROJECT-MANAGEMENT TECHNIQUES?

The discussion of environmental changes in Chapter 8 and of project dynamics in this chapter naturally leads to the above question: Just what are those conditions under which project management is essential or advisable?

The most obvious answers to this question come in the form of familiar illustrations such as those in the aerospace and construction industries (as discussed in Chapter 8).

In some situations, the decision to use project management techniques is made by the customer. Companies who bid on Department of Defense contracts will find that they are expected to establish a project-management system as a precursor to winning a contract for a weapons system. The Department of Energy also expects its contractors to have a project manager appointed to act as a focal point for the management of the project.

In general, project management can be applied effectively to any ad hoc undertaking. If such an undertaking is unique or unfamiliar, the need for project management is intensified. In some cases, such as that of an undertaking whose successful accomplishment involves complex and interdependent activities, a project manager can be appointed to "pull everything together."

A key reason for electing to manage things on a project basis is the fragmentation of functions and skills throughout the organizational structure. When an activity is introduced into the organization that is too large for any one functional department to manage, a single focal point is required to integrate the functional efforts. A good illustration of this need is the case of the Alaska gas pipeline from northern Alaska to the U.S. Midwest. Companies bidding on the construction of the Alaska gas pipeline claim they have learned about holding down costs through a process of streamlining management. "We plan to have a single point of management responsibility, not management by committee as Alyeska did, says John C. McMillian, Chairman of Northwest Energy!"[15]

A manufacturer established the following guideline for the use of project management:

> The program/project concept *may* be appropriate in those instances where management specifically intends to focus added attention, to highlight, or to emphasize the importance of a particular activity which can be isolated, specified, tracked and measured separately (either within or outside of the existing organization). Such activity might be functional or multi-functional in nature, may or may not be expected to be ultimately integrated into an ongoing operation, and might occur at any organizational echelon.[16]

Davis and Lawrence[17] insist that one should turn to a project "matrix" only when the following three conditions exist simultaneously: (1) when outside pressures require that intensive attention be focused on two or more different kinds of organizational tasks simultaneously—e.g., functional groupings around technical specialties and project groupings around unique customer needs; (2) when tasks become so uncertain, complex, and interdependent that the information-processing load threatens to overwhelm competent managers; (3)

[15]Quoted in article, "Costly Conduit," *The Wall Street Journal,* Mar. 6, 1978.

[16]Unpublished corporate document.

[17]Stanley M. Davis and Paul R. Lawrence, *Matrix,* Addison-Wesley Publishing Company, Inc., Reading, Mass., 1977.

when the organization must achieve economies of scale and high performance through the shared and flexible use of scarce human resources.

Basic to successful project management is recognizing when the project is needed—in other words, when to form a project, as opposed to when to use the regular functional organization to do the job. At what point in time do the changes in the organization add up to project management? The senior executives must have a basis for identifying undertakings which the regular functional groups cannot manage successfully. Of course, there are no simple rules to follow, but several general criteria, discussed below, can be applied.

Magnitude of the Effort

Project management is appropriate for ad hoc undertakings concerned with a single specific end product, such as a complex system for a customer, a move to a new plant site, a corporate acquisition, or the placing of a new product in the market.

The question of size is difficult to pin down because size is a relative matter. When an undertaking requires substantially more resources (people, money, equipment) than are normally employed in the business, project techniques are clearly indicated. Even though the functional elements for the end product are discernible in the organization, a function can be easily overwhelmed by the diversity and complexity of the task. In these cases, project management provides a logical approach to the organizational relationships and problems encountered in the integration of the work. For example, let us consider the move of a company from an eastern city to one in a southern state. This may appear to be a simple operation, but the complex development and correlation of plans, the coordination required in constructing the new site, and the task of answering numerous inquiries about the new site can easily swamp the existing organizational structure. These difficulties are compounded by the fact that the company must continue its normal operations during the period of the move. In such a situation, managing the move along traditional lines would be difficult, if not impossible.

The magnitude of an effort which is project-oriented depends on the basic strategy of the organization. A company engaged in routine manufacturing probably does not require much project management. However, if the company were to go through a major redesign of its product line which dictated significant special tooling and facility changes, a project manager could be set up to manage the change.

In the development and production of a weapons system, an awesome inventory of human and nonhuman resources has to be synchronized and integrated in an operable system. Initially, the Department of Defense requests organizations throughout the defense establishment to conduct feasibility studies and advanced planning. As the studies progress, the cost, technology, schedule, supportability, maintainability, and many other areas related to the project are added. Key Department of Defense officials enter the planning activities, funds are allocated, and contracts are awarded. As the project matures, widening circles of people throughout the defense-industry complex enter the project to identify the resources and organizational changes needed to satisfy it. Once a prototype is developed, the user (e.g., the Strategic Air Command), the producer (the aerospace company), and many other government and industrial representatives test the entire system. Throughout the life of the project, therefore, there is a continuous meshing of the philosophies, policies, structures, procedures, and resources of all the participating organizations, which requires the active involvement of thousands of people. Eventually the project is completed: a weapons system capable of performing an operational mission has been created. In one sense, the project has reached an end; in another, it has only begun, for the system still has to be supported, and this may also have to be done on a project basis.

Sometimes the size of the organization motivates the need to develop a "suborganization" as a self-contained entity to provide a suitable organizational ambience for success. For example, in the innovation process, some form of team or task force is required to provide the professional people an "organization" from which to motivate them in their work. The challenge, according to Hammerton,

> . . . is to create within a large organization the type of environment offered by a small company. The positive aspects of the latter are close cooperation among all participants, ease of identification with the production activity, absence of make-work tasks, minimum managerial restriction, little need to sell and resell tasks and projects to upper levels of management.[18]

The project team or task force (the small organization within the large organization) helps to foster an environment which facilitates establishing and maintaining good communications and respect within a professional community. In such a task force or project team, a "minicompany" cultural ambience can develop.

Unfamiliarity

An undertaking is not a project unless it is something out of the ordinary, different from a normal, routine affair in the organization. For example, an engineering change to an existing product could be conducted without setting up a project, although there would probably be a loss of overall efficiency in accomplishing the objective. The redesign of a major product, on the other hand, would probably require project management. In the first instance, each of the functional managers could draw on past experience to accomplish the work.

[18]James C. Hammerton, "Management and Motivation," *California Management Review,* vol. 13, no. 2, p. 52, Winter 1970.

In the latter case, however, the changes in cost, schedule, and technology would require a central management office (a project office) to bring together the functional activities required and relate them for compatibility.

Unique, one-of-a-kind opportunities (or problems) can be handled through the use of project management. Such one-of-a-kind situations are usually project-oriented. Work on these opportunities is usually scattered in the organization, yet it is interrelated. Various functional groups have to provide different disciplines to support the undertaking.

Environmental Change

Many firms and organizations operate in a "turbulent" environment that is characterized by continually changing products, rapid technological innovations, and rapid changes in the values and behavior of customers and markets. To be successful in such conditions places a premium on innovation, creativity, rapid response, and flexibility. Heterogeneous, changing environment requires a management system that can flourish in the ambiguity of changing objectives and goals where the life cycle of many projects places varying managerial and professional support.

Interrelatedness

Another decisive criterion for establishing a project is the degree of interdependence existing between the tasks of the effort. If the effort calls for many functionally separated activities to be pulled together, and if these activities are so closely related that moving one affects the others, project techniques are clearly needed. Consider the development and introduction of a new product. The early planning would require sales forecasts to be completed before plans for manufacturing processes, industrial facilities, special tooling, and marketing strategy could be developed. Sales promotions cannot be completed until the marketing research points the direction for the promotions. Performance and technical specifications must be resolved, as well as the many interdependencies between the production, marketing, finance, advertising, and administration groups. Provincialism cannot be tolerated. If no one agency can pull all the separate parts together, if the functional groups fail to make credible estimates, or if the plans submitted by the different departments cannot be reconciled, the activity needs the singleness of purpose of project management.

Projects are characterized by strong lateral working relationships requiring continuing coordination and decisions by many individuals, both within the parent organization and in outside companies. During the development of a major product, there will be close collaboration between the process and design engineers, and perhaps even closer collaboration between the individuals of a single department. These horizontal relationships do not function to the exclusion of the vertical relationships. The technicians and managers who make project decisions must also seek guidance from their supervisors.

Organizational activities which are interrelated, yet dispersed within an organization, and requiring the integration of multiple clientele resources, require the existence of an integrator-generalist. This is made clear by an employment advertisement for an individual to fill such a position. The advertisement utilizes a checklist for a prospective candidate to use in determining if he or she qualifies for a project-manager position with the corporation. Figure 9-9 shows the checklist.

Organizational Reputation

The overall organizational stake in the undertaking is another crucial determinant in the decision of whether or not to use the project techniques. For instance, if a failure to complete a contract on time and within the cost and performance limits would seriously damage the company's image and result in stockholder dissatisfaction, the case for using project management is strong. In the final analysis, a company's financial position can be seriously damaged if its performance on a contract fails to meet standards. In the case of government contracting, the company faces a single, knowledgeable customer, and failure to perform successfully can be catastrophic in terms of obtaining further contracts with the government.

Project management is no panacea, but it does provide a means for controlling the undertaking. A project manager who sees the role as that of an

"Our checklist enables you to assess your compatibility to our requirements.

- ☐ You require a position in which you control contracts of between $1 million and $10 million.
- ☐ You are confident of your ability to interface with the customer (Military) in terms of scheduling, marketing and resolving technical and production related problems.
- ☐ You are capable of making presentations to top management which are clear, concise and accurate.
- ☐ You possess the ability to cohere a team of engineering and manufacturing managers.
- ☐ Your resource management objectives include skillful budgeting of support services.
- ☐ You control cost factors with bottom-line cognizance.
- ☐ You appreciate the requirement for effective time management in a tight schedule environment.

Only those with a perfect score should forward detailed resumes (informal is fine) including present salary."

FIGURE 9-9
Prospective employee checklist. (*The New York Times*, January 28, 1979.)

integrator-generalist, responsible for meeting time, cost, and performance objectives, can do much to lessen the dangers inherent in a large undertaking. The project-management techniques concentrate into one person the attention demanded by a complex and unique undertaking. Before any recommendations are made concerning whether to use project techniques or not, the effects of the unique environment on the project must be weighed and evaluated. First, the objective of the undertaking must be considered. Possible improvements in methods that might take some time to implement would require considerable thought. The size and complexity of the project must be considered, since too much sophistication is also an ever-present danger. Other factors which merit consideration are the number of current projects in the company, the number in prospect for the organization, and the length of time remaining to complete the project. For example, establishing project management would be more appropriate at the start of an undertaking or at least early in its life, before large expenditures of work-hours and resources are made. Each situation is unique, and whether to manage by a project or a functional approach should be resolved on the basis of specific problems found as well as present concepts of organization.

No company management takes a purely project or a purely functional approach. All companies combine the two, although one form may predominate. If the company's undertakings are small, there would be a series of small, one-person projects. If the undertakings are large, as in the case of a new product development or an acquisition of a major system, the project group would be large and might exist over a period of several years. A project group in a development laboratory may be responsible for production and marketing planning. In some cases, a project takes on the characteristics of a permanent functional organization.

There are many other circumstances under which project techniques are desirable:

A multilateral objective exists, toward which many people and many relatively independent organizations work together.

There are pressures to improve the product and advance the state of the art.

Plans are subject to change, requiring organizational flexibility.

The risks are high, and the uncertainty factors make prediction of the future difficult.

Project integration requires the concurrent contribution by two or more functional elements and/or independent organizations.

The project is a type requiring advanced feasibility studies and development.

The government procurement agency requires a project-oriented approach.

A management climate exists which permits the temporary "shorting" of reporting relationships within the organization.

Frequently, the traditional vertical relationships of an organization prove inadequate to cope with the technology for a singular undertaking, and the horizontal relationships arising out of that technology must be considered. Project management requires new organizational setups and permits horizontal and diagonal relationships which are not patterned or clearly defined. Some sort of balance between the functional and the project organizations is required, however, even when the need for a project approach is clear.

SUMMARY

This chapter deals with the dynamic nature of systems, projects, and project management. As such, it is the linking mechanism between the environment, which creates the need for project management, and the specific organizational and management techniques of project management that are briefly introduced here and covered extensively in later chapters.

Project management is an approach for responding to the dynamic nature of systems. Since complex systems are the entities which will comprise the solutions to most of the complex problems of society, there is a real need to develop management techniques and devices which address themselves to the dynamic nature of systems.

One may think of systems dynamics in terms of such concepts as the *sales life cycle* for a product or a *systems development life cycle* for an emerging idea or concept. In either case, the dynamics of the system impose a requirement for a corresponding dynamic nature for the projects which are created to direct and control the system through its various phases.

Of course, project management is not universally applicable. The utility of the idea depends on the magnitude of the effort, the complexity, the degree of unfamiliarity and interrelatedness, and the concern with the organization's reputation.

DISCUSSION QUESTIONS

1 Project management is dynamic. What makes this dynamism manifest in projects?
2 Implementation of project management is difficult because of the dynamic nature of the systems involved. What is the life cycle, and how does it contribute to this dynamism?
3 Take Larry Greiner's description (p. 237) of the evolution and revolution as an organization grows, and apply the idea to an organization of your choosing. Does an appreciation of this idea have value for contemporary managers?
4 What are some of the key factors to consider in viewing the sales life cycle of a product? What are some of the key decision points involved during a product's life cycle? If so, what are some of these key decisions?
5 Each system goes through a life cycle. Take a system of your choosing (other than one mentioned in the book) and trace the stages of its life cycle and some of the key questions that should be asked about the system during its life cycle.
6 What is a work breakdown structure for a project? Why is it important to understand the concept of a work breakdown structure in the management of a project?

7 A work breakdown structure is both vertical and horizontal within an organization. Explain what is meant by this. Give some examples.
8 What is a work package? Take an organization, such as a functional organization, and develop the key work packages that make up that organization.
9 Develop a work breakdown structure with accompanying packages for a family camping trip.
10 What are the purposes for which a work breakdown structure can be used?
11 When should project-management techniques be used in contemporary organizations?
12 A project manager is much like a general manager. Defend or refute this statement.
13 What is the key focal point around which the project-management process must be carried out? Give some specific examples on how the management process can be carried out in this context.
14 The larger the project, the more important it becomes to have a single point of management responsibility. Do you agree with this? Why or why not?
15 In the early 1960s, project management was required by the U.S. Department of Defense for all major weapons system contracts. Some aerospace companies resisted developing project-management systems in their organizations to support these contracts. Why do you think this happened? Today, one finds that many of these aerospace companies are using project-management systems in the commercial side of their business. Why did they first resist and then later embrace project management?
16 If an organization is in a market environment where technological and social innovation is changing rapidly, a form of project management is necessary for survival. Do you agree with this statement? Why or why not?

RECOMMENDED READINGS

Abramson, Bertram N., and Robert D. Kennedy: *Managing Small Projects,* TRW Systems Group, 1969.

Archibald, Russell D.: *Managing High-Technology Programs and Projects,* John Wiley & Sons, Inc., New York, 1976.

Avots, Ivars: "Why Does Project Management Fail?," *California Management Review,* Fall 1969.

Bobrowski, Thomas M.: "A Basic Philosophy of Project Management," *Journal of Systems Management,* May 1974.

Davis, Stanley M., and Paul R. Lawrence: *Matrix,* Addison-Wesley Publishing Company, Inc., Reading, Mass., 1977.

Fox, Harold W.: "A Framework for Functional Coordination," *Atlanta Economic Review,* vol. 23, no. 6, 1973.

Greiner, Larry E.: "Evolution and Revolution as Organizations Grow," *Harvard Business Review,* July–August 1972.

Hammerton, James C.: "Management and Motivation," *California Management Review,* vol. 13, no. 2, Winter 1970.

Hilton, Peter J.: *New Product Introduction for Small Business Owners,* Small Business Management Series no. 17, U.S. Government Printing Office, Washington, D.C.

Kimberly, John R., et al.: *The Organizational Life Cycle,* Jossey-Bass, Inc., Publishers, San Francisco, 1980.

King, W. R.: "Strategic Planning for Management Information Systems," *MIS Quarterly,* March 1978.

Levitt, Theodore: "Exploit the Product Life Cycle," *Harvard Business Review,* November–December 1965.
Mantell, Leroy H.: "The Systems Approach and Good Management," *Business Horizons,* October 1972.
Miller, J. Wade, Jr., and Robert J. Wolf: "The 'Micro-Company,'" *Personnel,* July–August 1968.
Nolan, R. L.: "Managing the Crises in Data Processing," *Harvard Business Review,* March–April 1979.
Seneker, Harold: "If You Gotta Borrow Money, You Gotta," *Forbes,* April 28, 1980.
Spengler, Oswald: *The Decline of the West,* Alfred A. Knopf, Inc., New York, 1945.
Thamhain, Hans J., and David L. Wilemon: "Leadership, Conflict, and Program Management Effectiveness," *Sloan Management Review,* Fall 1977.
Ullman, P. E.: "Project Management Teams—What's the Score?," *Process Engineering,* September 1978.
The Initiation and Implementation of Industrial Projects in Developing Countries: A Systematic Approach, United Nations Industrial Development Organization, United Nations, New York, 1975.
"Costly Conduit," *The Wall Street Journal,* March 6, 1978.
"How GM 'Project Centers' Create Cars," Customer Information Sheet, General Motors Corporation.
Management Guide, Pratt & Whitney Aircraft Group, Government Products Division, United Technologies, undated.
"Texas Instruments Shows U.S. Business How to Survive in the 1980s," *Business Week,* September 18, 1978.

CASE 9-1: Project Management Start-Up—How to do it!

Situation You manage one of several engineering subsections of an Engineering Department. As the result of the "need for enhanced effectiveness and productivity," project management is being introduced in your Engineering Department for the first time. At a recent all-employee meeting, your Engineering Department vice president has outlined the need for project management, and explained its value. At that meeting, it was stated that the Manager of Projects Section, which has been partially in place for some time now, "will increase in size and take on full responsibilities beginning immediately."

Several of the project engineers who report to you have asked to meet with you to discuss project management. The engineers have each had total responsibility for "their" own projects and they are asking specifically, "What should we do differently when project management takes over?"

As a dedicated manager, you have a strong interest in ensuring that the transition to project management works effectively. You therefore want to have a meeting and use it not only to answer the specific question, but also to encourage your project engineers to (1) support the project-management concept, and (2) take whatever actions they can to help make it work.

Your Task As a team, outline the meeting which you would conduct with your people in such a way as to contribute to a speedy and successful start-up.

Specify all questions, issues, and probable concerns which your people might have, and outline how you would plan to handle each during the meeting. If there are any questions, issues, or concerns which will require additional action on your part, specify and outline your plan for resolving each.

Choose a spokesperson to review the outline of your meeting for the conference participants (as you might for your own manager).

CHAPTER 10

ORGANIZATIONAL CONCEPTS OF PROJECT MANAGEMENT

. . . sound organization structure requires both (a) hierarchical structure of authority, and (b) a capacity to organize task forces, teams, and individuals for work on both a permanent and a temporary basis.[1]

In this chapter we examine an organization from a systems viewpoint. First, we discuss the concept of an organizational system. From this vantage, we review some generic organizational alternatives as a preliminary to analyzing the matrix organizational form. Finally, the chapter offers some suggestions on how to align matrix organizational elements.

The systems view of an organization emphasizes the interrelatedness of organizational forces and stresses an integrated totality rather than a parochial component view. A systems-oriented organization is a dynamic, purposeful goal-seeking entity, since a systems viewpoint is based on the notion of the interdependency of subsystems and components. Implicit in the matrix interpretation of this systems notion is the idea of the *temporary nature of organizations:* constantly changing, fluid, and varied. It is the emergence of such temporary

[1]Peter F. Drucker, "New Templates for Today's Organizations," *Harvard Business Review,* January–February 1974, p. 53.

organizations that give rise to the need to understand the organizational concepts of project management.[2]

Often the term "project organization" is used to denote an interdisciplinary and interorganizational team pulled together for a specific purpose. Personnel are drawn from the hierarchically arrayed units to perform a specific task; the organization is temporary in nature, built around the purpose to be accomplished rather than on the basis of functional homogeneity, process, product, or other traditional bases. When such a team is assembled and superimposed on the existing structure, a *matrix organization* is formed. This matrix organization includes the *functional units* and the *project teams* assembled to accomplish the specific purpose. The matrix organization encompasses the complementary *functional and project* units.

Management theory associated with project management is still evolving. This is to be expected from a concept which has grown to fulfill a practical need. Theoretical developments are most often well defined and logically interrelated but often may not be useful in the real world. Pragmatic concepts such as project management grow from real-world need. If they succeed, as project management clearly has, management theorists try to build a framework of definitions and logic around existing ideas. It is therefore natural that such a process would, at some stage of development, have to rely on tentative definitions. One of the purposes of this chapter is to build such a framework within the context of the practical knowledge of the project and the need for a facilitative matrix organizational subsystem.

ORGANIZATIONS

"Organization" is a deceptively simple label; in common usage it has three interrelated meanings. According to Gilman, it is (1) a set of understandings of how human and other resources are to be marshaled toward the achievement of an objective; (2) the structure of understandings ranging from policies and procedures to personnel assignments; and (3) the acting agency that is formed when organization as process and structure relates the efforts of a number of individuals in joint accomplishment.[3]

When we speak of organizations, we must do so in operational terms. Rather than conceiving of the organization as a static structural entity, it is useful to consider it as a dynamic system having:

1 An explicit or implicit objective toward which the participants are working
2 A formal and an informal pattern of authority and responsibility among the participants
3 A given quality and quantity of resources, both human and nonhuman

[2]The permanent matrix organization is one in which an organization is structurally defined on two or more bases simultaneously. See, for example, S. M. Davis, and P. R. Lawrence, *Matrix*, Addison-Wesley Publishing Company, Inc., Reading, Mass., 1977.

[3]Glen Gilman, "The Manager and the Systems Concept," *Business Horizons*, August 1969, pp. 19–28.

4 A constant interaction between subsystems, as decisions are made, as strategies are designed for the implementation of decisions, and as decisions are themselves implemented

In this context, an organization is created when two or more people agree to cooperate in seeking a common goal. The integration of these organizational elements is carried out through plans, policies, procedures, and rules, which formally prescribe how the elements *are* to relate. On the human side, the *informal organization* prescribes how the people *want* to relate.

A "reorganization" of an organization can be effected by varying one or more of the components. For example, an organization is changed by adding or taking away groups of certain skilled people. An organization is also changed by a new policy which prescribes how people will work together to pursue a particular objective. When a project team is superimposed on an existing structure, a matrix organization has been created *and* an important reorganization has taken place.

Evolving Principles of Organization

Few "principles" of organization can be applied universally. In the 1950s and early 1960s, organizational theorists set forth various organizing techniques. *Centralization, decentralization, functional, departmental, product, process, geographical*—these were representative of the numerous techniques and patterns for structuring the organization. These techniques were well documented and explained in all the standard texts. *Line and staff* concepts and the vertical, hierarchical, chain-of-command beliefs provided basic points of departure from which to organize activities.

In classical theory, the organization was typically structured from some generic model; the process of adapting the particular structure to the requirements of the organization received only secondary consideration. In the late 1950s and early 1960s, some organizations were wrestling with *product and project management* forms which did not fit into the traditional patterns of organization. Such developments cast serious doubt on the universality of certain of the most venerated of management principles. Experimentation with alternative forms of organization structure began to solve some of the problems that did not fit into the traditional theories of organization. According to Mockler: "The traditional organization structures, with their rigid divisions of responsibility and authority and their mechanistic chains of command, were too inflexible to meet the needs of the dynamic business environment of the 1960s."[4]

In the late 1960s, a general philosophy of "no best way" to organize caused a shift from traditional organizational patterns to development of individualized and flexible approaches to meet the particular situation. Delbecq summarized the thinking about organization methodology by stating:

[4]Robert J. Mockler, "Situational Theory of Management," *Harvard Business Review,* May–June 1971, p. 147.

> One cannot use a single stereotyped organizational model and meaningfully understand the rich variety of Task and Administrative units within modern complex organizations. One must necessarily speak of variety of organizational designs, and a variety of administrative systems for coping with different mixes of these model forms.[5]

Many other writers have commented on a trend away from simple, unitary concepts of formal organization and toward increasingly complex models.[6]

A Systems View of an Organization

An understanding of the systems concept is necessary to a discussion of the project-management organization. The concepts of *organization* and *system* have been associated together since the beginning of the scientific management era. Kendall, writing in 1912, perceived this relationship and spoke of the need to be organized so that "separate processes and unit members are brought *into systematic* connection and operation as efficient parts of the whole."[7] In a more recent period, Barnard built his theme around the equivalency of systems and organizations by stating:

> It is the central hypothesis of this book that the most useful concept for the analysis of experience of cooperative systems is embodied in the definition of a formal organization as a *system of consciously coordinated activities or forces of two or more persons.*[8]

Barnard's view of the organization as a system related to larger systems in the environment implied reciprocal activities between the organizations in a particular sphere of activity. In Barnard's view, the organization has multidimensional characteristics with interrelated goal-seeking subsystems.

The systems concept appeared in management literature approximately sixty years ago, but it has only recently gained any degree of acceptance in the study of organizations. One can very easily question the usefulness of the concept. What can be done to further show the usefulness of the systems approach to a manager in managing internal and external organizational affairs?

One reason why practitioners have hesitated for so long to accept the systems approach is that it is based on analogical reasoning rather than on empirical verification. In addition, there was no real need for it in the past, and now that the need has arisen, the development of techniques has lagged far behind, as usual.

[5]André L. Delbecq, "Matrix Organization—An Evolution beyond Bureaucracy," *Paper,* The University of Wisconsin Press, Madison, p. 142.

[6]For example: See W. G. Bennis, "Organization Developments and the Fate of Bureaucracy," paper presented at the annual meeting of the American Psychological Association, Los Angeles, Sept. 4, 1964.

[7]Henry P. Kendall, *Scientific Management: First Conference at the Amos Tuck School,* The Plimpton Press, 1912, p. 113.

[8]Chester I. Barnard, *The Functions of the Executive,* Harvard University Press, Cambridge, Mass., 1962, p. 73.

The Component View In the past, management education and practices have emphasized a component view; accounting, production, marketing, finance, and human relations were taught as unrelated subjects. It was assumed, perhaps naïvely, that experience would give top-level managers the ability to integrate these disciplines in a practical management situation. The systems approach to management has gained in favor because it considers the interdependencies of today's industrial and public systems. The interconnections and interactions between the components of the system are often more important than the separate components themselves.

Not only large and complex companies need a systems view; some small companies with relatively stable environments need it too. Small, stable companies that have an unchanging product or service can operate very well on a vertical management philosophy, except that they have a role in the larger scheme of things. All companies, both large and small, are part of their environment, and the extent to which a company is integrated into its environment can be a good measure of its success.

The boundaries of a company are not simple. The operation of a company from a systems view requires people and groups that never appear on the organizational charts. Seymour Tilles has written of the systems approach:

> Many organizations have a management team that includes individuals—auditors, lawyers, bankers, brokers, and a variety of other specialists—who never appear on the organization chart. In some cases, these outside experts are consulted with such regularity that they are really a part of the management system. In fact, the extent of the management system is frequently an indication of the manager's ability.[9]

Viewing the manager's job as Tilles has done may lead one to conclude that managers have been using the systems concept all along. Perhaps the use of coordination as a managerial technique qualifies as the systems approach; coordination—the synchronizing of activities with respect to time and place—entails working with many individuals outside the scalar chain.

Coordination is necessary to the management process—it can do much to assure the timely accomplishment of objectives. But coordination per se lacks the inherent integration implied by the systems approach, which requires interrelatedness and organizational reciprocity.

An Integrated System The organization must be aligned so that the individual functions are brought together into an integrated, organized system with the parts related to the common goal. The systems approach to the organization of human and nonhuman resources includes the basic ideas that:

1 An organization is composed of many subsystems within the parent company and overlapping with the general environment.

[9]Seymour Tilles, "The Manager's Job: A Systems Approach," *Harvard Business Review*, January–February 1963, p. 75.

2 The organization is not an entity in itself, but must interface with other groups (public organizations, customers, suppliers, unions, creditors, bankers, stockholders, governments, etc.) to survive in its environment.

3 Subordinates are only part of the personnel groups that "work" for the supervisor.

4 The legitimacy of the task may sometimes dominate the vertical hierarchies.

5 Each person in an organization has a mix of roles in the total system, not a narrow specialization.

Individual managers do not occupy a single, central fixed position in the organization; rather, managers must have considerable organizational mobility to maintain the necessary relationships in the environment.

6 The manager must understand the relationships between the parts of the system as well as the specific requirements of his job.

Systems theory has had significant impact on how we view organizations. The systems approach provides a basis for understanding the nature of the organization, their problems, and their opportunities. Systems theory enables us to look at an organization as an input/output system. According to Chappie and Sayles:

> Most systems theorists conceive of the organization as a complex input-throughput-output system. The organization is separated from its environment by a permeable boundary. Through this boundary, transactions occur which enable the organization to secure human, financial, and material inputs. Within organization boundaries, a number of interacting subsystems transform these inputs into a final product suitable to the environment. Output passes through the organization's boundaries, reactivating the input-throughput-output cycle.[10]

VARIOUS FORMS OF ORGANIZATION

Although the focus of this book is on horizontal, or project, organizations, in this section we shall explore a variety of organizational forms. At one extreme is the pure project organization, where the project manager is given full authority to run a project as if it were a one-product company; at the other is the pure functional organization departmented on a traditional basis, reflecting the traditional hierarchy. In the middle lie an infinite variety of project-functional combinations—the *matrix organization*. Each of these forms has certain advantages and disadvantages; no one form is best for all projects, or even best for one throughout its entire life cycle. The essence of project organization is versatility —the project can be built around the objective; as the objective changes, so must the scope of the organization.

[10]Elliot O. Chappie and Leonard R. Sayles, *The Measure of Management,* The Macmillan Company, New York, 1961, p. 78.

Vertical Organizational Forms

There are several basic principles commonly used as organizational bases: *functional, product, process, geographic,* and *customer.* The most common form is the *functional;* it exists where long-lived products are found. In the functional form, people and resources having a common activity to support are grouped together. In a typical manufacturing company, one would find human and nonhuman resources grouped together to support major activities such as production, finance, marketing, and research and development.

In the *product* type, substantial product differences exist; within a product section, there may be functional or other organizational structure used. Usually integration between product lines is minimal or nonexistent, depending on the degree of interdependence of these product lines. For example, the technologies required to produce auxiliary equipment and turbines are different, yet both types of equipment may be required to meet customers' needs.

An organization aligned on a *process* basis facilitates a product passing through the organization in some logical fashion, e.g., fabrication, assembly, finishing, packaging, and shipping. Using a process form requires a close integration between the individual processes. A process type of organization is often used within a functional organization, e.g., manufacturing.

The geographic arrangement of the market may dictate the use of the *geographic* form. Frequently, this organization is used where there is a single product or range of products and where it is appropriate for physical reasons to segregate responsibilities in accordance with location . Each manager in each area is responsible for customers, the market, and keeping the rest of the organization informed about what is happening in the area.

In some situations, there are singular characteristics of the customer that make it appropriate to organize on the basis of *customer* or class of customer. An example of the customer form is found in companies that do business with the U.S. government. In case of each customer, the peculiarities of that customer dictate a need for customer specialization.

Pure Functional Organization

Of the various vertical forms of organization, the function is probably the most common. It is also an integral element of the project-based matrix organization, so we shall devote some attention to its characteristics.

In the pure functional form, people are grouped according to their similarities —in terms of activity, expertise, etc. Thus, all marketing people are in one department, all finance people in another. The continuity of knowledge and experience is therefore great. Expertise in functional areas is retained and developed, and clear career paths exist in each function.

One disadvantage of the functional organization acting alone, however, is that it does not provide the emphasis necessary to accomplish project objectives. No one individual is responsible for the total objective except the general manager who is in charge of all of the functions; there is no customer focal point. Since no one person functions as the champion of the product or customer, responsibility may be difficult to pinpoint, coordination unduly complex, response to customer needs slow, and motivation and innovation decreased. Ideas may tend to be functionally oriented, and approaches to the management process will tend to perpetuate the functional organization without regard for ongoing projects.

Pure Project Organization

In this approach, the project is truly like a minicompany. The project team is independent of support from any major functional units or departments. Minor functional support—in such matters as industrial relations, payroll, and public relations—are provided by a functional element that takes care of this for the entire organization. The major advantage of the pure project organization is that it provides complete line authority over the project; the project participants work directly for the project manager, with the chief executive (or some general manager) in the main line of authority. One of the strongest disadvantages to this type of organization is that the cost in a multiproject company would be prohibitive because a duplication of effort and facilities would be required among the projects. Since there would be no reservoir of specialists in a functional element, there might be a tendency to retain personnel on the project long after they were needed. Then, too, there would be no functional group to look toward the future and work to improve the company's capability for new projects.

A mixed project and functional structure, or *matrix* organization, is desirable for producing large projects within desired cost, schedule, and performance standards. The mixture can lie anywhere between the two extremes, the exact structure being determined by the particular task requirements.

The Matrix Organization

The *matrix* form of organizational alternative demands our greatest attention, for not many managers have a clear, consistent concept of what it means. Although the matrix is used in a wide variety of different organizations, there is a limited understanding of its structure, processes, and impact on the greater organizational system of which it is a part. A *matrix organization* is a network of intersections between a project team and the functional elements of an organization. As additional project teams are arrayed across an organization's functional structure, more intersections come into existence. In its most elemental form, a matrix organization looks like the model in Figure 10-1, where the confluence of the *project* and *functional* elements comes about.

The confluence of these elements centers on the project work packages. The underlying concept of the work package is simply that of management by

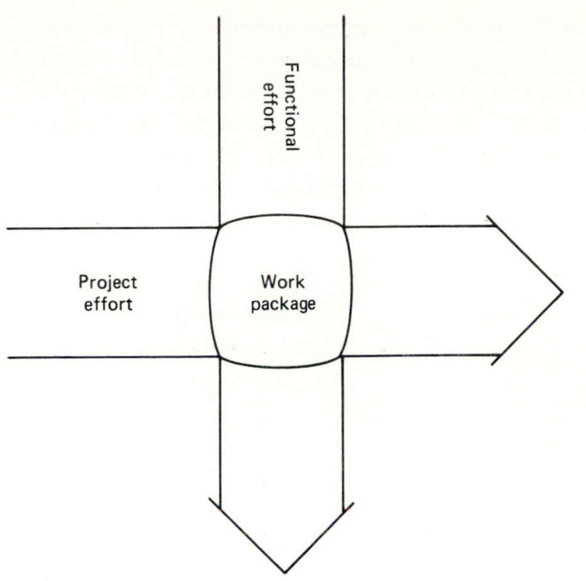

FIGURE 10-1
Confluence of the project and functional effort around the project work package.

objectives[11] and the decentralization of authority and responsibility. Implementation of the work package requires that the total job be broken down into components (hardware, software, and services) and then that these components be further broken down into assignable work packages. Each work package is basically a "bundle of skills" that an individual or individuals have to perform in the organization. A work package is negotiated with, and assigned to, a specific functional manager. The functional manager who accepts the work package agrees to specific objectives which are measurable, and to detailed task descriptions, specifications, milestones, budget for the work package, etc. This work package manager is then held fully responsible for the work package in terms of the package's meeting its objective on time and within budget.

"Matrix organization" is the term that is used to describe the policy, procedure, and work relationships resulting when project terms are superimposed on an existing hierarchical structure. It is the theoretical opposite of the bureaucratic design. The fundamental premises of the matrix and the bureaucratic design are opposite, yet complementary. The bureaucratic structure tends to be rigid and predictable; the matrix structure, flexible and unpredictable, since each project tends to have its own particular attributes and may require a somewhat different management philosophy and process.

The underlying premise of the matrix structure is simply that objectives can

[11]As discussed in Chap. 7.

best be obtained if the organization resources can be directly oriented toward those objectives without regard to traditional organizational structures and constraints. The organization form of the matrix is used as a means to an end; it can be readily adapted to a changing environment. As the organizational need for new projects changes, the matrix structure tends to be fluid. Since organizations are organized around specific projects, the matrix is in a constant state of flux as projects are completed and resources are deployed to new or other current projects.

The justification for applying matrix organizational techniques to the delivery of a product or service lies in the fundamentals of open-systems theory. We can view the matrix organization as a system of interrelated activities, all of which can have some impact on the goal. The purpose of the matrix organization is to provide a mechanism for different disciplines to bear on a common problem or opportunity—the project objective.

In mathematical terms, a matrix is a rectangular array of numerical or algebraic quantities treated as an algebraic entity. In organizational terms, a matrix is a

> . . . cross-hatch of structural elements, with disciplines or functional units forming the vertical dimensions, and programmatic or project units providing the horizontal dimensions.[12]

A Prototype Matrix Organization The dynamic nature of complex systems has led to the institution of a project framework on organizations which has come to be referred to as the matrix organization.

Consider, for example, a company made up of two divisions. Division A is an operating entity which produces a standardized product in high volume. Within Division A are the functional departments through which the standardized work can flow. Functional departmentalization is the traditional way of organizing in such an instance. A major functional department such as finance would normally be comprised of a number of minor functional departments such as credit, disbursements, fund control, and accounting. Division A may be thought of in terms of four major functional departments—production, engineering, personnel, and finance. The managerial emphasis in these departments would, because of the nature of the activities which they perform, be primarily on improving operating efficiency. Figure 10-2 shows Division A to be traditionally organized to achieve such efficiency.

Division B of the company, on the other hand, is a "job shop" operation which performs contract work for the government. Its work load is composed of various projects, each with rather specific objectives and a well-defined point of completion. The managerial emphasis here needs to be on the timely completion of these projects.

[12] F. A. Shull and R. J. Judd, "Matrix Organizations and Control Systems," *Management International Review,* vol. 11, 1971, p. 65.

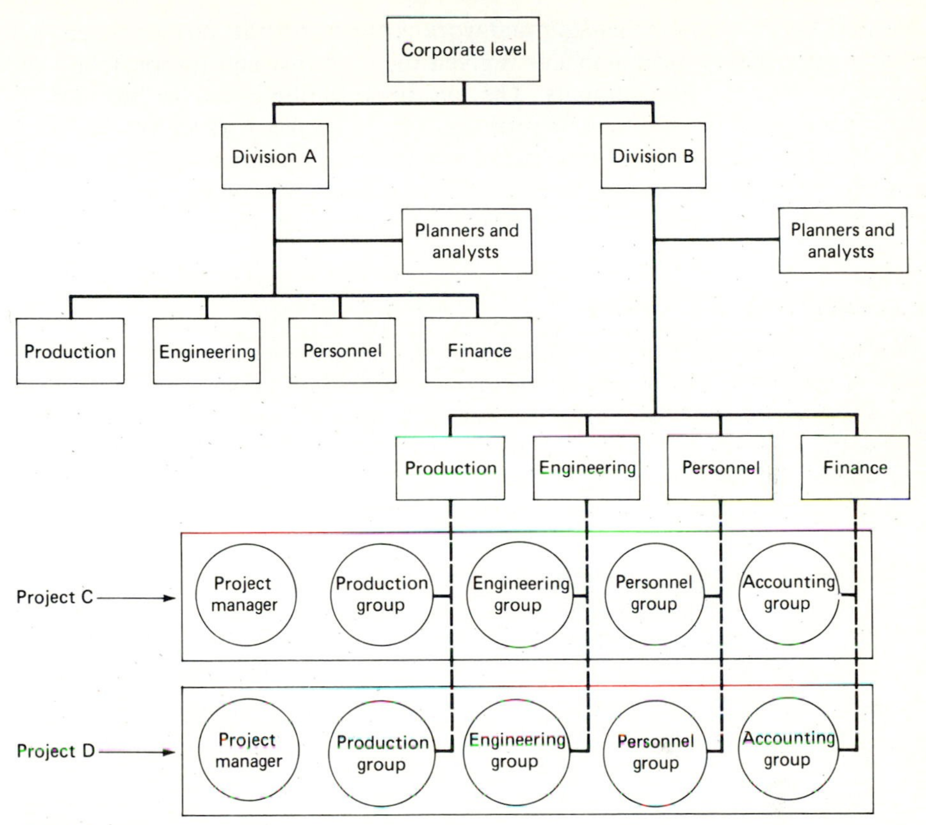

FIGURE 10-2
Illustrative matrix organization.

The level of participation by various functional entities in each of Division B's projects is highly variable as each project progresses toward completion. In the early stages of a project, the emphasis may be on the choice of materials and components, while at a later stage this facet will be deemphasized. This is the case because of the natural "life cycle" of a project. Thus, the combination of people and resources that can best jointly pursue project goals changes from time to time in each of Division B's projects. Here, then, the primary management emphasis must be on organizing and controlling the projects, rather than on the seeking of greater efficiency.

The organizational chart of the hypothetical two-division company is illustrated in Figure 10-2. Division A needs no matrix concept because of its continuing standardized work load. Division B has all the same functional departments as does Division A, but each of them provides facilities and functional support to the two major activities of the division—Projects C and D.

Each of the project organizations depicted in the lower portion of Figure 10-2 is comprised of a project manager and work groups from the various functional departments. The project manager is given the authority and responsibility for the achievement of project goals. The line organization is left to function by providing support for the various projects. Typically, the work groups for each project are assigned on a full-time basis from their various functional units to the project. When the project is completed, or when their services are no longer essential to it, they are assigned back to their functional units.

Matrix Management Concepts

The matrix concept of organization has important implications for the myriad tasks of management. In matrix organizations, we find a management philosophy which dictates that the organization shall reflect major work relationships rather than traditional work alignments. This new organizational structure contains four generic elements: *functional support, project management, routine administration,* and *research and development (strategic planning)*.

Functional Support Functional support consists of facilitative technology provided for the company by various groups. In a manufacturing organization, this element would be supplied by four groups, designated "production," "marketing," "R & D," and "finance." Functional support is provided for all projects in the organization as well as for the advancement of the state-of-the-art in a particular discipline.

Project Management Project management is carried out by a set of managers acting as *unifying agents* for particular projects in respect to the current resources of time, funds, materials, people, and technology. The project managers act as focal points for their project activities through an organization superimposed on the traditional functional organization structure. *The project managers are, in effect, the general managers of the company for their particular projects*. They actively participate in planning, organizing, and controlling those major organizational and extraorganizational activities.

Routine Administration Routine administration involves the accommodating services provided for mission-related activities. These services include the centralized activities required to keep score on the business as a whole, as well as the routine administration and accounting of funds, people, material, and ideas. Examples are the personnel function, repetitive business data processing, and recurring logistic support.

Research and Development (Strategic Planning) Research and development activities are those concerned with advancing the *strategic* state-of-the-art in the functional areas and with developing a system of plans and products/services for the company's future. This group is less concerned with accomplishing current work than with obtaining future work and finding new uses for existing resources; consequently, its work is more conceptual and abstract than that of other elements.

Contemporary managers face the problem of keeping themselves up to date on the many ad hoc activities needed to complete a project within time, cost, and technological boundaries. These projects may be in varying degrees of maturity; some may be merely concepts, while others are nearing completion and phase-out. This stream of projects, encompassing a number of different products, each of which requires considerable effort and attention, often first manifests itself in the R&D function. The R&D functional manager cannot stay on top of all the development efforts for which he or she is responsible. To assist in these developments, a project manager is assigned to each of the major projects to achieve, first of all, unity of communication and coordination across the disciplines of the R&D organization, as well as within the total organization. Project managers become a source of integrated information concerning their particular projects and interaction points for coordinating the diverse organizational and extraorganizational activities involved. This communication function, coupled with the coordinative function, enables them to exercise control over many aspects of their projects.

The Basis for the Matrix Approach

The ability to attain goals and to solve problems in large, complex organizations is often fragmented and diffused throughout the structure of the establishment. Such a diffusion makes it difficult to marshal organizational forces to deal with a problem of opportunity. If one functional unit studies one aspect of a project and other parts are studied elsewhere, who provides the synergy? Who decides which element is more important? Can a functional unit provide the answers to these questions—and through the vertical structure achieve a project goal?

Unfortunately, the answers to these questions are not positive, particularly in large, complex bureaucracies. The very nature of assigning organizational work to functional units carries with it the opportunity for suboptimization and a failure to assume a total-system approach. According to Mantell:

> There exists an inevitable tendency for hierarchically arrayed units to seek solutions and to identify problems in terms of the scope of duties of particular units rather than looking beyond them. This phenomenon exists without regard to the competence of the executive concerned. It comes about because of authority delegation and functionalization.[13]

Within the pyramid type of organization, there is also the ever-present concern that a person reporting to an immediate superior may pay more attention to the nuances of that relationship than to getting the work done. Instead of fragmenting the problem among functional groups, one may establish what has been

[13]Leroy H. Mantell, "The Systems Approach and Good Management," *Business Horizons*, October 1972, p. 50.

referred to as a "microcompany" for project work representing all the action elements of a conventional company.[14]

All these questions and problems suggest that an alternative to the purely functional organization is both viable and potentially effective. That alternative —the matrix organization, or whatever else (venture, task force, program) it may be called—has grown to be a significant force in modern management.

PLANNING THE MATRIX ORGANIZATION

Any undertaking requires some degree of organization. In some instances, the organization may be informal but unique to the situation. As the number of individuals engaged in an enterprise increases, however, and interorganizational and interpersonal complexities increase, more formality is required to achieve the desired objectives. Project management is not unlike other management in this respect. Organization is necessary to establish a framework, not only to produce the desired results, but also to clarify individual responsibilities, privileges, and authority.

Why Matrix Organization?

The project manager accomplishes the project objectives by working with functional groups of the company and with outside organizations. The total project organization has no discrete boundaries; it is a complex structure that facilitates the coordination and integration of many project activities. While project managers use many traditional organizational principles in planning their structure, they must be guided by some considerations that go beyond traditional theory, such as:

How shall the parent and outside organizations be aligned to accomplish the multilateral objectives of a project?

How applicable are traditional principles of organization such as span of management, the scalar principle, unity of command, parity of authority and responsibility, unity of direction, and functional homogeneity?

Are the authority and responsibility relationships subject to alignment in a scalar chain, or will the flow of authority and responsibility form a web of relationships in the total project environment?

As project manager, will the first responsibility be to plan and to organize and control subordinates, or to provide the environment in which others can accomplish the project objectives successfully?

How should the organization be aligned to give contributors due recognition?

What will the organization consist of—the blocks on the organizational chart or something greater?

Are conditions such that a simple bureaucratic organization will not suffice for

[14]J. Wade Miller, Jr., and Robert J. Wolf, "The 'Micro-Company,'" *Personnel,* July–August 1968, pp. 35–42.

the technological progress and the interdependencies between complex organizations?

What effect will technology have on the project organization structure?

What alternative matrix forms are available to choose as a model around which to build the organization?

Alternative Matrix Forms

Matrix organizations, representing the complementary interface of the project and the functional elements, vary considerably. While the traditional line-and-staff organization emphasizes production, matrix organizations stress project completion.

The different types of matrix organization represent varying degrees of authority and responsibility which are assumed by the project manager and which have been splintered from, or are shared with, the functional manager. Project managers are located at different levels in the organization and direct varying numbers of people temporarily assigned to them; the amount of the work done in the functional areas also varies.

One arrangement is to have a traditional functional organization, with project managers who report to the company president or general manager in a staff capacity (Figure 10-3). In this setup, project managers act more or less as assistants to the chief executive in matters involving the project, relieving him or

FIGURE 10-3
Functional organization with project manager acting as an assistant in a staff capacity.

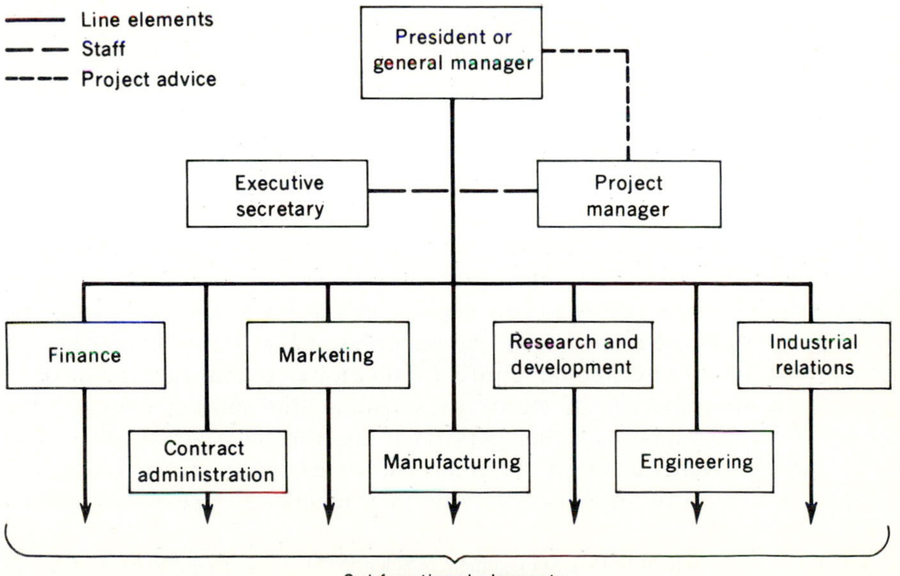

her of some of the burdensome details. Project managers may be provided with some nontechnical (clerical) personnel, but usually have no direct control over the work which is done in the functional units. They play the role of staff officers: they investigate, research, analyze, recommend, and coordinate matters relative to the project. They also serve as expediters of the project activities by dealing directly with individuals in the coordination of project affairs. Keith Davis has described this type of project manager well:[15]

> This reflects his (the project manager's) second function, that of serving as a center of communication to be able instantly to report to general management on the *whole* of the project and thus relieve general management of the tedious task of keeping up with all the details. Accordingly, he accomplishes unity of command, a key necessity in the complex world of advanced technology.

Most decisions are made by the chief executive since project managers have no real authority except to persuade and report results to their superior. They are monitors who can "view with alarm" the adverse machinations of a project, but they are something less than a true coordinator or integrator of the project. Although they do not function in a line capacity, they usually have wide functional authority, and since they are in close organizational proximity to the chief executive, they have significant influence.

Placing project managers in a staff capacity as assistants restricts their freedom to function as true integrators and decision makers. In management literature, and in practice, assistants are not considered to have the authority to act alone. Instead, they merely furnish their superiors with information and recommendations. In this arrangement, the project managers' responsibilities are likely to exceed their authority. In a staff position, their ability to act decisively would depend almost solely upon personal persuasiveness and superior knowledge. While superior knowledge and persuasiveness are powerful additives to authority, their position would probably be weak because the role of staff members, according to traditional theory, is restricted to giving advice and assistance.

The second organizational pattern is a true *matrix* organization. In this type of structure (Figure 10-4), a functional organization exists in which the project manager reports to the chief executive in a line capacity. The staff of the project manager's office may vary in number from only the manager to hundreds of people, depending on the size of the effort and the degree to which the project activities are centralized. As the project manager's responsibilities increase and more facets of the project are centralized under his or her control, the company may establish an organizational entity (a division) to manage the project independently. In this type of matrix organization, the project manager has authority over the functional managers regarding the *what* and *when* of the activities; the functional manager determines *how* the support will be given. The functional managers are responsible to both their functional supervisors and the

[15]Keith Davis, "The Role of Project Management in Scientific Manufacturing," *Arizona Business Bulletin,* May 1965, p. 2.

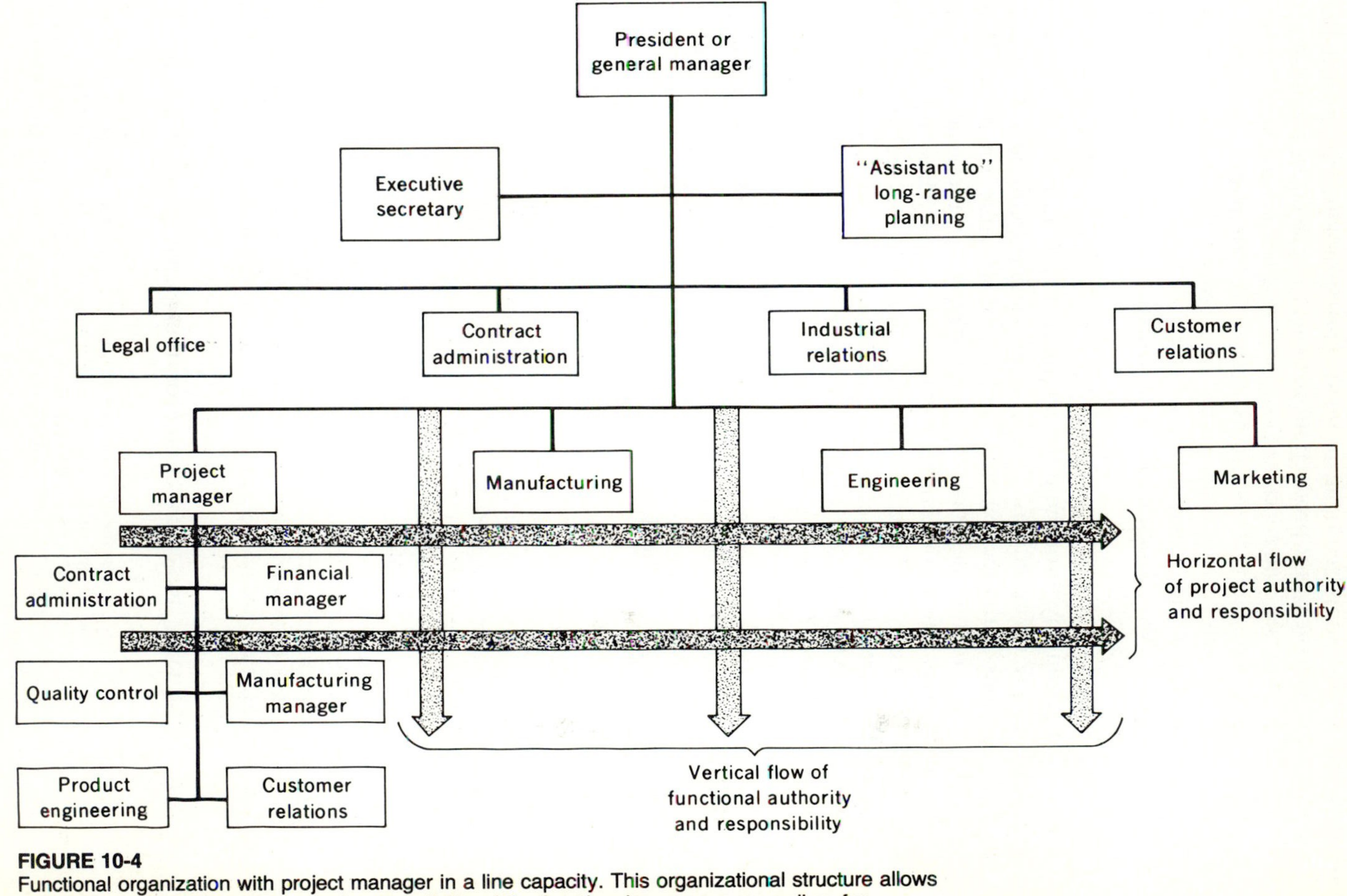

FIGURE 10-4
Functional organization with project manager in a line capacity. This organizational structure allows for vertical flow of the authority and responsibility of the functional managers, as well as for horizontal flow of project authority and responsibility.

project manager for support of the project. This situation, in which a line functional manager (such as a production manager) is placed in a position of providing advice, counsel, and specialized support to a project manager, who is concerned with unifying project activities across the company, represents a change in the authority relationships. This is a radical departure from the line-staff organizational dichotomy that has been the mainstay of management theory for decades. Also, it seems to be a violation of the scalar principle described by Henri Fayol.[16]

Authority patterns in the organization shown in Figure 10-4 flow both vertically and horizontally throughout the company. In addition, there is flow to outside participating organizations. Project organization frequently disregards levels and functions and superimposes the project structure on the existing organization. The structure depends, to a large degree, on the location of the project clientele, regardless of where the clientele is located. Thus, at times it is difficult, if not impossible, to chart the relationship; they will be discussed in connection with charting the project organization in a subsequent chapter.

At TRW, Matrix Management is a variation on the "project" organization of the aerospace industry. Under the Matrix Management of TRW, each project office is not self-sufficient—but only a minimum core team for overall planning, budgeting, coordination, and systems engineering. Design and technical work is performed by professionals from the specialized departments—a team is formed for each special project—employees may be on several teams at once and report to several bosses. Such a system is shown in Figure 10-5 and described by *Business Week* as:

> At TRW Systems, Matrix involves both permanent and temporary lines of authority. When these two kinds of authority meet, there are inherent conflicts. Ideally, organization development (OD)—a potpourri of behavioral science techniques—keeps the communications channels open, minimizing the difficulties that result.
>
> The manager of Laboratory B-1 may agree, for example, at the start of Project Y that the project manager is the "boss" of Depts. III, IV, and V as far as work on his Project Y is concerned. The laboratory manager also loans people from his lab to be subproject managers, who report to Project Y's manager and recruit expertise from any number of departments—including Depts. II or VI in other divisions.
>
> Under Matrix management, an engineer in Dept. III may work on several project teams at once and have any number of bosses. For instance, he may work for subproject manager Y-1 and project manager Y as well as for the head of Project X in another division.[17]

The Laramie Energy Research Center of the Department of Energy is

[16]Fayol envisions the scalar chain as the chain of superiors ranging from the ultimate authority to the lowest rank, with the line of authority following every link in the chain. In today's large organizations, where lines of authority cross functional lines and extend into outside organizations, this principle requires some modification. See Henri Fayol, *General and Industrial Management,* Sir Isaac Pitman & Sons, Ltd., London.

[17]"Teamwork Through Conflict," *Business Week,* Mar. 20, 1971, pp. 44–47.

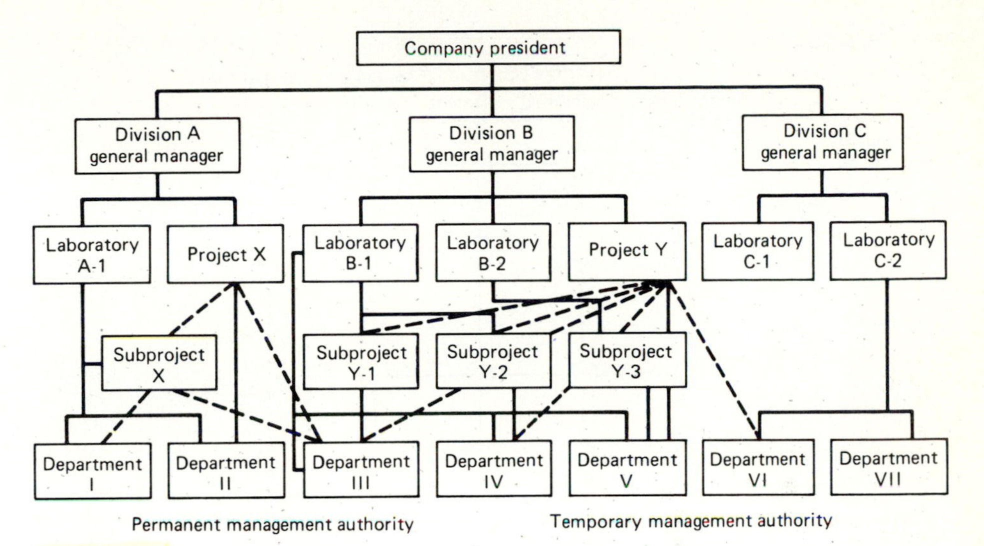

FIGURE 10-5
TRW Matrix Management. (Adapted with permission from "Teamwork Through Conflict," *Business Week*, Mar. 20, 1971, p. 44.)

organized under a matrix management system. Within this research center this is interpreted to mean that R&D investigations in each of the research target areas—oil shale, coal, and petroleum (tar sands)—are each supported by resource evaluation, product characterization, analysis, contract management, and administrative services through a pool of multidisciplinary personnel. This organizational form at the research center is one designed for minimum fixed structure and maximum adaptability and change. The organization is directed toward the development of an expert staff of federal scientists and engineers to accomplish the Department of Energy (DOE) mission in the research center's areas of expertise. The management plan calls for a matrix by which both internal work and external work are brought into focus to accomplish the two major roles of this U.S. government federal in-house laboratory. The roles are to act as an expert internal advisor to the U.S. government and to ensure that federal government dollars and resources are well spent and placed. A two-dimensional diagram showing an abbreviated concept of matrix management of the Laramie Energy Research Center's programs is shown in Figure 10-6.

Texas Instruments has developed a structure which could be described as a three-dimensional matrix. Within Texas Instruments, a product structure is overlaid on a functional structure. This has been complemented by a time-related structure known as "Objectives, Strategies, and Tactics" (OST). This program has its own chain of command and its own budget. This organizational element is charged with developing a future sense of direction for the firm.

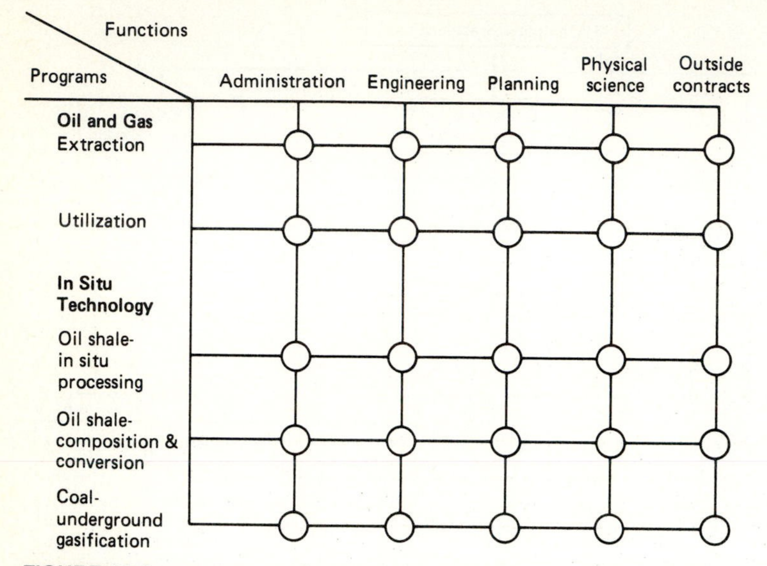

FIGURE 10-6
Abbreviated management matrix of LERC programs. (From Annual Report for 1975, Laramie Energy Research Center, U.S. Energy Research & Development Administration, Laramie, Wyo., p. 26.)

Individuals within this structure are appointed as objective managers, strategy managers, or tactical managers. Some of these managers are full-time, while others carry on their current responsibilities in functional or product hierarchies. These managers are provided with budgets, and their progress toward agreed-upon objectives is regularly reviewed by top management. Texas Instruments places a lot of credit for its impressive growth on the use of this system.[18]

Another form of a facilitative matrix organization may be found in the Dow Corning Corporation. In addition to the functional, product, and time structures used by Texas Instruments, Dow Corning adds geography as a fourth dimension. The complexity of such an organization is apparent.[19]

Project managers may be differentiated on the basis of the degree of organizational involvement of the projects they manage. The size of the project is one distinction; i.e., does the project manager act as the focal point for a project that is being carried out only within the boundaries of a major functional unit, or does the project require cutting across the lines of several functional

[18]See "Texas Instruments Shows U.S. Business How to Survive in the 1980's," *Business Week*, Sept. 18, 1978, pp. 66–92. Reprinted with permission. Copyright © 1978 by McGraw-Hill, Inc. All rights reserved.

[19]William C. Goggin, "How the Multidimensional Structure Works at Dow Corning," *Harvard Business Review*, January–February 1974.

departments? Or, is the project so small that its affairs can be conducted within the bounds of a major functional department? The organizational alignment of such a project is reflected in Figure 10-7. This project manager has direct responsibility over the units doing the main work of the project. For the most part, the project affairs are conducted without any negotiation and coordination with outside organizations. This type of project organization is typical of many small affairs conducted in an organization which do not fit into any specific department, but require some degree of formality and central location to facilitate their performance.

Another type of a project organization functions within a given department, but has direct authority over the units doing the work. Such an organization is portrayed in Figure 10-8, where the project manager acts as an assistant to the department head. In this organization, the subordinate managers are, in effect, working for two people, i.e., the head of the engineering department and the project manager. People in this dual relationship report first through the traditional line organization and then through the project organization. The sphere of influence of the project manager is limited to the functional boundaries of the organization.

Another type of project organization is portrayed in Figure 10-9, where the project manager is placed in a line capacity with direct responsibility over the project affairs. In this case, the project manager would be a senior executive in the company and might bear the title of vice president or assistant general manager. The range of the project manager's jurisdiction in this case would

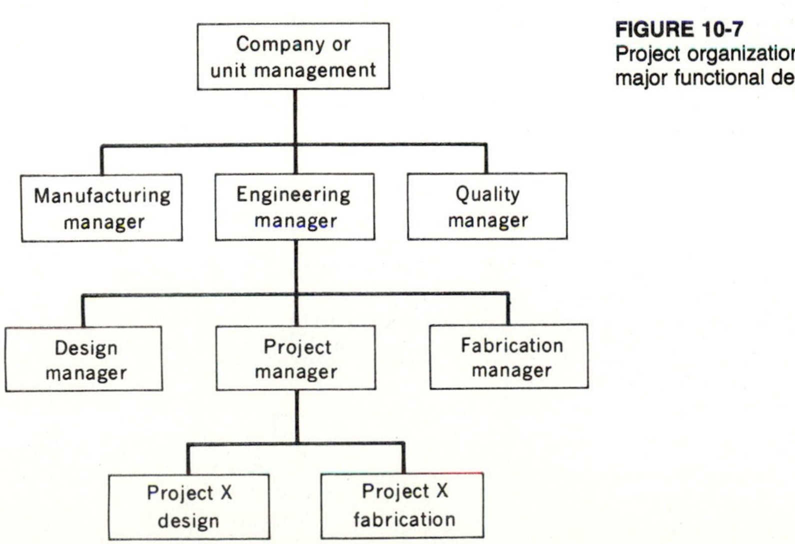

FIGURE 10-7
Project organization within a major functional department.

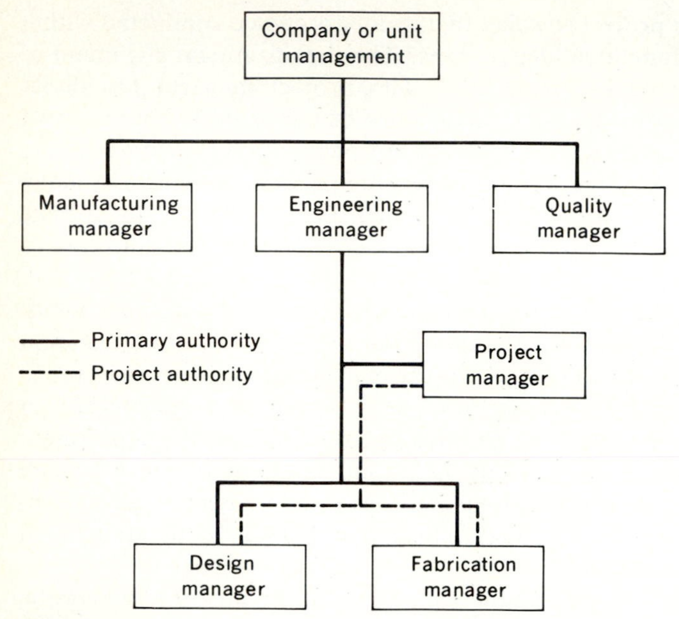

FIGURE 10-8
Project organization within a major functional department with the project manager acting in the capacity of "assistant."

FIGURE 10-9
Project organization in which project authority and line authority are the same.

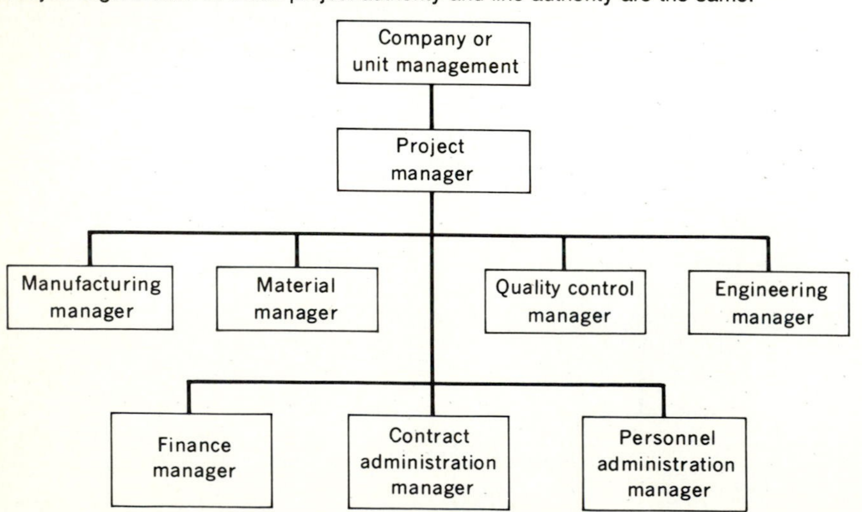

encompass most of the functional activities of the company. The project organization is similar to the traditional vertical model, with project authority and line authority being one. This project manager normally has authority and responsibility that cut across functional lines, although the chart does not depict a matrix relationship.

One transnational corporation organized a projects profit center (a division) to draw on the resources of other profit centers to bring together the people, facilities, equipment, systems, and services for industrial construction customers around the world. This division acts as a "strike force" to develop project markets in a market segment for the overall benefit of the corporation. This division organizes, coordinates, delivers, and installs the electrical-equipment package for large industrial projects. Project management is used to the advantage of single-source responsibility for all stages of a project to include:

Up-front studies and analyses
A single, coordinated proposal
A single contract covering all products and services
Interface with customers, or with other contractors as required by the contract
Large and complex projects packaged from smaller pieces
Integrated equipment design and installation
Stringent control of scheduling, shipping, and installation
Single point of contact for problem resolution
Centralized invoicing

Project management is carried out by a team composed of representatives from the participating profit centers. The team is formed during proposal development; team responsibility runs through the warranty period. In addition to the design, development, and production phases of a project, the team manages installation, erection, and equipment start-ups (all integrated into the overall project) and support activities such as personnel training and start-up engineering. In some projects, field service and contract maintenance and repairs are provided.

The mission of this Projects Division is to provide corporate-wide teamwork and synergy for electrical equipment–related industry/construction projects that cut across many corporate profit centers (divisions) sales organizations and staffs that produce product components. Figure 10-10 shows the coordination of a number of important functions necessary to provide a successful project capability within the electrical-equipment market segment of this corporation. The functions on the left of Figure 10-10 are provided by the Projects Division. Those functions on the right are handled by other divisions or staffs of the corporation.

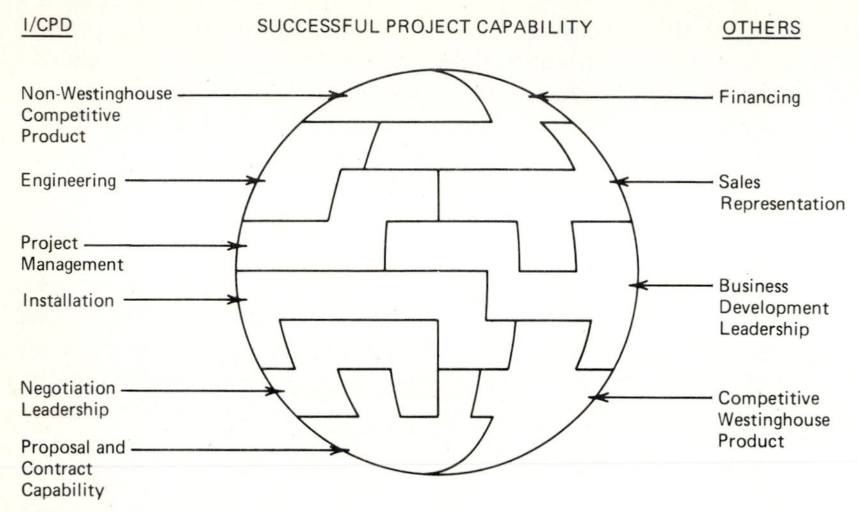

FIGURE 10-10
Successful project capability. (*Westinghouse International News,* vol. 2, no. 1, February–March 1981, p. 8.)

INTERORGANIZATIONAL PROJECT MANAGEMENT

The interactions between a government project office and industry agencies can be appreciated by reviewing Figure 10-11. The interactions suggested by the figure are only a partial illustration of the number, size, and intensity of the project interrelationships. On a major government project, the project manager and office personnel interact with the highest levels of government and industry. Contractors doing business with these organizations tend to develop project offices which mirror the skills of the government project office. The relationship of the two organizations—the military department and the defense contractor—revolves around the two project managers, as illustrated.

Although we use an example drawn from the defense industry, the same model could be used to describe any interorganizational project management situation.

The establishment of a special project office in both the buyer's organization (e.g., the Air Force) and the seller's organization (e.g. an aerospace company) permits a focal point for concentration of attention on the major problems of the project or program. This point of concentration forces the channeling of major program considerations through a project manager who has the perspective to integrate relative matters of cost, time, technology, and systems compatibility.

This managerial model is not meant to stifle the interfunctional lines of communication or the necessary and frequent lateral staff contacts between the functional organizations of the defense contractor and the military organization.

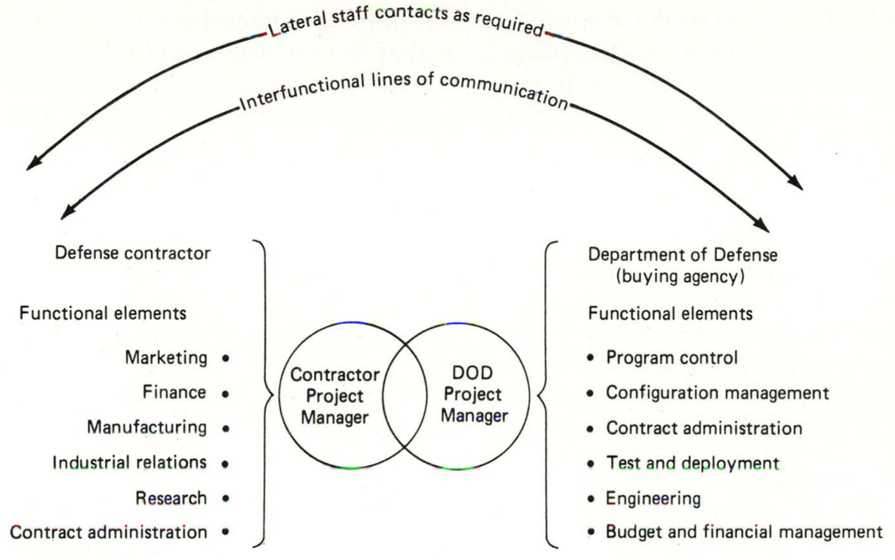

FIGURE 10-11
Interorganizational project manager relationships. Critical decisions involving policy and managerial prerogatives are directed through the central focal point. Decisions involve cost and cost estimating, schedules, product performance (quality, reliability, maintainability), resource commitment, project tasking, trade-offs, contract performance, and total system integration. (Adapted from David I. Cleland, "Project Management—An Innovation in Management Thought and Theory," *Air University Review,* January–February 1965, p. 19.)

Rather, what is intended is the establishment of a focal point for critical decisions, policy making, and key managerial prerogatives relating to the project manager when trade-offs between the key elements of the research or production activity are involved. By being in a face-to-face relationship, the two project managers can control and resolve both interfunctional and interorganizational conflict.

SUMMARY

People could probably devise more varied organizational arrangements than there are companies or projects. Varied as these arrangements might be, they would all be based on the concept of pulling together technical and managerial talents into a team, to operate without limits of discipline or organizational lines in undertaking a project. The organization form that finally evolves out of the project requirements will undoubtedly be a compromise between pure project management and standard functional alignment—some form of the matrix model.

In the matrix organization, lines of authority and responsibility must be set out clearly. These lines will be crisscrossing; fuzziness is bound to emerge even with the best of policy documentation. Thus, policy documents are not enough. A major educational effort is needed, so that the managers concerned fully understand the new relationships that emerge and how some of the traditional principles of management, such as unity of command and parity of authority and responsibility, are modified.

DISCUSSION QUESTIONS

1 What is meant by a systems view of an organization? How does this differ from the more classical view of an organization?

2 How is an organization defined in operational terms? Why is such a definition useful? How can an organization be changed?

3 In designing the structure of an organization, several approaches are commonly used. What are these approaches? How do they differ? Take an organization of your choosing and determine where and how the various approaches are used.

4 An understanding of the functional organization in project management is an essential prerequisite to success as a project manager. Why is this so? Would it be desirable for a project manager to have an experience as a functional manager before taking over the management of a project? Why or why not? Defend your answer.

5 Describe the nature of a project-driven matrix organization. How does this organization relate to the work packages of the project?

6 The matrix organization and the bureaucratic organization can be considered complementary. Describe the nature of this reciprocal relationship.

7 What are some of the considerations to be evaluated in planning for the structure of the project-driven matrix organization?

8 What are some of the alternative ways by which the project-driven matrix can be structured? What else is important in the design of the matrix organization besides the structural part thereof?

9 Describe the nature of the Texas Instruments three-dimensional matrix. Why does this matrix suit that company?

10 There is an old law of management: "Structure follows strategy." What relevance does this have in the context of an organization initiating a project-management system?

11 Project management is often used in the management of international projects in a transnational corporation. What are some of the reasons that its use is particularly applicable there? What are some of the problems and constraints that might be encountered in setting up such a management system?

12 Project management can be used within a corporation in the centralized research and development laboratory of the company. What are some of the reasons for using this approach? How might a profit-center manager in a large corporation be concerned about the management of development projects in the central R&D laboratory of the corporation?

13 What is the nature of the pure project organization? What are some of the advantages and disadvantages of using a pure project organizational approach?

14 Peter Drucker has called the matrix organization "fiendishly difficult" to work with. What do you think he meant by this statement? Do you agree?

15 Why is it important to deal explicitly with the allocation of authority and responsibility in the matrix organization?

6- takes advant of one and overlay on the disadvan of the other.

RECOMMENDED READINGS

Barnard, Chester I.: *The Functions of the Executive,* Harvard University Press, Cambridge, Mass, 1962.

Bennis, W. G.: "Evolving Organization Obsoleting Pyramid, Etc." *Steel,* April 11, 1966.

Carlisle, Howard M.: "Are Functional Organizations Becoming Obsolete?" *Management Review,* January 1969.

Chappie, Elliot O., and Leonard R. Sayles: *The Measure of Management,* The Macmillan Company, New York, 1961.

Cleland, David I.: "Project Management—An Innovation in Management Thought and Theory," *Air University Review,* January–February 1965.

Davies, C.; A. Demb, and R. Espejo: *Organization for Program Management,* John Wiley & Sons, Inc., New York, 1979.

Davis, Keith: "The Role of Project Management in Scientific Manufacturing," *Arizona Business Bulletin,* May 1965.

Fayol, Henri: *General and Industrial Management,* Sir Isaac Pitman & Sons, Ltd., London.

Gilman, Glen: "The Manager and the Systems Concept," *Business Horizons,* August 1969.

Kelly, A. J. (ed.), *New Dimensions of Project Management,* Lexington Books, Lexington, Mass., 1982.

Kendall, Henry P.:*Scientific Management: First Conference at the Amos Tuck School,* The Plimpton Press, 1912.

Kolodny, Harvey F.: "Evolution to a Matrix Organization," *Academy of Management Review,* vol. 4, 1979.

Mantell, Leroy H.: "The Systems Approach and Good Management," *Business Horizons,* October 1972.

Mee, John F.: "Ideational Items: Matrix Organization," *Business Horizons,* vol. 7, no. 2, Summer 1964.

Miller, J. Wade, Jr., and Robert J. Wolf: "The 'Micro-Company,'" *Personnel,* July–August 1968.

Mintzberg, Henry: "Organization Design: Fashion or Fit?," *Harvard Business Review,* January–February 1981.

Mockler, Robert J.: "Situational Theory of Mangement," *Harvard Business Review,* May–June 1971.

Shull, F. A., and R. J. Judd: "Matrix Organizations and Control Systems," *Management International Review,* vol. 11, 1971.

Tilles, Seymour: "The Manager's Job: A Systems Approach," *Harvard Business Review,* January–February 1963.

CASE 10-1: Matrix Management Roles

A series of roles with supporting definitions that can be used to describe what the manager does in a matrix management situation are shown in Table 10-1.These roles can be classified as:

1 Primary (P)
2 Supportive (S)

TABLE 10-1
ROLES OF THE KEY MANAGERS IN THE MATRIX MANAGEMENT

Role and definition	Business results manager	Resource facilitation manager	General manager	Work package manager	Chief executive
Technologist Provides leadership to establish technical (engineering, production, finance, etc.) objectives, goals, and strategies.					
Agent Represents and acts for another in a transaction.					
Negotiator Arranges an agreement on an issue.					
Logistician Procures, distributes, maintains, and replaces human and nonhuman resources for the organization.					
Strategist Uses science and art in the development of a sense of future direction for an organization.					
Counselor Participates in an exchange of opinions and ideas and provides advice and guidance.					
Disciplinarian Enforces organizational policies and procedures and keeps the organization on the most promising path toward objectives.					
Motivator Provides an environment whereby people attain social, psychological, and economic satisfaction in their work.					
Organizer Organizes human and nonhuman resources.					
Decision maker Chooses the alternative that the organization will follow.					
Figurehead Represents the organization in all matters of formality.					
Leader Leads people along a way.					
Liaison officer Interacts with peers and "systems" community to gain favors for the organization.					
Monitor Receives and collects information which permits an understanding of the organization.					

TABLE 10-1 (*Continued*)
ROLES OF THE KEY MANAGERS IN THE MATRIX MANAGEMENT

Role and definition	Business results manager	Resource facilitation manager	General manager	Work package manager	Chief executive
Disseminator Distributes information.					
Spokesperson Disseminates organization information in the environment.					
Entrepreneur/innovator Initiates change in the organization.					
Disturbance handler Takes charge when the organization is threatened.					
Subordinate Receives direction from a superior in the organization.					
Controller Establishes standards and judges results.					
Coordinator Synchronizes activities with respect to time and place.					
Teacher One whose purpose is to instruct.					
Other ?					

Your Task Place a *P* or *S* in each column segment to describe appropriately the role(s) each of the managers carries out. If you think that a manager does not carry out a particular role, leave that column blank. Be prepared to defend your team's choice of role(s). Have the team leader present your findings.

CHAPTER 11

CHARTING ORGANIZATIONAL RELATIONSHIPS[1]

This one fact—that the chart is essentially a device for clarifying authority relationships—goes a long way toward explaining its usefulness.[2]

The organizational model which is commonly called the *organization chart* is much derided in the satirical literature and in the day-to-day discussions among organizational participants. However, organizational charts can be of great help in both the planning and implementation phases of management.

In this chapter, we shall explore a systems-oriented version of the traditional organizational chart that is quite useful in implementing the project management and matrix organizational ideas of the previous chapters.

THE PYRAMIDAL ORGANIZATIONAL CHART

The traditional organizational chart is of the pyramidal variety; it represents, or models, the organization as it is *supposed* to exist at a given point in time. Many charts of this variety have already appeared in previous chapters, and even

[1]This chapter includes certain material from the articles by David I. Cleland and Wallace Munsey, "Who Works with Whom," *Harvard Business Review,* September–October 1967; and "The Organization Chart: A Systems View," *University of Washington Business Review,* Autumn 1967. Used by permission.

[2]Allen R. Janger, "Charting Authority Relationships," *The Conference Board Record,* December 1964.

though these charts have departed from the traditional by depicting matrix organizations, they have not differed generically in their composition and implicit assumptions about organizational forms and the way that they operate.

At best, such a chart is an oversimplification of the organization and its underlying concepts which may be used as an aid in grasping the concept of the organization. Management literature indicates various feelings about the value of the chart as an organization tool. For example, Cyert and March say:

> Traditionally, organizations are described by organization charts. An organization chart specifies the authority or reportorial structure of the system. Although it is subject to frequent private jokes, considerable scorn on the part of sophisticated observers, and dubious championing by archaic organizational architects, the organization chart communicates some of the most important attributes of the system. It usually errs by not reflecting the nuances of relationships within the organization; it usually deals poorly with informal control and informal authority, usually underestimates the significance of personality variables in molding the actual system and usually exaggerates the isomorphism between the authority system and the communication system. Nevertheless, the organization chart still provides a lot of information conveniently—partly because the organization usually has come to consider relationships in terms of the dimensions of the chart.[3]

Jasinski is critical of the traditional, pyramidal organizational chart because it fails to display the nonvertical relations between the participants in the organization. He says:

> Necessary as these horizontal and diagonal relations may be to the smooth functioning of the technology or work flow, they are seldom defined or charted formally. Nonetheless, wherever or whenever modern technology does operate effectively, these relations do exist, if only on a nonformal basis.[4]

Unfortunately, too often the policy documentation describing the role of a project manager will describe this manager's relationship with the functional organizations as a "dotted line" relationship which is ambiguous and confusing and can mean anything one wishes it to mean. In this respect, Davis and Lawrence note:

> For generations, managers lived with the happy fiction of dotted lines, indicating that a second reporting line was necessary if not formal. The result had always been a sort of executive ménage à trois, a triangular arrangement where the manager had one legitimate relationship (the reporting line) and one that existed but was not granted equal privileges (the dotted line).[5]

One suspects that managers use a dotted line on an organization chart because at the time the chart was developed the relationship had not been

[3]Richard M. Cyert and James G. March, *A Behavioral Theory of the Firm,* Prentice-Hall, Inc., Englewood Cliffs, N.J., 1963, p. 289.

[4]Frank J. Jasinski, "Adapting Organization to New Technology," *Harvard Business Review,* January-February 1959, p. 80.

[5]Stanley M. Davis, and Paul R. Lawrence, "Problems of Matrix Organization," *Harvard Business Review,* May–June 1978, p. 142.

completely defined. The use of a dotted-line technique in depicting authority and responsibility gives a manager a great deal of flexibility. The price of this flexibility is confusion and unclear understandings of reciprocal authority and responsibility.

Charting Horizontal Relationships

The desire to retain a simple idealistic *concept* of an organization probably accounts for the durability and venerability of the pyramidal chart. The growing realization that organizations have internal and external modes which are neither simple nor idealistic probably accounts for much of the criticism of these charts. This observation leads to the generalization that the vertical organizational chart is a graphic portrayal of the traditional school of organization theory.

Today's organizations are considerably more complex than those of earlier times; this implies that the organizational chart should go beyond the limits of classical or traditional doctrines. A method of charting is needed which recognizes the role of the many organizations, in addition to the parent organization, that play a role in company fortunes. As quoted previously from Tilles:

> Many organizations have a management team that includes individuals—auditors, lawyers, bankers, brokers, and a variety of other specialists—who never appear on the organization chart. In some cases, these outside experts are consulted with such regularity that they are really a part of the management system. In fact, the extent of the management system is frequently an indication of the manager's ability.[6]

Such an organization is a complex of clientele relationships existing in the total organizational systems environment and having tying bonds of reciprocity.

The lack of an organizational chart which can be used to analyze the environment can be a matter of frustration and consternation to managers. One project manager, when asked to diagram his project organization on the blackboard, replied, "It would be impossible to show a meaningful diagram of the organization. The work load weaves back and forth between the two branches of the division and other agencies in such a way that there is no practical way to tell anyone just how we go about getting our work done." To this successful project manager, the principal usefulness of the vertical organizational chart was in establishing superior-subordinate responsibilities in his organization and in showing the relative alignment of the various departments under his jurisdiction. Although this project manager could draw from memory numerous systems schematics of the system with which he worked, he could not draw a schematic of the organization he managed. He readily acknowledged that conceptually his organization was a system, but he had not developed any sort of schematic to analyze the interrelatedness of the project.

[6]Seymour Tilles, "The Manager's Job: A Systems Approach," *Harvard Business Review*, January-February 1963, p. 75.

Usefulness of the Traditional Chart

The organizational chart is a means of visualizing many of the abstract features of an organization. The possible value of the organizational chart as a way of depicting organizational relationships has not been fully exploited in contemporary literature.[7]

In summary, the organizational chart is useful in that:

It provides a general framework of the organization.

It can be used to acquaint the employees and outsiders with the nature of the organizational structure.

It can be used to identify how the people tie into the organization; it shows the skeleton of the organization, depicting the basic relationships and the groupings of positions and functions.

It shows formal lines of authority and responsibility, and it outlines the hierarchy—who fills each formal position, who reports to whom, and so on.

Limitations of the Traditional Chart

The organizational chart is something like a photograph. It shows the basic outline of the subject, but tells little about how individuals function or relate to others in their environment. The organizational chart is limited in that:

It fails to show the nature and limits of the activities required to attain the objectives.

It does not reflect the myriad of reciprocal relationships that exist between peers, associates, and many others with a common interest in some purpose.

It is a static, formal portrayal of the organizational structure; most charts are out of date by the time they are published.

It shows the relationships that are supposed to exist, but neglects the informal, dynamic relationships that are constantly at play in the environment.

It may confuse organizational position with status and prestige; it overemphasizes the vertical role of managers and causes parochialism, a result of the blocks and lines of the chart and the neat, orderly flow they imply.

Alternatives to Traditional Organizational Charting

The usefulness of organizational charts, together with their limitations, suggests that new means for graphically displaying the interactions between people and activities in an organized activity should be sought. The charts should be constructed in such a way that their limitations, which seem to be universally recognized, are reduced, thus making the charts more useful.

Introducing new techniques of charting will be difficult because people are so resistant to change. The popularity of the pyramidal organizational chart is

[7]George R. Terry, *Principles of Management,* Richard D. Irwin, Inc., Homewood Ill., 1964, pp. 443–462.

probably due to its simplicity and ease of preparation. For many executives, reorganizations are brought about by changing the alignment of the blocks and lines on the chart. Too often, a reorganization is rushed into when the realignment that is required could be accomplished by a careful thinking through of policies and procedures which prescribe how people are to relate in their work.

The realignment of blocks and lines on the organization chart is the most drastic form of reorganization because it breaks up patterns of formal and informal authority and responsibility and disrupts the prevailing procedures and policies. Dalton E. McFarland's statement illustrates the problem.

> Innovations in organization charting can be expected from time to time, but so far none has appealed widely to executives. Principal reliance must continue to be placed on standard top-to-bottom charts. Custom and habit support their use. These represent the basis upon which other types of charts are made. The standard charting procedure is realistic in the sense that most business organizations actually are set up so that authority resides at the top and flows in a downward direction. Charting procedures that are in the nature of euphemism provide no enduring practical value.[8]

Perhaps the greatest threat to new methods of charting is the feeling of some executives that they will lose prestige if their relative positions on a hierarchical chart are changed. There is a tendency to equate the ascending levels of an organization with gradations of talent and competence.[9]

In selecting a new means of charting an organization, one should consider the purpose for which the charts will used: (1) information display and (2) organizational analysis. Both are required, but can one chart satisfy both needs? If the organization is simple in concept, structure, and environmental relationships (i.e., if there is very little interdependency between functions and organizations), perhaps the traditional chart will adequately portray organizational information. However, if the organization is a complex system, that is highly interrelated with a larger system, the organizational chart that displays the skeletal formal structure is far from being adequate for organizational analysis.

If the organizational chart is intended as a means of communicating information, then its use as an analytical tool is an unintentional by-product. The creation of the chart itself is a way of forcing planning and analysis; its greatest value may be the analysis required for its preparation. In this respect, it is like systems analysis and the network planning and control of techniques to be discussed in a later chapter.

Our discussion has centered around the assumption that the chart, derived as a result of some organizational analysis, has been subsequently used for communications purposes. If the chart is intended primarily as a vehicle for the dissemination of information, it must of necessity be simple in format and devoid

[8]Dalton E. McFarland, *Management Principles and Practices,* The Macmillan Company, New York, 1958, p. 288.

[9]See William H. Read, "The Decline of the Hierarchy in Industrial Organizations," *Business Horizons,* Fall 1965.

of detail. On the other hand, if it is to be used for detailed fact gathering, analysis, and correlation, something more than conventional techniques is required. This reasoning might well lead to the conclusion that two organizational charts are needed in certain organizations, one to show the intended legal or formal overall structure of the organization (the picture) and the other to portray the manner in which the day-to-day details of the organization are carried out (the schematic diagram).

LINEAR RESPONSIBILITY CHARTS (LRCs)

The linear responsibility chart (LRC) is an innovation in management theory that goes beyond the simple display of formal lines of communication, gradations of organizational level, departmentation, and line-staff relationships. In addition to the simple display, the LRC reveals the task-job position couplings that are of an advisory, informational, technical, and specialty nature.

The LRC has been called the "linear organization chart," the "matrix responsibility chart," the "linear chart," and the "functional chart." None of these names adequately describes the device. The LRC (or the table or grid, as Janger calls it)[10] shows who participates, and to what degree, when an activity is performed or a decision made. It shows the extent or type of authority exercised by each executive in performing an activity in which two or more executives have overlapping authority and responsibility. It clarifies the authority relationships that arise when executives share common work. The need for a device to clarify the authority relationships is evident from the relative unity of the traditional pyramidal chart, which (1) is merely a simple portrayal of overall functional and authority models and (2) must be combined with detailed position descriptions and organizational manuals to delineate authority relationships and work-performance duties.

The typical pyramidal organizational chart is not adequate as a tool of organizational analysis since it does not display systems interfaces. It is because of this inadequacy that a complementary array of position descriptions and organizational manuals has come into being. As organizations have grown larger, personnel interrelationships have increased in complexity, and job descriptions and organizational manuals have grown more detailed. Typical organizational manuals and position descriptions have become so verbose that an organizational analysis can be lost in semantics.

Position descriptions do serve the purpose of describing a single position, but executives are also concerned with how the people under their jusrisdiction relate to one another. On many occasions, executives are confronted with the task of examining and explaining relationships. Project management, corporate staff organization, concepts of product planning, the development of a corporate plan—all these lead to highly complex working relationships. A dynamic organization is often—even continually—redefining large numbers of positions and establishing new responsibility and authority patterns.

[10]Allen R. Janger, op. cit.

Structure and Philosophy of the LRC

Typically, the LRC shows these characteristics:

1 Core information from conventional organizational charts and associated manuals displayed in a matrix format[11]

2 A series of position titles listed along the top of the table (columns)

3 A listing of responsibilities, authorities, activities, functions, and projects down the side of the chart (rows)

4 An array of symbols indicating degree or extent of authority and explaining the relationship between the columns and the rows

Such an arrangement shows in one horizontal row all persons involved in a function and the extent and nature of their involvement. Furthermore, the one vertical column shows all functions that a person is responsible for and the nature of this person's responsibility. A vertical column represents an individual's job description; a horizontal row shows the breakout of a function or task by job position.

Figure 11-1 shows the salient structure of an LRC, in terms of (1) an organizational position and (2) a *work package,* in this case "Conduct Design Review"; the symbol Δ indicates that the Director of Systems Engineering has the primary responsibility for conducting the system design review. When many work packages are involved, one of the key advantages of using an LRC to define organizational relationships becomes apparent; i.e., various people are brought into a dialogue about their specific authority and responsibility to a work package and to other individuals in the organization.

Figure 11-2 shows an LRC for an organization involved in the design, manufacture, and delivery of an industrial system from the time of receipt of a firm order until the system is accepted by the customer. In this example, the organizational departments of *Information Systems, Manager of Programs,* and *Strategic Planning* have been added to the *Systems Engineering* position. The LRC in Figure 11-2 shows how these four positions relate to the work packages, to each other, and to other departments that are not represented as columns in the array.

THE ROLE OF THE LRC IN DESCRIBING ORGANIZATIONAL RELATIONSHIPS

The LRC can be viewed as a potentially important element in both planning and control functions of management. The LRC can be thought of as a plan specifying how the organization *should* work; subsequently, the plan can be used as a standard against which to assess whether responsibilities are being properly carried out.

Moreover, the LRC can also have motivational value because it clarifies

[11]For example, one writer proclaimed: "On one pocket-size chart it shows the facts buried in all the dusty organizational manuals—plus a lot more."

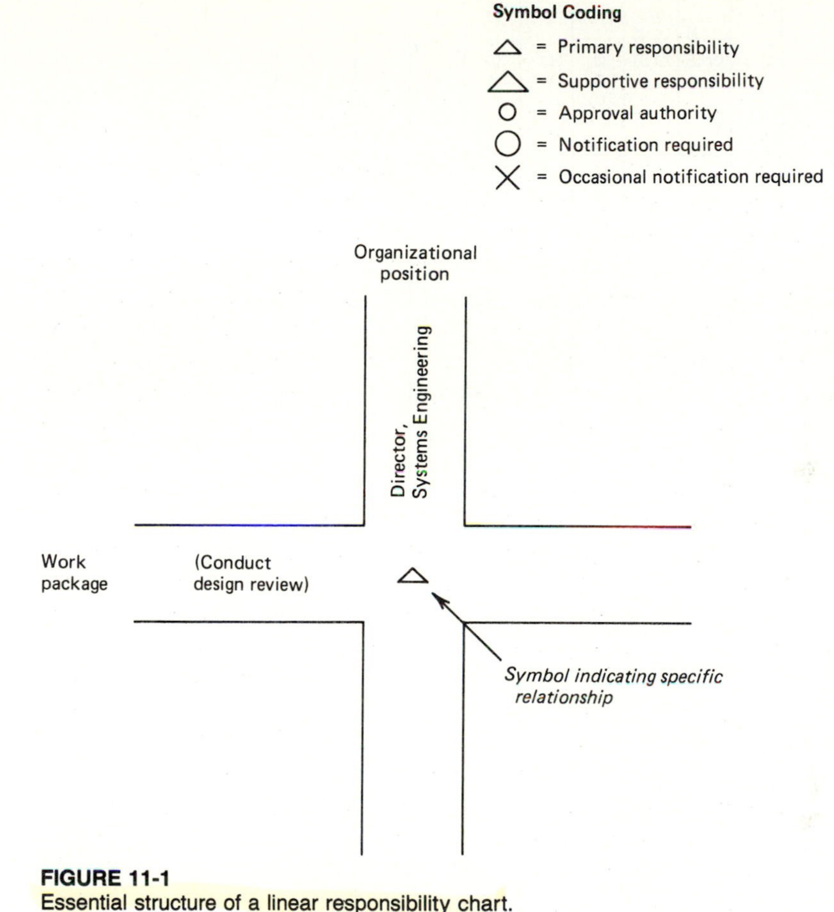

FIGURE 11-1
Essential structure of a linear responsibility chart.

relationships that can, under less clear circumstances, lead to conflict; with an LRC, individuals have a clear picture of what is expected of them and what role they are to play in a wide variety of functions and activities.

Perhaps the greatest potential value of the LRC to the overall organization is through a participatory process of development. Through a process of getting together the people who occupy key positions to discuss how each position relates to other positions and to the work packages, much can be gained in substantive terms, in learning, and in "mutual understanding."[12]

This variety of LRC development is an excellent illustration of *participative management,* since various individuals are given the greatest practical latitude to

[12]See C. W. Churchman, and A. H. Schanblatt, "The Researcher and the Manager: A Dialectic of Implementation," *Management Science,* vol. 11, no. 4, February 1965.

WORK PACKAGE	STRATEGIC PLANNING	INFORMATION SYSTEMS	SYSTEMS ENGINEERING	MANAGER OF PROGRAMS (PROGRAM MANAGERS)	OTHER
1 Program definition	●	●	▲	△	▲ Sales
2 Review and define systems design	▲	✷	△	▲ ○	
3 Master schedule	✷	▲	▲	△	▲ CA ● Field sales
4 Develop program budget		△	▲	▲ ○	● CA ▲ Sales
5 Funds allocation		▲		▲ ○	△ CA ▲ Sales
6 Define hardware/software work packages	✷		△	▲	● CA
7 Define outside hardware/ software work packages			△	▲	● CA
8 Schedule work packages		●	△	▲ ○	● CA
9 Schedule other work packages		●	▲	△	● CA
10 Define customer information requirements		●	▲	△	● ▲ CA
11 Conduct on-going design review		✷	△	▲	
12 Operate schedule tracking system		△			▲ CA
13 Schedule review		▲	▲	△	▲ CA

WORK PACKAGE	STRATEGIC PLANNING	INFORMATION SYSTEMS	SYSTEMS ENGINEERING	MANAGER OF PROGRAMS (PROGRAM MANAGERS)	OTHER
14 Manage customer-initiated changes		▲	▲	△	▲ CA
15 Manage corporate-initiated changes		▲	▲	△	▲ CA
16 Develop start-up plan		●	▲	▲ ○	△ Field service ▲ CA ▲ Field sales
17 Start-up implementation		●	▲	△	▲ Field service ▲ Field sales
18 Warranty problems	*	●	△	▲	▲ Field service
19 Customer acceptance			▲	△	▲ CA
20 Out-of-warranty problems			▲	△	▲ Field sales

CODES:
- △ Primary responsibility
- ▲ Support responsibility
- ● Notification
- ○ Approval authority
- * Occasional notification

OTHER:
- Contract administration (CA)
- Sales
- Field sales
- Field service

FIGURE 11-2
Linear responsibility chart: design, manufacture, and delivery of an industrial system.

influence their jobs and their relationships to other members of the organization through:

Acceptance of responsibility and accountability Individuals have a major voice in determining job scope and objectives and goals which must be achieved.

Planning Team members do their own planning.

Authority Individuals have a voice in defining the nature of and limitations of their authority.

Organizing and Controlling People see the basis for aligning human and nonhuman resources and the basic unit of control (work package) for meeting project schedule, cost, and performance objectives.

Decision Making Decisions can be delegated to the lowest possible level (the work package), where information and competence come together.

Supervision This provides a basis for individual and team self-supervision.

Communication and Feedback These are tailored to fit individual needs on the known basis of the work package.

An Organizational Design Illustration

An illustration of the use of the LRC in this fashion is one that we developed in the context of the development of a management information system for a police department.[13]

The first few steps in that information-systems design project involved determining precisely how things were done in the organization and how they might be better done. The LRC was used as the basic "model" for accomplishing both of these tasks.

There has been a good deal of controversy in the field of systems design over how large a role should be played by the existing system in the design of a new system. Proponents of a fresh approach, or idealized-systems design,[14] argue that the new system should be designed quite independently of any system which may currently exist in the organization. Others take the position that this is inefficient and violates basic information principles by ignoring the information which is already present and which has been previously evaluated to be relevant. The approach to system design that was used melds these apparently disparate points of view. This general-level design phase of the overall system design process initially involved the definition, by the analysts, in close cooperation with the organization's managers, of "user sets," "interfacing organizations," and "decision areas."

Identification of User Sets and Interfacing Organizations The relevant user sets for a system—consisting of those managers who are designated to be the primary users of the system's output—are often specified by the stated objec-

[13] W. R. King, and D. I. Cleland, "The Design of Management Information Systems: An Information Analysis Approach," *Management Science,* November 1975, pp. 286–297.

[14] See R. L. Ackoff, *Creating the Corporate Future,* John Wiley & Sons, Inc., New York, 1981.

tives of the system. Using the statement of objectives, traditional organization charts, job descriptions, and other documents as guides, the system designer designated, on a preliminary basis, the relevant "user sets." These were defined to permit the system design to be oriented toward a reasonable number of user groups, each of which was treated as a homogeneous entity.

Often, the system objectives reflect the recognition that past information inadequacies—which usually precede the decision to develop a new system—are largely due to a lack of relevant information. Most of the information or data processed by the organization's existing system will be descriptive of the past history of the internal organization, usually much of it is outdated and inward-directed. To be useful, such information must be more prospective and more focused toward those environmental and competitive elements of the organization that will most critically affect its future.

Because of this, the analyst must consider external interfacing organizations as well as internal users. These organizations are defined in terms of specific informational inputs and outputs, that is, those organizations with which information is communicated in support of, or as a result of, the functions which the system is to support.

Identification of Decision Areas The next step in the general-design phase of the process involved the identification of decision areas. This step was initiated by the analysts on the basis of existing knowledge and refined through discussion with the appropriate managers who were to be the users of the system.

Table 11-1 shows such a "decision inventory" related to the planning function. It represents one broad delineation of some of the critical decision areas involved in the planning process in any organization.

Other categorizations could be used as well. In fact, during the process of developing such an inventory, the analyst will usually begin from a theoretical point of view based on a highly abstracted view of the organization and then

TABLE 11-1
DECISION AREAS

Policy formation—internal
Policy formation—external
Direction of operations
Organizing activities
Budget
Tactical planning
New programs
Training
Personnel selection
Allocation of resources
Research
External coordination
Internal coordination

proceed to revise the "inventory," based on discussions with executives and members of the user sets. This process serves to provide a good theoretical foundation and to make sure major omissions are avoided, while at the same time avoiding the problems of confusing terminology and overlapping decision areas.

Definition of Decision Areas After decision areas have been identified, they must be specified in detail. Discussions held with executives and the members of the user sets, as outlined in the previous step, assist greatly in achieving this greater level of specificity. These discussions serve a secondary purpose as well—the obtaining of support and acceptance by the people on whom ultimate system success will depend.

The row labels at the left side of Table 11-2 illustrate various decision elements associated with one of the broad decision areas such as those in Table 11-1. Thus, *all* of the tasks listed at the left side of Table 11-2 have been specified as the detailed functions that must be performed within *one* of the broad decision areas which the system is to support.

Developing the Basic LRC Models of the Organization After the user sets, interfacing organizations, and decision areas have been identified and specified, they are incorporated into an overall model of the organization. To address both the problem of how the organization currently operates and how it should operate, three LRCs were developed:

a A descriptive LRC model which describes existing system characteristics
b A normative LRC model which defines the ideal system
c A consensus LRC model which integrates elements of both (a) and (b)

Development of a Descriptive LRC Model The user sets, interfacing organizations, and decision areas may be used to develop a descriptive LRC. Table 11-2 shows an LRC which describes the relationships among various task elements (listed down the rows of the chart) and various user groups: positions and interfacing organizations as listed at the head of the columns. The entries in Table 11-2 represent a number of organizational characteristics with regard to the single decision area described:

1 Authority and responsibility relationships
2 Initiation characteristics
3 Input-output characteristics

The codes used to describe these characteristics for internal positions are:

E Execution
A Approval
C Consultation
S Supervision

Numbered subscripts on these role descriptors serve to identify the specific

TABLE 11-2

	Marketing VP 1	Executive VP 2	Comptroller 3	Department manager 4	Sales manager 5	Product managers 6	Marketing analyst 7	Salespeople 8	Field sales department 9	Legal department 10	Production department 11
Analysis of routine complaints				A	C_4	S	E	C_7	7^{io}		
Observation of field practices				A	S	E	C_6				
Complaint analysis				A	S	E	C_6		6^i	6^i	
Warranty analysis				A	E	C			5^i	5^i	5^i
Call-back analysis				A	S	E	C_6		6^i	6^i	6^i
Analysis of new legislation				A	S	E	C_6		6^i	6^i	6^i
Issue clarification and definition				A	S	E	C_6				
Selection of alternatives				A	E	C_5					
Obtaining of relevant facts				A	S	E	C_6	C_6	6^i	6^i	
Analysis of facts				A	E	C_5					
Review	A	A		E					4	4^{io}	
Formation				E	C						
Articulation		A	A	A	C_4	S	E	6°	6°		
Training for implementation					E	S	E				
Execution and control	A	A		A	S	E	C_6		6°	6°	

relationship. For instance, the first row of the table tells us that the marketing analyst (7) performs the analysis of routine product complaints, since he is designated as the executor (E). He consults with salespeople in doing this as denoted by C_7. He is supervised (S) by the product manager, and the department manager has responsibility for approving his work (A). The department manager consults with the sales manager in determining this approval (C_4). Also, the field sales department provides both input (^{i}) to and receives out-put (^{o}) from the marketing analyst (7) to permit him to perform this task.

The model depicted in Table 11-2 is an abstract description of the way the organization actually operates with regard to the single decision area. While descriptive models such as this are often developed by analysts to provide a basis for understanding the functioning of a system, the purpose for this descriptive model in this methodology is, in fact, prescriptive. However, the use of the model in this fashion first requires that a comparable normative model be developed.

Development of a Normative LRC Model The descriptive model of the organizational environmental system provides insights into who actually does what, the interactions among organizational units and between internal and external units, the general nature of information required, and the direction of information flow.

However, the use of a model of this variety as the sole basis for the design of systems would represent an abrogation of the analyst's proper role. Rather than creating a system to serve an existing organizational system, the analyst should attempt to influence the restructuring of the decision-making process so that the system may be oriented toward the support of a more nearly optimal process. To do this, the analyst may call on the best knowledge and theory of management to construct a normative model of the organization which is consistent with, and comparable to, the descriptive one previously developed.

A normative model looks much like the descriptive model of Table 11-2 except that it will normally involve more entries. This is because the normative model is an exercise in idealized-system design,[15] which will normally reflect many activities that the organization should be performing which are not depicted in the descriptive model.

However, it is just as inappropriate for the analyst to impose his concept of the way the organization should operate as it is to design slavishly the system to conform to the way in which the organization does operate. To achieve broad optimality based on improved effectiveness and usability, the designer must seek some consensus model which can serve as a realistic basis for system design.

Development of Consensus LRC Model Although few organizations desiring a new information system would be willing to restructure their organization's authority and responsibility patterns and relationships to suit the needs of the system, it is generally recognized that procedural improvement is a valid by-product of system design. Therefore, organizations are normally willing to consider some elements of a normative organizational model rather than to insist merely that the system service only existing procedure, functions, and authorities.

The development of a consensus model hinges on an objective comparison of a descriptive model, such as that of Table 11-2, with a comparable normative model. This comparison and evaluation must be done by managers with the aid and advice of the analyst. Thus, the consensus model "looks like" the LRC in Table 11-2. However, it is developed as an organizational consensus on "how we want the organization to operate." Since a consensus model emanates from a comparison of a descriptive model ("how we operate now") with a normative model ("how we theoretically should operate"), it integrates the best of current practice with those aspects of theory which are considered by the organization's managers to have practical applicability.

The descriptive model was developed by analysts on the basis of observation

[15]Ibid.

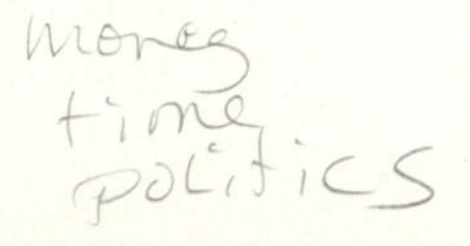

and discussion with operatives in the system, much as in the traditional system design model. So, too, is the normative model developed in quite traditional fashion by analysts using criteria of "technical" optimality.

However, the different and crucial aspect of this design process is the consensus model. This model is developed using the descriptive ("how we do it") and normative ("how we *should* do it") models as the basis for organizational discussion and design decision making. One format which has been successfully used for this is that of a "participative executive development program" in which the descriptive and normative models were explained in detail to managers in lecture/discussion sessions, and manager/analyst workshops were used to develop the consensus model—the one which is the "best" blend of theory and practical limitations to theory.

The considerations which go into the development of a consensus model are many and varied. "We can't do it the way the books say we should because of some unique feature of our organization" is a frequently heard comment in the consensus-seeking process. Sometimes such a comment is entirely valid. The recognition of this validity and its incorporation into the consensus model serves to make the system design more useful, since the system is being designed in terms of this practical consideration, which otherwise might have provided a basis for rejecting a "more optimal" design.

This design process also serves to make the managers feel that they have, in fact, designed their own system—a not insignificant element in gaining their acceptance of it.

In the police department application of this design use of LRCs, the consensus design for how the organization was to operate was then used as a basis for the development of the informative system.[16] Here, our interest is in the potential use of LRC for *organizational redesign*—whether in a project context or in a more general one.

THE LRC IN PROJECT MANAGEMENT

The LRC can be very directly useful for project managers to use to understand their authority relationships with their clientele. For a simple project, these relationships may be easy to depict; for more complex projects, a series of descending charts, from the macro-level of the project to successively lower levels, may be necessary.

The LRC in a Large-Scale National Development Project

An example of increasing complexity of the project clientele may be gained by considering project management situations in developing countries. These countries have become increasingly aware of the need to evolve a systematic

[16]King and Cleland, op. cit.

approach to the management of their large development projects. These projects are very large and complex in that they require the synchronization and cooperation of many different agencies and bodies.

Figure 11-3 shows a typical project environment in a developing project where the manager has to relate to all of the appropriate clientele. As the project moves from initiation to implementation, the clientele changes and the decision-making focal points change. A central planning council may do the preliminary project analysis, and the state planning board may make the evaluation and selection. An implementation agency may manage the designing and engineering. It is important in such a situtation that the responsibility be tracked appropriately. A *responsibility matrix* is helpful in this sense, and such a matrix is reflected in Figure 11-4.

FIGURE 11-3
Project management environment. (Hamed K. Eldin and Ivars Avots, "Guidelines for Successful Management of Projects in the Middle East: The Client Point of View," *1978 Proceedings of the Project Management Institute,* 10th Annual Symposium, Los Angeles, October 8–11, 1978.)

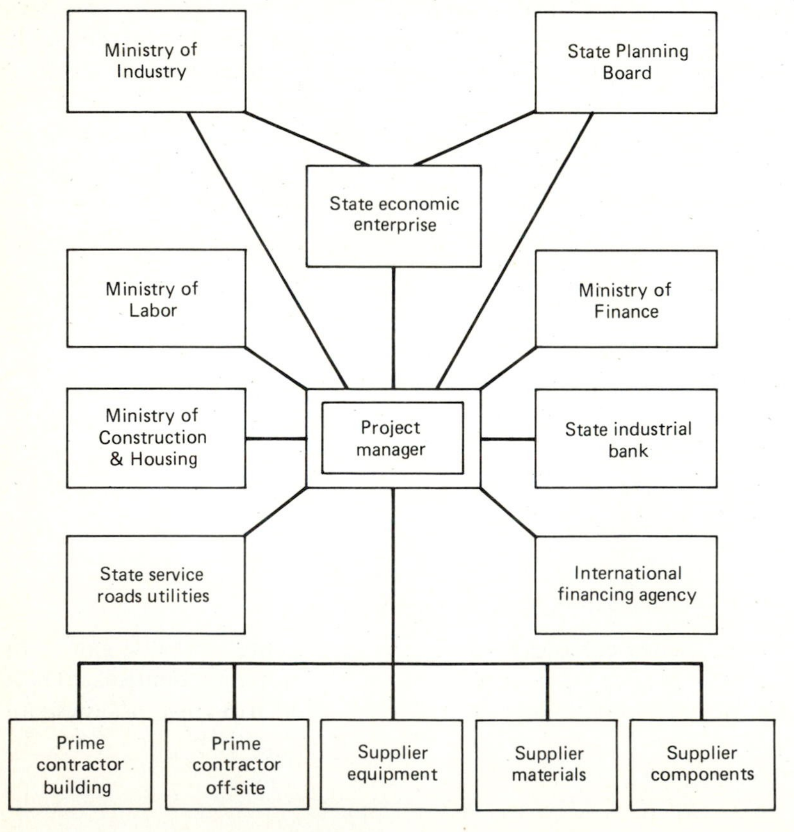

	Stages of Industrial Development Project	Decision-making bodies							
		Higher authorities council for planning[1]	State planning board	Project implementation agency	Project manager[2]	Assistant project manager	Contractors[3]	Subcontractors	Suppliers
1	Identification of project idea (preliminary analysis)	a	b	c					
2	Preliminary selection	b	a	b	c				
3	Feasibility (formulation)	c	b/c	a	b				
4	Post-feasibility evaluation and decision	a	a	b	b				
5	Detailed project design and engineering; and initial project implementation, scheduling		c	a	b	b	b/c	c	c
6	Contracting and purchase			a/b	a	b	b	b	c
7	Construction and pre-operations (system implementation, start-up)			b	a	b	b	b	b

Involvement:
- *a* *Ultimate responsibility*
- *b* *Assigned to project*
- *c* *Peripheral activities*

[1] *Development strategy and policy, target approval*
[2] *Or, prime contractor*
[3] *Or, project coordinator*

FIGURE 11-4
Types of decision-making bodies involved with project management. (Hamed K. Eldin and Ivars Avots, "Guidelines for Successful Management of Projects in the Middle East: The Client Point of View," *1978 Proceedings of the Project Management Institute,* 10th Annual Symposium, Los Angeles, October 8–11, 1978.)

The LRC in Establishing Project Relationships

Figure 11-5 shows an LRC for project-functional management relationships *within* a matrix organization. The development of such a chart, combined with the discussions that usually accompany such a development, can help greatly to facilitate an understanding of project management and how it will affect the day-to-day lives and activities of the various participants.

Developing an LRC in the Project Environment

The development of the project LRC is inherently a group activity—a getting together with the key people in the organization who have a vested interest in the work that is to be done. The following plan for the development of an LRC has proved useful:

1 Distribute copies of the current traditional organization chart and position descriptions to the key people.

2 Develop and distribute blank copies of the LRC.

3 At the first opportunity, get the people together to discuss:

- **a** The advantages and shortcomings of the traditional organization chart
- **b** The concept of a project work-breakdown structure and the resulting work packages
- **c** The nature of the linear responsibility chart, how it is developed, and how it is used
- **d** A simple way of establishing a code to show the work package–organizational position relationship (getting a meeting of the minds on this code is very important because individuals who believe the code to be either "too fine" or "too coarse" will find it difficult to accept)
- **e** The makeup of the actual work-breakdown structure with accompanying work packages
- **f** The fitting of the symbols into the proper relationship in the LRC

4 Encourage an intensive dialogue during the actual making of the LRC. In such a meeting, people will tend to be protective of their organizational "territory." The LRC by its nature requires a commitment to support and share the allocation of organizational resources applied to work packages. This commitment requires the ability to communicate and decide. This process takes time, but when the LRC is completed, the people are much more knowledgeable about what is expected of them.

Even though the LRC can be of great assistance in helping to understand personal and organizational interfaces, the *commitment* of people to support the project is vital. In the planning stages of a project, this commitment becomes very important. One manager emphasized its importance in the context of project planning by noting:

> . . . the project plan is simply the summation of all of the individual commitments reached between the project manager and the functional managers. If the project

ACTIVITY	General Manager	Manager of Projects	Project Manager	Functional Manager
Establish department policies & objectives	1	3	3	3
Integration of projects	2	1	3	3
Project direction	4	2	1	3
Project charter	6	2	1	5
Project planning	4	2	1	3
Project – functional conflict resolution	1	3	3	3
Functional planning	2	4	3	1
Functional direction	2	4	5	1
Project budget	4	6	1	3
Project WBS	4	6	1	3
Project control	4	2	1	3
Functional control	2	4	3	1
Overhead management	2	4	3	1
Strategic programs	6	3	4	1

CODE

1: Actual responsibility
2: General supervision
3: Must be consulted
4: May be consulted
5: Must be notified
6: Approval authority

FIGURE 11-5
Linear responsibility chart of project-management relationships.

> manager . . . does not have these commitments, he is in no position to make a commitment on behalf of the division. Remember, no one can commit you but yourself. Likewise, if you ever serve as project manager, *you* cannot commit anyone else.[17]

Those professionals who aspire to be project managers should note the importance of this commitment idea. A good way to start is to ask yourself the questions: Am I fully committed to those projects I am currently supporting? Have I provided the kind of personal commitment to the project manager that I would like if I were a project manager? Have I helped to develop an environment in the project work which is conducive to accomplishing the project objectives on time and within budget?

[17]Personal communication with the authors.

Much of the success of project management depends on how effectively people work together to accomplish project objectives and gain personal satisfaction. The development of a project LRC can greatly contribute to achieving this.

Limitations of LRCs

Charts such as those shown in Figures 11-4 and 11-5 are not a panacea for all organizational difficulties. The LRC is a pictorial representation, and it is subject to some of the characteristic limitations and shortcomings of pyramidal organizational charts. The LRC does reveal the task breakdown of the work to be done and the interrelationships between the tasks and job positions; it does not, however, show how *people* act and interact.

It is doubtful that any contemporary management theorists would deny that organizational effectiveness is as dependent on the informal organization of human actions and relations as it is on the structured, formal organization. The LRC, as we have so far discussed it, is limited to showing the task-position relationships that constitute the formal organization; it does not purport to reveal the infinite number of variations in human relations arising out of the informal organization. The LRC technique simply extends the scope of charting formal organizations wherever they are located in the hierarchical order. Thus, a note of caution is in order about the LRC. But, as Karger and Murdick have implied, we still must give it a vote of confidence:

> Obviously, the chart has weaknesses, of which one of the larger ones is that it is a mechanical aid. Just because it says something is a fact does not make it true. It is very difficult to discover, except generally, exactly what occurs in a company—and with whom. The chart tries to express in specific terms relationships that cannot always be delineated so clearly; moreover, the degree to which it can be done depends on the specific situation. This is the difference between the formal and informal organizations mentioned. Despite this, the Linear Responsibility Chart is one of the best devices for organization analysis known to the authors.[18]

SUMMARY

This chapter demonstrates the values and limitations of traditional organizational charts and introduces a variety of charts—all based on the *linear responsibility chart* (LRC)—which can aid in both the planning and implementing phases of management.

One of us in an earlier writing, took the position that project management is a "theoretical framework for viewing the internal and external environmental

[18]Delmar W. Karger, and Robert G. Murdick, *Managering Engineering and Research,* The Industrial Press, New York, 1963.

factors as integrated into the whole. . . . Explicit in this concept is the interdependency of decisions between all parts of components of the management problem."[19] By adding people to the concept of the organization as a system, we can define an organization as an assemblage of persons, each of whom carries out assigned tasks and all of whom are interrelated in such a way as to obtain a particular goal.

Systems-oriented organizations are always concerned with the underlying structure of human interrelationships. The use of linear responsibility charts is a step forward in management, therefore, since it provides a means of charting the interrelationships among human beings.

DISCUSSION QUESTIONS

1. What useful purpose does the traditional organizational chart serve?
2. What do the solid lines depict on the pyramidal organizational chart? Many organization charts have dotted lines to describe an organizational relationship. What do these dotted lines mean? Is there any advantage in using a dotted line on an organizational chart?
3. What are some of the alternatives to traditional organizational charting?
4. What is the nature of the linear responsibility chart (LRC)? What are its advantages over the traditional pyramidal chart?
5. What role does the work package play in the development of an LRC?
6. For what purpose are the symbols or legend in the LRC used? Why is it important to get agreement on the legend by the people concerned with developing the chart?
7. The process of developing an LRC may be more important than the end product. What is meant by this statement? Do you agree?
8. What is the relationship of the LRC with respect to the management functions carried out by the executive?
9. Participative management and LRCs have something in common. What is this?
10. The concept of the LRC can be applied in an information-systems context. Why is this application possible and practicable?
11. What strategy should be used in setting out to develop an LRC for the management of a project? What are some of the pitfalls to be avoided in developing an LRC in this context?
12. Commitment is important in the management of a project, particularly on the part of the key members of the project team. How can the project manager go about developing a sense of commitment on the part of such key project team members?
13. The development of an LRC for an ongoing project tends to "unclothe the project." What is meant by this statement?
14. The development of an LRC for an ongoing project can be very threatening to the manager and professionals if it is not done properly. Why might the process be threatening? What might be done to reduce the intensity of this threat?
15. LRCs can be used effectively to replace detailed policy and procedural manuals. Do you agree or disagree with this statement?

[19] David I. Cleland, and David C. Dellinger, "Changing Patterns in Management Theory," *Aerospace Management,* General Electric Company, vol. 1, Spring 1966, p. 4.

RECOMMENDED READINGS

Ackoff, R. L.: *Creating the Corporate Future,* John Wiley & Sons, Inc., New York, 1981.

Allen, Louis A.: *Charting the Company Organization Structure,* Studies in Personnel Policy, no. 168, National Industrial Conference Board, Inc., New York, 1959.

Churchman, C. W., and A. H. Schanblatt: "The Researcher and the Manager: A Dialectic of Implementation," *Management Science,* vol. 11, no. 4,, February 1965.

Cleland, David I., and David C. Dellinger: "Changing Patterns in Management Theory," *Aerospace Management,* General Electric Company, vol. 1, Spring 1966.

———and Wallace Munsey: "The Organization Chart: A Systems View," *University of Washington Business Review,* Autumn 1967.

———and Wallace Munsey: "Who Works with Whom," *Harvard Business Review,* September–October 1967.

Cyert, Richard M., and James G. March: *A Behavioral Theory of the Firm,* Prentice-Hall, Inc. Englewood Cliffs, N.J., 1963

Davis, Stanley M., and Paul R. Lawrence: "Problems of Matrix Organizations," *Harvard Business Review,* May–June 1978.

Donnelly, John F.: "Participative Management at Work," *Harvard Business Review,* January–February 1977.

Duncan, Robert: "What is the Right Organization Structure?," *Organizational Dynamics,* Winter 1979.

Eldin, Hamed K., and Ivar Avots: "Guidelines for Successful Management of Projects in the Middle East: The Client Point of View," *1978 Proceedings of the Project Management Institute, 10th Annual Symposium,* Los Angeles, October 8–11, 1978.

Galbraith, Jay R.: *Organization Design,* Addison-Wesley Publishing Co., Inc., Reading, Mass., 1977.

Handy, Charles C.: "Through the Organizational Looking Glass,"*Harvard Business Review,* January–February 1980.

Higgans, Carter C.: "The Organization Chart: Its Theory and Practice," *Management Review,* October 1956.

Janger, Allen R.: "Charting Authority Relationships," *The Conference Board Record,* December 1964.

Jasinski, Frank J.: "Adapting Organization to New Technology," *Harvard Business Review,* January–February 1959.

Karger, Delmar W., and Robert G. Murdick: *Managing Engineering and Research,* The Industrial Press, New York, 1963.

King, W. R., and D. I. Cleland: "The Design of Management Information Systems: An Information Analysis Approach," *Management Science,* November 1975.

Labovitz, George H.: "Managing Conflict," *Business Horizons,* June 1980.

Landsberger, Henry A.: "The Horizontal Dimension in Bureaucracy," *Administrative Science Quarterly,* December 1961.

McConkey, Dale D.: "Participative Management: What it Really Means in Practice," *Business Horizons,* October 1980.

Mintzberg, Henry: *The Structuring of Organizations,* Prentice-Hall, Inc., Englewood Cliffs, N.J., 1979.

Randall, Clarence B.: "The Myth of the Organization Chart," *Dun's Review and Modern Industry,* February 1960.

Read, William H.: "The Decline of the Hierarchy in Industrial Organizations," *Business Horizons,* Fall 1965.

Reeser, Clayton: "Some Potential Human Problems of the Project Form of Organization," *Academy of Management Journal,* vol. 12, December 1969.
Rogers, Lloyd A.: "Guidelines for Project Management Teams," *Industrial Engineering,* December 1974.
Shannon, Robert E.: "Matrix Management Structure," *Industrial Engineering,* March 1972.
Stuckenbruck, Linn C.: "Implementing Project Management—A Communication Problem," PMI 1980 Proceedings, Phoenix, October 27–29, 1980.
———: *The Integration Function in the Matrix,* paper presented at the PMI Seminar/Symposium, Atlanta, Oct. 17–20, 1979.
Tilles, Seymour: "The Manager's Job: A Systems Approach," *Harvard Business Review,* January–February 1963.
Weisbrod, M.: *Organizational Diagnosis,* Addison-Wesley Publishing Company Inc., Reading Mass., 1978.
Youker, Robert: *A New Look at WBS (Work Breakdown Structure),* CN-851 Course Note Series, July 1980, International Bank for Reconstruction and Development.
"Changing the Company Organization Chart," *Management Record,* November 1959.
"Linear Responsibility Charting," *Factory,* vol. 121, March 1963.
Understanding and Designing Formal Organizational Structure, ICC Case no. 9-478-034 (prepared by Prof. Cyrus F. Gibson).

CASE 11-1: Linear Responsibility Chart

Situation A project will be initiated within an aerospace firm for the construction of a new plant.

Your Task Using this situation under whatever assumptions you wish, develop an LRC for this project within the organization's environment. In doing this, consider the following:

An appropriate work breakdown structure for the project—select some representative "work packages" to use in the LRC

- The appropriate summary position descriptions of the managers/profession(s) in the organization
- Appropriate symbols required
- Need for additional information to develop the LRC

What do you think should go into this LRC? What are the advantages in developing and using an LRC? Have the team leader present your findings.

CHAPTER **12**

PROJECT AUTHORITY

The project manager who fails to build and maintain his alliances will soon find indifference or opposition to his project requirements.[1]

Authority is necessary if one is to get a project completed on time and within the cost and performance requirements. However, a degree of personal freedom is required in the project environment—particularly for the professional people. Balancing these two conditions of freedom and authority, which are both contradictory and complementary, is one of the more challenging problems facing the project manager.

In this chapter, we will outline a conceptual framework for authority, its limits, and its needs. One might say that technically oriented project managers have no concern for the philosophical niceties of authority, that their greatest concern is the management of the technical affairs of the project. This is truly not the case. However complex the technology, and however great the magnitude of the material resources, project management is still a function of executive leadership acting through organized groups.

WHAT IS AUTHORITY?

Authority is a conceptual framework and, at the same time, an enigma in the study of organizations. The authority patterns in an organization, most commen-

[1]David I. Cleland, "Understanding Project Authority," *Business Horizons,* Spring 1967. The present chapter is an extension of this article.

tators agree, serve as both a motivating and a tempering influence. This agreement, however, does not extend to the emphasis that the different commentators place on a given authority concept. Early theories of management regarded authority more or less as a gravitational force that flowed from the top down. Recent theories view authority more as a force which is to be accepted voluntarily and which acts both vertically and horizontally.

Increasingly, managers find themselves having to rely on some technical expert in making decisions, whether a solid-state physicist, or an information-systems designer. These professionals are assuming decision-making functions within the organization and managers are increasingly dependent on them. They, in effect, make decisions simply because their work is so technical and their supervisors, particularly higher-level executives, do not understand enough to be able to make many judgments in the technical areas of an organization.

Defining Authority

Although authority is the key to the management process, the term is not always used in the same way. Authority is usually defined as a "legal or rightful power to command or act." As applied to the manager, authority is the power to command others to act or not to act. The manager's authority provides the cohesive force for any group; it comes into being because of the group effort. In the traditional theory of management, authority is a right granted from a superior to a subordinate.

Authority, at least formal authority, is the right of a person to be listened to and obeyed. But where does this right originate? Every manager obtains formal authority as a delegation from the next higher level. In this view, therefore, the ultimate source of authority (i.e., hierarchical authority) is the right to private property[2] in our society or in the charismatic power of hierarchical role.[3] In theory, authority is still concentrated at the top of the organization and is delegated in the scalar chain to subordinate organizational elements. This hierarchical authority exists primarily as a contingent force, to be used in resolving intraorganizational disputes, in making basic strategic decisions affecting the whole organization, and in establishing overall policy for the organization.

Barnard tempers the traditional view of authority by recognizing the right of the contributors to, or members of, a formal organization to accept or reject an order given by a higher official.[4] Traditional theory did not adequately consider that the *sources* and *uses* of authority are manifested outside the boundaries of the parent organization. This traditional viewpoint therefore ignores the authority patterns that exist between managers and technicians in different organizations. Nor does the traditional view recognize the impact of the reciprocal

[2]Ralph C. Davis, *The Fundamentals of Top Management,* Harper & Row, Publishers, Incorporated, New York, pp. 281–322.

[3]Victor A. Thompson, *Modern Organization,* Alfred A. Knopf, Inc., New York, 1966, p. 77.

[4]Chester I. Barnard, *The Functions of the Executive,* Harvard Business Press, Cambridge, Mass., 1938, p. 163.

authority relationships existing between peers and associates. With the exception of functional authority, the traditional view presupposes some superior-subordinate relationship in the organizational arrangement.

Power

Power is a concept frequently associated with authority and is defined as the ability to unilaterally determine the behavior of others, regardless of the basis for that ability.[5]

Authority provides the power that is attached to the organizational position; it is delegated through the media of position descriptions, organizational titles, standard operating procedures, and related policies. Authority as influence, however, may be assumed by individuals without the legitimacy of an organizational position. Individuals can influence their environment simply because they have knowledge and expertise, even without documented authority. There is little doubt that a duly appointed superior has power over subordinates in matters involving pay, promotion, and effectiveness reports. This delegated power functions unilaterally, from the top down. A manager's authority, however, is a combination of power and influence, such that subordinates, peers, and associates alike willingly accept his or her superior judgment. To conceive of this combination of power and influence is to take an *integrative* approach to a discussion of the project manager's authority, emphasizing both the legal grants and the personal effectiveness of this organizational position. Fayol may be taken to favor this integrative approach. He explained that:

> Authority is the right to give orders and the power to exact obedience. Distinction must be made between a manager's official authority deriving from office and personal authority, compounded of intelligence, experience, moral worth, ability to lead, past services, so forth. . . . [Personal authority] is the indispensable complement of official authority.[6]

In the traditional bureaucratic sense, authority (the power to act) depends on power inherent in the structure of the organization rather than on the quality of the interpersonal relationships within the peer group. Power arises within the organization through attachment to a legitimate organizational position; reliance is based on the workings of the formal structure, operating through defined policies and procedures, rather than on the initiative of individuals, operating through peer group interrelationships. Project managers who have been newly appointed will often be uncomfortable about their power in the organization because they are not located in the hierarchical structure. If such project managers perceive a superior-subordinate relationship as a prerequisite to the exercise of authority, they are missing a fundamental point about power in organizations. Drucker provides relevant thoughts about this power:

[5]For example, see James D. Thompson, "Authority and Power in 'Identical' Organizations," *The American Journal of Sociology*, vol. 60, November 1956.

[6]Henri Fayol, *General and Industrial Management*, Sir Isaac Pitman & Sons, Ltd., London, 1949, p. 21.

> The hierarchy does not, as the critics allege, make the superior more powerful. On the contrary, the first effect of hierarchical organization is the protection of the subordinate against arbitrary authority from above. A scalar or hierarchical organization does this by defining carefully the sphere within which the subordinate has the authority, the sphere within which the superior cannot interfere. It protects the subordinate by making it possible for him to say, "This is *my* assigned job." Protection of the subordinate underlies also the scalar principle's insistence that a man have only one superior. Otherwise the subordinate is likely to find himself caught between conflicting demands, conflicting commands, and conflicts of interest as well as of loyalty.[7]

Power structures in contemporary organizations are much different from those found in earlier organizations. Power today springs from the legitimacy of an organizational position *and* from knowledge, expertise, personal effectiveness, and influence. The granting of formal authority—providing a manager with an element of power—is necessary, if only for the preservation of a sense of discipline in the organization. When all else fails—persuasion, knowledge, interpersonal influence—there are times when a manager simply has to resort to the power granted from a higher echelon. But even in these circumstances, influence and knowledge play a role in making a formal power structure endure.

How Much Project Authority?

The project manager manages across functional (intraorganizational) and organizational (interorganizational) lines to bring together activities required to accomplish the objectives of the specific project. In the traditional bureaucratic organization, business is conducted up and down the vertical hierarchy. The project manager, on the other hand, is more concerned with the flow of work in horizontal and diagonal directions than with flows in the scalar chain. Problems of motivation exist for the traditional vertical manager, but these problems are compounded for project managers because the traditional leverages of hierarchical authority are not at their disposal. They must act as the focal point for major project decisions and considerations, however, so they must be given adequate authority to accomplish these objectives. All too frequently inadequate attention is given to the matter of authority in setting up a matrix organization. The experiences of a company in the construction industry illustrate the point.

Ebasco Services Incorporated, an old line builder of power plants, began experimenting with the matrix organization in the late 1970s. In Ebasco, a 4,500-person work force was responsible for "bottom-line" results in engineering excellence. The introduction of a matrix organization into the company created new managerial problems. Lines of authority became blurred. A construction manager, for example, who used to choose the equipment only in terms of what worked best from an engineering viewpoint, now had to give consideration to the project manager's budget concerns. According to John A. Scarola, "Suddenly these people are expected not only to fulfill their discipline

[7]Peter F. Drucker, *Management: Tasks, Responsibilities, Practices,* Harper and Row, Publishers, Incorporated, New York, 1974, p. 525.

needs but to look at what they're doing in terms of the entire project on which they are working."[8]

Other examples are plentiful. For instance, at some universities in the 1960s the formal power structure broke down—students seized property, evicted faculty and administrators from their offices, and disrupted classes. Power was not regained by the legitimate officeholders until negotiation was carried on with the students.[9] The new power structure was a modified version of the old, reflecting the participative right of the students in the governance of the universities.

The emergence of the matrix organization has necessitated a reassessment and redefinition of authority relationships—indeed, of "power." *Project* managers, *functional* managers, and *general* managers all must operationally redefine their formal authority requirements and reassess their ability to bring technical and managerial knowledge to bear in attaining organizational objectives.

The Objective of Project Authority

Although the specific "amount" and nature of project authority certainly varies from organization to organization and from one situation to another, the essential objective of project authority is captured by Goodman:

> Because of the complex interdependencies, the project manager must be vested with some authority to enable him to prevent sub-optimization on the part of the work units while maximizing the optimization of the total task.[10]

Thus, the objective of the authority that is granted to the project manager is the resolution of the essential dilemma of the systems approach to management that was discussed in Chapter 2—the *optimization of overall effectiveness in the light of the conflicting objectives of organizational subunits.*[11]

A Brief Review of Contemporary Theory

Contemporary management theory identifies three basic kinds of authority: formal line authority, staff authority, and functional authority.

Formal Line Authority This is the right, derived from a legitimate source, to command, act, or direct. This formal authority flows from the right of private

[8]"When Bosses Look Back to See Ahead," *Business Week,* Jan. 15, 1979, p. 60. Reprinted with permission. Copyright © 1979 by McGraw-Hill, Inc. All rights reserved.

[9]One might speculate about the ultimate example of the use of power carried out at Kent State University when troops fired on assembled students. The question of the legitimacy and necessity of the use of this type of power will undoubtedly be debated in our society on a continuing basis.

[10]Richard Alan Goodman, "Ambiguous Authority Definition in Project Management," *Academy of Management Journal,* p. 396, December 1967.

[11]W. R. King and D. I. Cleland, "A New Method for Strategic Systems Planning," *Business Horizons,* vol. 18, no. 4, August 1975.

property, conferred by the society, through the owners of the business and their delegates. The formal line authority is based on some enforceable contract. The use of formal authority in contemporary organizations is backed by long-standing custom and certain basic values in our society.

Staff Authority The legal authority of staff officials is derived from their appointment as staff members to assist, advise, and counsel the line official to whom they report. Staff officials may have *line* authority over their subordinates, but they do not have command authority over other people in the organization except as it operates through their superior knowledge.

Functional Authority This is the legal right to act with respect to specific activities or processes.[12]

These three concepts of authority—line, staff, and functional—and their relationship to *project* authority can be understood better by thinking of them in terms of organizational parameters. This is illustrated in Figure 12-1, where the

[12]One derivative of formal authority is functional authority. At first glance, it might appear that functional authority and project authority are one and the same. Functional authority is defined as the manager's legal right to act or command with respect to specific activities or processes in departments other than his or her parent department. It is a small slice of the authority of some line manager and relates to particular phenomena in the organization, for example, the authority of the personnel officer to prescribe certain grievance procedures. The project manager's authority vastly exceeds any that could be delegated under the concept of functional authority.

FIGURE 12-1
The project manager's authority.

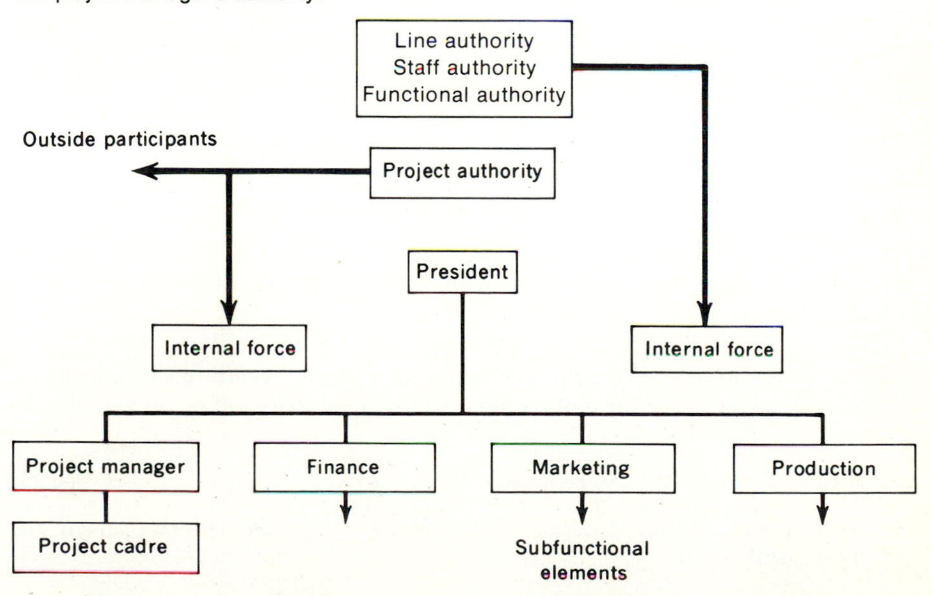

four types of authority are depicted as having an *internal* force in the organization, while only project authority operates outside the boundaries of the parent organization.

What Is Project Authority?

Managers must have authority in order to accomplish their work. No philosophy of authority, however, is going to tell them how to exercise their authority in specific cases; what a philosophy can do is give them a conceptual framework as a base for their thinking about it. Authority has not, says Golembiewski, enjoyed "conceptual unanimity."[13] The concept of authority is in a period of transformation, from the bureaucratic hierarchical force to a participative and persuasive one. The elements of participation and persuasion in the authority relationships are products of our modern organizations and reflect the influence of the democratic and scientific revolution in contemporary society.

De jure project authority is the legal or rightful power to command or act in the management of a project. Inherent in this authority is the legal right to commit or withdraw resources supporting the project. The legal authority of a project manager is usually contained in some form of documentation; such documentation must of necessity contain, in addition, the complementary roles of other managers (e.g., functional managers, work package managers, general managers) associated with the project.

De facto project authority is that influence brought to the management of a project by reason of a particular person's knowledge, expertise, interpersonal skills, or personal effectiveness. De facto project authority may be exercised by any of the project clientele, managers, or professional people.

Typical project managers lament their "lack of authority." In this respect, O'Brien captures the essence of their views on their authority by noting:

> Almost without exception, the project manager sees himself as an individual with great responsibility and limited authority. Generally, he is quite correct. For a variety of reasons, management chooses to assign responsibility to the project manager, for a different variety of reasons, management is always reluctant to delegate authority on a broad basis.[14]

Even with a grant of legal authority, project managers may still have problems. O'Brien cautions:

> The project manager who expects to have his authority clearly documented in a written charter or in the form of a crisp organizational chart will be disappointed in

[13]Robert T. Golembiewski, "Authority as a Problem in Overlays: A Concept for Action and Analysis," *Administrative Science Quarterly,* June 1964, p. 24.

[14]James J. O'Brien, "Project Management: An Overview," *Project Management Quarterly,* vol. 7, no. 3, September 1977, p. 30.

either of two ways. Generally, he will not get the clear, crisp charter which he desires. Should he be fortunate (or unfortunate) not to achieve his apparent clear authority he will often find it difficult to enforce.[15]

Thus, effective project managers must develop de facto authority to complement their limited de jure authority.

The Basis for Project Authority There is a clear basis for the desirability and practicality of such authority patterns in project management positions. For instance, Marquis and Straight found, after a study of 50 different projects in 50 different firms, that the authority granted to the project manager was not related to the organizational structure within the firm.[16] Steiner and Ryan, in a study of how successful industrial project managers conduct their programs, concluded that, under certain circumstances, "greater authority placed in the hands of competent project managers will result in a superior technical product developed at a time and cost less than one can normally expect."[17]

Goodman[18] has conducted research dealing with corporations that have tried to cope with the "dual authority" situation. The research was carried out in six defense aerospace corporations located in the midwestern and far western areas of the United States. The results suggest that the complex interdependencies for an R&D project require a person to be designated as a project manager and authority to be delegated to this person which establishes the "dual authority network."

The research found that the major adaptation to the dual authority network was

> . . . the use of an ambiguous authority definition between the general management of the company and the project management of the company. It is the conclusion of this author [Goodman] that ambiguous authority definition is beneficial to an organizational climate that has evolved a concern for the overall impact of each individual on the company objectives. Even in organizations which do not have such a climate the use of ambiguous authority will make certain, at least, of double coverage of every problem. The major drawback to the use of ambiguous authority is the conflict situation and the uncertainty it provides for many of the organizational members.[19]

The "Authority Gap" Issue Project managers are often perceived as having an "authority gap" because they do not possess the authority to reward or

[15]Ibid.

[16]Donald G. Marquis and David M. Straight, Jr., "Organizational Factors in Project Management," Working Paper no. 133-65, Cambridge, Mass., Alfred P. Sloan School of Management, Massachusetts Institute of Technology, August 1965.

[17]George A. Steiner and William G. Ryan, *Industrial Project Management,* The Macmillan Company, New York, 1968, p. 127.

[18]Richard Alan Goodman, op. cit.

[19]Ibid.

promote their personnel. They lack the traditional line of authority over the team and possess what is called "project authority." Hodgetts conducted research into the question: "What ways do project managers find of increasing their authority and minimizing their 'authority gap'?"[20]

In his research, he employed two steps to determine the way in which project managers cope with the authority gap. First, four project organizations in different industries and undertakings were visited, and interviews were conducted with appropriate project managers. Second, a list of firms using project management was sent questionnaires. The firms interviewed were located in the fields of aerospace, construction, chemicals, and state governments.

Hodgetts' results indicate that there are numerous techniques for overcoming the authority gap. An interesting aspect of his study is the determination that the authority gap was handled differently in projects of varying magnitudes. In small projects, it was overcome by the project manager, who used different persuasive techniques. In large projects, the authority gap was mitigated through relying principally on the definition of the authority of the project manager in an organization charter.

Interfaces in Project Authority

In its total sense, project authority is the legal and personal influence that the project manager exercises over the scheduling, cost, and technical considerations of the project. Project authority manifests itself within the legitimacy of the project; it extends horizontally, diagonally, and vertically within the parent organization and radiates to outside participating organizations. The traditional line-staff relationships are modified in the project environment since, in the project environment, a line functional manager (such as a production manager) gives advice, counsel, and specialized support to the project manager.

Functional managers may very well perceive that their power and status within the organization tend to be diluted as they go through a role reversal brought about by the project manager "calling the plays."

Project authority provides the way of thinking required to unify all organizational activities toward accomplishment of the project end, regardless of where they are located. Project authority also determines how project requirements are to be met within the planned scheduling, technological, and cost restraints. The work of project managers varies in accordance with the type and degree of authority vested in them. At one extreme, project managers may serve as assistants to the general manager and, as such, play the role of project coordinators. At the other extreme, they may run their programs with a line-type degree of authority that denies the functional executive any significant right of appeal.

[20]See Richard M. Hodgetts, "Leadership Techniques in the Project Organization," *Academy of Management Journal,* June 1968.

One requirement in defining the role of project managers is to delineate their role vis-à-vis the role of the functional managers with whom they must work. This typically takes a form along the lines of the *project-functional interface* described in Table 12-1.[21]

The project-functional interface may be described as the "deliberate conflict" in the matrix organization. This deliberate conflict arises out of the competition for scarce resources in the organization among project managers and functional managers. The conflict is also manifested in the question "Who works for whom?" in the matrix organization.

When project management is introduced in an organization, it is essential that these project-functional interfaces be *understood* and *accepted* by general

[21]These interfaces are defined in the context of an organization chart in Chapter 11.

TABLE 12-1
THE PROJECT-FUNCTIONAL INTERFACE

Project manager	Functional manager
Project direction Determines *what* effort will be accomplished and *when* it will be performed through the development and issuance of master program plans to include the project work breakdown structure and work statements with accompanying budgets and schedules. Ensures the accomplishment of the technical objectives, schedule requirements, and effective cost management of the project	*Operational direction* Determines *who* will perform specific tasks, *where* they will be done, and *how* they are to be accomplished Provides a stable base for the development of talent and skills to ensure the maintenance of technical capability Provides necessary facilities and services to support program requirements
Project control Monitors cost, schedule, and technical results against master program plans; initiates any necessary corrective action to ensure accomplishment of project objectives; and monitors contractual reporting	*Operational control* Ensures that the technical excellence and quality requirements of assigned tasks are met and that the tasks are accomplished on schedule and within the budget
Configuration management Controls changes and ensures configuration accountability affecting the project	*Administration* Performs administrative services in support of personnel assigned to a project Initiates merit increases for all personnel within their organization
Customer coordination Provides the prime contact with the customer of project activities	
Administration Approves the assignment and concurs in merit increases of key functional personnel assigned to the project	

Adapted from David I. Cleland, "The Deliberate Conflict," *Business Horizons*, February 1968, pp. 78–79.

managers, project managers, and functional managers. This understanding can be facilitated if all the managers concerned jointly participate in the development and publication of a policy document containing a description of the intended authority and responsibility relationships characterized by Table 12-1.

The relationships depicted in that table are the major interfaces with which the project manager must deal. Other interfaces are important as well, such as:

With line managers
Between the project manager and other project managers in the company
With managers in the customer's organization
With staff managers
With supportive technical staff agencies
With associate and/or subcontractors
With technical and managerial peers

Figure 12-2 portrays a typical overall organizational complex of authority and responsibility interrelationships found when a matrix organization is established. While project managers may be described as the "general managers" of the company as far as the project is concerned, they are nevertheless dependent on the technical and managerial support of many people throughout the organiza-

FIGURE 12-2
Overall company authority and responsibilities for accomplishing project effort.

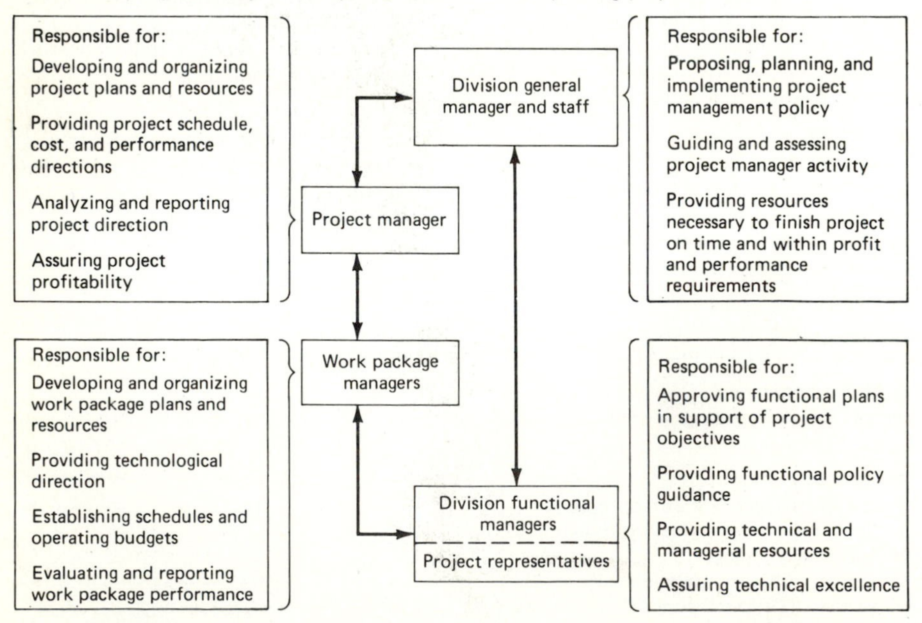

tion. Project managers who try to establish a fiefdom for their projects soon find lack of support from the many managers with whom they must deal. As more and more active projects are added to the organizational work, the interrelationships reflected in Figure 12-2 become more complex and varied; the opportunities for conflict in the application of resources are intensified. In a large company, there may be hundreds of active projects, all competing for the scarce resources available to the functional managers for application to a particular project. Aggressive project managers will want top priority for their projects; inevitably, a higher level of management will have to become involved in establishing project priorities.

Project managers typically exercise authority granted to them by the appropriate general manager and perform in behalf of the general manager on those matters pertaining to the assigned project. There may be exceptions to a general grant of authority. For example, one company enumerated specific exceptions as follows:

1 Financial functions for which there must be acceptable, consistent, and accurate accounting systems and techniques in accordance with Division and Corporate requirements

2 Direct management authority over the activities of the Quality and Reliability Assurance Department and the certification of quality

3 Specific acts in the areas of contracting and subcontracting for which there must be designated legal authority and responsibility [22]

Of course, there is the situation where a general manager may find it necessary to preempt the authority of the project manager, as in the case of final, critical negotiations with the customer. These preemptions should be done, however, only with the full knowledge of the project manager.

The exercise of authority in the execution phase of a project is far removed from the organic power of the chief executive. Decisions are made constantly by the functional and the project managers, and the success of these decisions depends upon the successful integration of these managers' delegated and assumed authority. In the project environment, the real basis of a person's authority is the professional reputation enjoyed among peers and associates. People gain this type of reputation only through recognition of their accomplishments by the other members of their environment and not by policy documentation, however extensive that may be.

A significant measure of the authority of project managers springs from their function and the style with which they perform it. Project managers' authority is neither all de jure (having specific legal foundations) nor all de facto (actual influence exercised and accepted in the environment). Rather, their authority is

[22]Division Policy Guide, Subject: *Responsibilities of the Program Manager,* Aerospace and Electronics Systems Division, Westinghouse Electric Corporation, Baltimore, Nov. 9, 1972.

a combination of de jure and de facto elements in the total project environment. Taken in this context, the authority of project managers has no organizational or functional constraints, but rather diffuses from their office throughout and beyond the organization, seeking out the things and the people it wishes and needs to influence and control.

The effectiveness with which project managers exercise authority depends to a large degree on their legal position as well as on their personal capabilities. But there are ways in which project managers can operate to enhance their basic authority.

The Work Package Commitment

How well project participants are committed to support the project can make the difference between the success and failure of a project. The importance of having individuals committed to support a work package of the project cannot be overemphasized. One engineering manager described the importance of the commitment to the work package in this manner:

> A work package is a commitment made and agreed to by two individuals "eyeball to eyeball" so to speak. It is a personal thing and has nothing to do with charts, graphs, or other inanimate pieces of paper. You don't have to make any commitment you don't agree to—normally. In very unusual circumstances your boss, and only your boss, will make a commitment for you, but this is the exception, not the rule. Only the parties to the commitment may agree to a change or cancellation of the commitment.[23]

Of course, this commitment presupposes that the necessary planning has been done; i.e., that the tasks have been adequately defined so that the personal commitment can be made. The same engineering manager noted:

> If you can't commit to what the other person is asking you for, then the two of you had better redefine the task in such a manner that a commitment can be made. "Time and Material" contracts cannot be successful without this agreement. If you don't have a clear understanding of what each party is committed to, the inevitable day will arrive when the "misunderstanding" surfaces and someone's reputation and integrity suffers. These "misunderstandings" are more typically "no understandings," i.e., the problems or definitions of tasks were difficult in the beginning so they were avoided or set aside like they didn't exist. A real commitment between the two parties eliminates the problem and allows a real project management to flourish.[24]

The work packages provide a common denominator or framework for monitoring performance, cost, and schedule throughout the life of a project. As an allied benefit, the work breakdown structure is a convenient method of

[23]Letter, J. C. McVickers, Director, Production Technology, Westinghouse Electric Corporation R&D Center, Nov. 25, 1975.

[24]Letter, J. C. McVickers, Mechanical R&D Division, Westinghouse Electric Corporation, Jan. 3, 1977.

managing by exception; that is, of providing a focus for highlighting critical items or areas of the project which must receive prompt attention by the managers concerned, working with those persons who are committed to do the work.

Authority through Negotiation

Project managers do not have unilateral authority in the project; they must frequently negotiate with the functional manager. These negotiations provide an opportunity to achieve trade-offs (checks and balances) among project performance, delivery, and cost objectives. In these negotiations, the flow of a project manager's authority departs from the vertical structure of the organization.

This flow of authority is more a network of alliances between the project participants than a recurring delegation of power between the chain of superiors and subordinates in the various hierarchies. This network of alliances depends heavily upon the reputation of individuals as reflected in their professional achievements. The character of project authority is supportive, with more reliance on peers, associates, and a consensus thereof than on the customary manifestations such as organizational position descriptions and policy documentation. The effect of the peer group in exerting influence is exemplified in the manner in which Texas Instruments manages innovation. A key part of the management of innovation at Texas Instruments is carried out through a procedure whereby an individual can submit a proposal for funding of a development project. The approval process for such a project proposal includes an evaluation by a peer group. After the development project is funded, a project team of peers is organized to provide the support required to accomplish the project. The use of such a peer team in Texas Instruments has helped to foster an organizational environment which facilitates built-in checks and balances. This results in a situation described as one in which:

> It's impossible to bury a mistake in this company; the grass roots of the corporation are visible from the top. There's no place to hide in TI—the people work in teams, and that results in a lot of peer pressure and peer recognition.[25]

Current literature in organizational theory has noted that informal and persuasive processes modify bureaucratic authority. Authority systems have become less arbitrary, less formal, and less direct. Under the bureaucratic theory, there was assumed to be one center of authority—the line. In the project environment, no such clear-cut line of authority exists except for nonprofessional support people. The professionals do not form into an authority structure in the usual sense of the term.[26] Various modifications of the formal authority structure

[25]"Texas Instruments Shows U.S. Business How to Survive in the 1980s," *Business Week*, Sept. 18, 1978, p. 81. Reprinted by permission. Copyright © 1978 by McGraw-Hill, Inc. All Rights Reserved.

[26]For a discussion of the authority structure in professional organizations, see Amitai Etzioni, "Authority Structure and Organizational Effectiveness," *Administrative Science Quarterly*, June 1959.

exist, even in those organizations considered to be bureaucratic. Hence, effective authority in the project environment depends on manifestations other than the legal ones. The effective authority of project managers is seldom autocratic. Their most meaningful authority may be based on their ability to build reciprocity in their environment, to create and maintain political alliances, and to resolve conflict between the functional managers. Unilateral decisions, dogmatic attitudes, and the *resort to the authority of a hierarchical position* are inconsistent with the development of this sort of authority. Instead, the project manager's job is to search for points of agreement, to examine the situation critically, to think reflectively, and only then to take an authoritative position based on the superiority of his or her knowledge. This, rather than their organizational position, is the most significant basis for the authority of project managers.

WHAT PLACE HIERARCHY?

Adequate authority to accomplish project objectives cannot be created by rearranging the compartments and shifting the lines on the organizational charts. Participants in the project organization at all levels must modify, negate, supplement, and reinforce the legal authority which emanates from a given arrangement of positions. Project managers accomplish their objectives through working with people who are largely professional. Consequently, their use of authority (both de jure and de facto) must be different from what we would expect to find in a simple superior-subordinate relationship. For professional people, project leadership must include explaining the rationale of the effort as well as the more obvious functions of planning, organizing, directing, and controlling.

Authority in project decisions must be indifferent to the hierarchical order of affairs. In many cases, the decisions that executives in the higher echelons reserve for themselves amount to nothing more than approving the proposals made by project managers. The role played by these line and staff managers can easily deteriorate to that of delayers, debaters, investigators, coordinators, and vetoers, and such parochial views may work to a manager's own disadvantage.

Upper-echelon executives may be in a more precarious position than they realize. One of the folklore figures of traditional management is the powerful executive who sits at the head of a highly organized, tightly run organizational pyramid, running things from the top down. In project management the vertical organization still plays an important role, but this role is concerned largely with facilitating project affairs and ensuring that the proper environment is provided for those participating in the project. George Weissman, president of Phillip Morris, Inc., has said that his role is "to carry the oil can."[27] Concerning the

[27]Quoted in *Business Week,* Mar. 4, 1967, p. 94.

operation of Phillip Morris brand (product) managers, *Business Week* has said that ". . . the basic moving parts are the brand managers. . . . They run the show. Within corporate budgetary limitations, they control packaging, promotion, sales effort, and advertising for their brands."[28]

Project management proves one thing about management theory, and that is that ". . . simply being in an executive hierarchy does not mean that one can freely direct those below him."[29] The higher-level officials in an organization are more dependent on their subordinates and peers than traditional theory will admit. The decisions made in the course of a large project are of considerable complexity; so many of them must be made that one individual, acting unilaterally, cannot hope to have sufficient time to make a thorough analysis of all the factors governing his or her decisions. The decision maker in project management must depend on many others to provide analysis, alternatives, and a recommended course of action.

Project authority depends heavily on the personality of the project managers and on how they see their role in relation to the project environment. Their authority is not necessarily weak because it is not thoroughly documented and because it functions outside the parent organization, between the participating organizations. A project manager is in a focal position in the project endeavors, and this focal position gives the opportunity to control the flow of information and to have superior knowledge of the project. The scope of power and control exercised by project managers may be virtually independent of their legal authority.

DOCUMENTING PROJECT AUTHORITY

Project managers should have broad authority over all elements of their projects. Although a considerable amount of their authority depends on their personal abilities, their position will be strengthened by the publication of documentation to establish their modus operandi and their legal authority. As a minimum, the documentation (expressed in a policy manual, policy letters, and standard operating procedures) should delineate their role and prerogatives in regard to:

1 The project manager's focal position in the project activities

2 The need for a defined authority and responsibility relationship among the project manager, functional managers, work-package managers, and general managers

3 The need for influence to cut across functional and organizational lines to achieve unanimity of the project objective

[28]Ibid.

[29]Herbert A. Simon, Donald W. Smithburg, and Victor A. Thompson, *Public Administration,* Alfred A. Knopf, Inc., New York, 1950, p. 404.

4 Active participation in major management and technical decisions to complete the project

5 Collaborating (with the personnel office and the functional supervisors) in staffing the project

6 Control over the allocation and expenditure of funds, and active participation in major budgeting and scheduling deliberations

7 Selection of subcontractors to support the project and the negotiation of contracts

8 Rights in resolving conflicts that jeopardize the project goals

9 Having a voice in maintaining the integrity of the project team during the complete life of the project

10 Establishing project plans through the coordinated efforts of the organizations involved in the project

11 Providing an information system for the project with sufficient data for the control of the project within allowable cost, schedule, and technical parameters

12 Providing leadership in the preparation of operational requirements, specifications, justifications, and the bid package

13 Maintaining prime customer liaison and contact on project matters

14 Promoting technological and managerial improvements throughout the life of the project

15 Establishing a project organization (a matrix organization) for the duration of the project

16 Participation in the merit evaluation of key project personnel assigned to the project

The publication of suitable policy media describing the project manager's modus operandi and legal authority will do much to strengthen his or her position in the client environment.

Examples of De Jure Authority

In practice, we find varying degrees of de jure (legal) authority delegated to project managers. In the U.S. Air Force, a policy governing the use and application of project management is contained in a publication. In this publication the project manager's de jure authority is defined as follows:

> The program manager is the individual, military or civilian, responsible for managing all activity concerned with planning and executing the program. He is responsible for the success or failure of the program. His functional responsibilities are those common to top-level executives everywhere; that is planning, organizing, coordinating, controlling and directing.[30]

Other responsibilities of the Air Force program manager in carrying out the above role include:

[30]Air Force Systems Command Pamphlet 800-3, April 9, 1976, p. 20-10.

1 Managing the collective actions of all participating organizations
2 Executing the program in accordance with the program management plan
3 Being the contact point for communications with contractors and Air Force organizations
4 Being the spokesman for the program
5 Initiating changes to the program consistent with the authority and directions from higher headquarters

In the final analysis, the program manager is responsible for the total system program while holding subordinates responsible for specific tasks or objectives.

The aerospace industry has contributed significantly to the evolving project-management thought and theory. One of the contributions that individual companies have made is in the creation of policy documentation which establishes the modus operandi and authority of the project manager. For example, in the military communications department of the General Electric Company, the project managers are appointed by the department general manager to manage a given program (or a group of closely related programs) on a full-time basis. Appointment of these individuals is done in the preproposal, proposal, or contract stage of a program. Although project managers may be located in various departments of the company, it is expressly intended that project managers represent the general manager on all facets of projects, regardless of where they may report. Each project manager has the following documented responsibilities and authority:

> **1** He will act for the Department General Manager in providing guidance to all elements of the Department on any and all aspects of the Program(s).
>
> **2** He is authorized to represent the Department to the customer on all matters concerning the Program (but with the exception of the negotiation of strictly contractual terms and conditions which will continue to be the full responsibility of Marketing). This is to be done through established Marketing channels including coordination . . . where appropriate.
>
> **3** He will define the requirements of the program (for contracts, these will be consistent with contractual terms) in relation to each element of the Department. He will define the work to be performed by each functional element in terms of cost, delivery, and performance. During the life of the contract, he will be responsible for auditing the relationship of these factors. If any of them deviate in such a way that the overall program commitments will be jeopardized, he will immediately make this fact known to the Functional Managers involved and the General Manager.
>
> **4** He will utilize the services of the Financial Section for all program financial information and will obtain all other services from appropriate functional components.
>
> **5** He will review manning of the program by each element of the Department. If in the Program Manager's opinion manning is not adequate or is excessive in relation to the task to be performed, he will immediately notify the Functional Managers involved and the General Manager.
>
> **6** He will determine types and details of various management controls and reports

to be used at program level, and insure that they are consistent with Department policy. (Functional elements will contribute to these controls and reports as requested; may utilize other techniques within their functions as are currently in use, but will not generate any other information at program level.)

It is the intent that individual Program Managers will integrate their Program reporting methods to achieve uniformity between programs to the maximum extent feasible and to facilitate aggregating.

7 He will schedule and hold a design concept review(s) covering all aspects of the design concept at a very early stage in the proposal or contract. He will determine the complete agenda and will determine who will make the presentations. He will recommend attendance by reviewers and will solicit additional recommendations for reviewers from Section Managers concerned. Those he selects will participate in the review and take any follow-up action as may be requested or appropriate.

8 For contracts, he will schedule and hold at least one design review at appropriate stages of the program covering all aspects of the design. He will determine the complete agenda and will determine who will make the presentations. He will recommend attendance by reviewers, and will solicit additional recommendations for reviewers from Section Managers concerned. Those he selects will participate in the review and take any follow-up action as may be requested or appropriate.

9 He is authorized to approve or disapprove all changes in the program, changes in planning, changes in specifications, major personnel changes, etc. He will normally establish change control procedures for administration through normal functional channels, and monitor such activity for compliance with policies, but may enter directly into the approval routine at his discretion.

10 For proposals, he will schedule and hold at least one pre-cost meeting with regard to costing and will establish guidelines for the program costing.

11 For contracts, he will control the dollar resources and will allocate appropriate amounts to each element of the Department through requisitions issued at his direction by Contract Administration and through Shop Orders issued at his direction. He will determine the degree to which the program is to be segregated by Shop Orders, consistent with the PERT [Program Evaluation and Review Technique][31] controls and with existing Department Policy and machine accounting capability. Cost performance of each functional manager will also be reported.

12 For contracts, he will schedule and hold expected cost-performance reviews at the initiation of each contract and actual and projected cost-performance reviews at significant stages of the contract. He will schedule these reviews at the beginning of the programs and publish the schedule to all concerned.

13 He will be chairman of a make-or-buy and source-selection board. He will constitute the(se) board(s) in close cooperation with functional managers and as approved by the General Manager.

14 He is authorized to make decisions consistent with Department Policies and Procedures for any aspect of this program including all functional areas. Where such decisions would appear to be in conflict with the overall objectives and goals of functional elements the Program Manager will pursue resolution of the matter up to and including the functional Section Managers. If concurrence cannot be achieved at this level, they will bring the matter to the attention of the General Manager. Pending

[31]PERT, a well-known project planning and control system, is discussed in Chapter 15.

resolution, however, the Program Manager's decision will be followed. The Program Manager will, of course, delay implementation of the decision until resolution is achieved in all cases where in his opinion the time schedule permits.

15 He will be responsible for the preparation of a monthly Report.

16 He will report regularly the status of the program to the Department General Manager and will bring any unusual items to his attention as they occur.[32]

The foregoing examples of project-authority documentation are examples of the care taken to delineate the legal position of the project manager. This constitutes an obvious source of power in the project environment. While this gives project managers the right to exercise that power, the significance of authority under the deliberate project-functional conflict cannot be understated. While project managers may have the final, unilateral right to order affairs in the project, it would be foolhardy for them to substitute their views without fully considering the "crystallization of thinking" of the other participants in their project. Project managers rarely hope to gain and build alliances in their environments by arbitrarily overruling the other managers who contribute to a project. They may not have the authority for such arbitrary action. Even if they did, the resolution of conflict would be better handled in a different manner.

Conflict Management

Project managers frequently indicate that the effective management of a project involves an ability to effectively *manage conflicts* and disagreements which arise in accomplishing the project ends. Thamhain and Wilemon have studied the opportunity for conflict in the management of a project. Approximately 150 managers from a variety of technology-oriented companies participated in their study.

The study involved three varieties of measure: first, the average intensity of seven *potential conflict sources;* second, the intensity of each of the seven conflict sources in the four project life-cycle phases; and third, the *modes used by the project managers for resolution of conflict.*

The seven potential *conflict sources* were:

1 Conflict over project priorities
2 Conflict over administrative procedures
3 Conflict over technical opinions and performance trade-offs
4 Conflict over personnel resources
5 Conflict over cost
6 Conflict over schedules
7 Personality conflict

The mean intensity for each of the potential conflict sources over the entire life

[32]General Electric Company, Military Communications Department Policy 51.1, Nov. 4, 1963. Subject: Program Manager.

of the project is presented in Figure 12-3. Thus, the three areas most likely to cause problems for the project manager over the entire project cycle are disagreements over schedules, project priorities, and personnel resources.

The prescriptions emanating from research results are summarized in Table 12-2. This table, according to Thamhain and Wilemon:.". . . provides an aid to project managers in recognizing some of the most important sources of conflict which are most likely to occur in various phases of projects."[33] It also incorporates strategies for minimizing the potential detrimental consequences of conflict.

[33]Hans J. Thamhain, and David I. Wilemon, "Conflict Management in Project Life Cycles," *Sloan Management Review*, vol. 16, no. 3, pp. 31–49, Spring 1975. Reprinted with permission.

FIGURE 12-3
Mean conflict intensity profile over project life cycle. (Hans J. Thamhain and David L. Wilemon, "Conflict Management in Project Life Cycles," *Sloan Management Review*, vol. 16, no. 3, pp. 31–49, Spring 1975. Used with permission.)

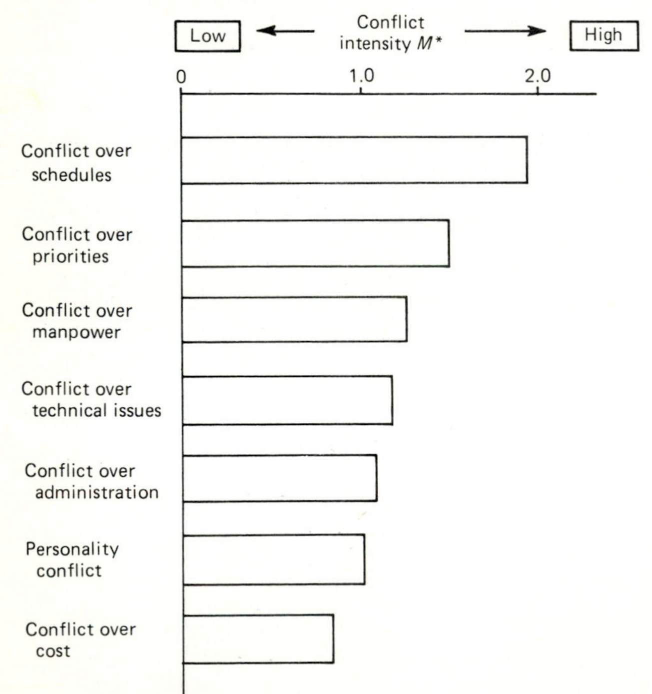

*M is the relative intensity of conflict perceived by project managers, measured on a four-point scale, 0-1-2-3, and averaged over the five sources (1) conflict with functional departments, (2) conflict with assigned personnel, (3) conflict between team members, (4) conflict with superiors, and (5) conflict with subordinates. Hence it follows $0 \leqslant M \leqslant 3$.

TABLE 12-2
MAJOR CONFLICT SOURCES BY PROJECT LIFE-CYCLE STAGE

Project life-cycle phase	Conflict source	Major conflict source and recommendations for minimizing dysfunctional consequences: Recommendations
Project formation	*Priorities:*	• Clearly defined plans. Joint decision making and/or consultation with affected parties. • Stress importance of project to goals of the organization.
	Procedures:	• Develop detailed administrative operating procedures to be followed in conduct of project. Secure approval from key administrators. Develop statement of understanding or charter.
	Schedules:	• Develop schedule commitments in advance of actual project commencement. • Forecast other departmental priorities and possible impact on project.
Build-up phase	*Priorities:*	• Provide effective feedback to support areas on forecasted project plans and needs via status review sessions.
	Schedules:	• Carefully schedule work breakdown packages (project subunits) in cooperation with functional groups.
	Procedures:	• Contingency planning on key administrative issues.
Main program	*Schedules:*	• Continually monitor work in progress. Communicate results to affected parties. • Forecast potential problems and consider alternatives. • Identify potential "trouble spots" needing closer surveillance.
	Technical:	• Early resolution of technical problems. • Communication of schedule and budget restraints to technical personnel. • Emphasize adequate, early technical testing. • Facilitate early agreement on final designs.
	Manpower:	• Forecast and communicate manpower requirements early. • Establish manpower requirements and priorities with functional and staff groups.
Phaseout	*Schedules:*	• Close monitoring of schedules throughout project life cycle. • Consider reallocation of available manpower to critical project areas prone to schedule slippages. • Attain prompt resolution of technical issues which may impact schedules.
	Personality and manpower:	• Develop plans for reallocation of manpower upon project completion. • Maintain harmonious working relationships with project team and support groups. Try to loosen up "high-stress" environment.

Source: Hans J. Thamhein and David L. Wilemon, "Conflict Management in Project Life Cycles," *Sloan Management Review,* vol. 16, no. 3, Spring 1975. Used with permission.

The conflict associated with the project management environment is not easily separated from the more generalized conflict found in any organization. One distinction in the matrix organization is the concentration of professionals with differing backgrounds on the project team. This team is a temporary organization; once the heterogeneous team members are working together, the project may go through a phase in its life cycle where new people enter the team, while some older members leave. This change may cultivate and amplify conflict.

The conflict is ubiquitous, but not necessarily dysfunctional. *What can be dysfunctional is an inadequate process for resolution of conflict.*

In process terms, while conflict can always (theoretically) be resolved by appeal to the common line supervisor, every effort should be made by the project and the functional managers to seek a successful resolution themselves. Management by exception is the rule! This resolution should consider (1) the impact on cost, schedule, and performance of the program and (2) the effect of the resolution on the total organization.

As might be expected in such an organizational setting, the opportunity for procrastination on controversial decisions is ever present, particularly if the decision might involve a showdown with the common supervisor. Yet timely resolution of significant differences is essential. *Suggested procedures* for effecting resolution should be based on the following:

1 Each project and functional manager will assume the initiative in taking actions which relate to his or her particular responsibilities and authorities. It should be the policy of the organization to emphasize the resolution of these disagreements through negotiation below the general-manager level. If, and only if, mutual resolution is not found quickly through such negotiation, and if the issue is a salient project-functional one, the project and functional managers should seek an audience with their mutual superior for resolution of the issue.

2 Before such an audience is given, each manager must clearly try to reach agreement on the following

a *The issue*

b *The impact* upon the project and the total organization

c *The alternatives*, examined together with the related costs and benefits of each

d *The recommendation* (as each sees it)

3 Further, each participant should be fully aware of the issues, their impact, and the alternatives and recommendations offered by the other. If this procedure is followed, it should reduce the opportunity for *personal* conflict and keep the conflict *organizational* and minimal in nature.

Conflict over issues and approaches in the matrix organization is to be expected, not as a detriment to organization efficiency, but, rather, as an anticipated matter, because people view problems and issues from the perspective of their loyalties. Deliberate conflict is not necessarily always bad or good; the conflict can become detrimental if it deteriorates to personal attacks and petty jealousies. "Deliberate conflict" is created by the *matrix* organizational structure and can be managed by the perceptive manager if it is put into proper organizational climate.

SUMMARY

Authority in the project environment flows horizontally, diagonally, and vertically. Technical competence, persuasion, negotiation, reciprocity, alliances, and the resolution of deliberate conflict are some of the means that project managers can use to augment their legal authority to accomplish project objectives. Thus, the effective authority of project managers is political as well as hierarchical.

Conflict in a matrix organization is intensified by the project-functional interface. Such conflict, properly managed, can facilitate the decision process in the management of the project.

Differences of opinion in the project-functional conflict can ultimately be resolved by the common line supervisor; however, the program manager's and the functional manager's credibility will ultimately be reduced by continued use of this alternative.

DISCUSSION QUESTIONS

1 The project manager who fails to build and maintain alliances with the project team and other project claimants will soon find indifference or opposition to the project requirements. Is this a valid statement? Why or why not?

2 Define authority. How does the concept of authority relate to the management of the family unit?

3 Each manager is in a dependent position with respect to the exercise of authority in the organization. What is the nature of this dependency?

4 Why is Chester I. Barnard's concept of authority so important in modern organizations?

5 What is the difference, if any, between power and authority?

6 The organizational position that the manager holds may be less important than the influence such an individual brings to the management process. Do you agree? If so, why do we not abandon the hierarchical form of organization and create a federation of coequals? Could we then manage the organization through the exercise of influence by those persons in charge?

7 Project managers often lament the lack of hierarchical authority in the management of a project. Is this a valid lamentation on the part of a project manager?

8 What is the objective of project authority? Is it important that all members of the project team support this objective?

9 A project manager may find it necessary to give direction to an executive who holds a higher organizational role. Is this not inconsistent with the idea of "unity of command" and "parity of authority and responsibility"?

10 In the exercise of authority, the project manager usually finds it necessary to manage several key interfaces in the organization. What are these interfaces? How can the project manager really manage these interfaces?

11 Describe the specific nature of the project-functional interface. Which managers in the organization should have a good understanding of this interface and a solid commitment to making it work?

12 What are some of the key factors to consider in delegating authority to a person who will serve as a project manager? Can all of the authority necessary for a project manager to function effectively be granted through the delegation process? Why or why not?

13 Purposeful conflict has been described as one of the cultural characteristics of the matrix organization. Describe what is meant by this statement.

14 Describe a process for reducing the number of times that a conflict between the project manager and the functional manager has to be resolved by the common line supervisor.

15 What are the differences, if any, between project authority and functional authority? From what source in the organization is the authority of the general manager and that of the project manager derived?

RECOMMENDED READINGS

Barnard, Chester I.: *The Functions of the Executive,* Harvard Business Press, Cambridge, Mass. 1938.

Barnes, Lewis B.: "Project Management and the Use of Authority: A Study of Structure, Role and Influence Relationships in Public and Private Organizations," doctoral dissertation, University of Southern California, 1971.

Cleland, David I.: "Understanding Project Authority," *Business Horizons,* Spring 1967.

———: "The Deliberate Conflict," *Business Horizons,* February 1968.

Drucker, Peter F.: *Management: Tasks, Responsibilities, Practices,* Harper & Row Publishers, Incorporated, New York, 1974.

Dunne, Edward J., Jr.; Michael J. Stahl, and Leonard J. Melhart, Jr.: "Influence Sources of Project and Functional Managers in Matrix Organizations," *Academy of Management Journal,* March 1968, pp. 135–140.

Etzioni, Amitai: "Authority Structure and Organizational Effectiveness," *Administrative Science Quarterly,* June 1959.

Gibbey, Lowell B.: *Project Management Authority in Matrix Organizations,* doctoral dissertation, University of California, Los Angeles, 1975.

Golembiewski, Robert T.: "Authority as a Problem in Overlays: A Concept for Action and Analysis," *Administrative Science Quarterly,* June 1964.

Goodman, Richard Alan: "Ambiguous Authority Definition in Project Management," *Academy of Management Journal,* vol. 10, December 1967.

Hodgetts, Richard M.: "Leadership Techniques in the Project Organization," *Academy of Management Journal,* June 1968.

King, W. R., and D. I. Cleland: "A New Method for Strategic Systems Planning," *Business Horizons,* vol. 18, no. 4, August 1975.

Kotter, John P.: "Power, Dependence and Effective Management," *Harvard Business Review,* July–August 1977.

O'Brien, James J.: "Project Management: An Overview," *Project Management Quarterly,* vol. 8, no. 3, September 1977.

Simon, Herbert A.; Donald W. Smithburg, and Victor A. Thompson: *Public Administration,* Alfred A. Knopf, Inc., New York, 1950.

Steiner, George A., and William G. Ryan: *Industrial Project Management,* The Macmillan Company, New York, 1968.
Stickney, Frank A.: *The Authority Perception of the Program Manager in the Aerospace Industry,* doctoral dissertation, The Ohio State University, Columbus, March 1969.
Thamhain, Hans J., and Gary R. Gemmill: "Influence Styles of Project Managers: Some Project Performance Correlates," *Academy of Management Journal,* vol. 17, no. 2, June 1974.
———, and David L. Wilemon: "Conflict Management in Project Life Cycles," *Sloan Management Review,* vol. 16, no. 3, Spring 1975.
Thompson, James D.: "Authority and Power in 'Identical' Organizations," *The American Journal of Sociology,* vol. 60, November 1956.
Thompson, Victor A.: *Modern Organization,* Alfred A. Knopf, Inc., New York.
"When Bosses Look Back to See Ahead," *Business Week,* Jan. 15, 1979.

CASE 12-1: Project Management Policies

Situation As an organization implements project management, certain policies will have to be developed to guide the operation of an industrial firm's *project management system.*

Your Task Develop a list of those policies that you feel are necessary to implement a *project management system* successfully within an industrial firm. In undertaking this task, you might want to consider such things as:

Need for clarification of existing policies
Application of life-cycle concept
Use of "matrix" organization
Definition of priorities
"Cost center" concept
Evaluation of project team members

What do you think these policy needs should include? Select a team leader to present your findings.

CASE 12-2: The Authority of the Project Manager

Situation Authority is the legal or rightful power to command or act. *Line* authority is the right, derived from a legitimate source, to act or direct. *Staff* authority is the legal authority of the staff official derived from that individual's appointment as a staff member to assist, advise, counsel, and support the line official to whom he reports. The exercise of staff authority is based on the philosophy of leading, within a framework of corporate mission and policy, by persuasion rather than by command. Responsibility is the obligation to act, direct, or command. Accountability is the condition of being answerable for one's actions or lack thereof in a position of authority and responsibility.

Authority, responsibility, and accountability are inseparable parts of the exercise of leadership in a project management position. When an organization implements the project-driven organizational matrix, certain actions must be taken to ensure that the project managers have an adequate understanding of and commitment to the effective discharge of authority, responsibility, and accountability in their project management roles.

Your Task Develop a list of those actions—or whatever—that are necessary to ensure the successful operation of project management in an organization from the viewpoint of authority, responsibility, and accountability. In doing this you might want to consider such things as existing lines/staff relationships, development of certain key policies/procedures, design of a strategy for changing the attitude of people, and so on. Select a team leader to present your findings.

CHAPTER **13**

PRESCRIBING ROLES IN THE MATRIX ORGANIZATION

Many matrix management problems can be traced to an important oversight—no one bothered to adequately clarify roles.

Individuals who hold managerial positions in the matrix organization operate through many organizational interfaces. Each project that is superimposed on the functional structure creates a series of such interfaces. A web of relationships is created as projects are initiated, pass through a life cycle, and are eventually terminated.

As more and more projects are added, general managers depend increasingly on the project managers to keep them informed, to manage their projects, and to coordinate action, as necessary, among the functional groups within the organization.

The general managers recognize, and even create, a "purposeful conflict" between project managers, on the one hand, and functional managers, on the other, as a means of evaluating relative trade-offs for the time, cost, and technical parameters of projects. The executive expects the project and functional managers to resolve daily operating problems among themselves and to bring only major unresolved questions to the executive level. Management by exception is the objective.

Rarely do project managers find that the project activities are limited to their own organizations; they must usually work with clients (or contributors) outside the company. They therefore have superior knowledge of the relative roles and functions of the individual parts of a project, which makes them logical persons to take part in major interorganizational decisions affecting the project.

The role of the project manager varies from one organization to another. In some cases, these managers have influence and responsibilities cutting across many of the organization's semi-independent units. In other cases, they operate merely as project representative for the several functions within their own facilities. In either case, their position is at the focus of an organizational relationship with unique work patterns.

Whatever their specific role may be, the essence of the matrix organization and the project manager's job is that of existing at, and exerting control over, *organizational interfaces.* The term "interface" is defined as: "A surface forming a common boundary between adjacent regions."[1] Here the word "interface" is used in an organizational systems context as "a coming together at common boundaries of interest." With the matrix organization, the project-functional interface deserves special consideration, although there are many such interfaces that are crucial to the success of the matrix organization.

Stuckenbruck comments on the complex organizational and project interfaces which must be managed in the matrix organization:

> These interfaces are a problem for the project manager, since whatever obstacles he encounters, they are usually the result of two organizational units going in different directions. . . . [2]

According to Stuckenbruck, three types of interfaces exist. These are:

> **1** Personal Interfaces—These are the people interfaces within the organization which are working to carry out his project.
>
> **2** Organizational Interfaces—Organization interfaces are the most troublesome since they involve not only people but also varied organizational goals, and conflicting managerial styles and aspirations . . . and
>
> **3** System Interfaces—System interfaces are the product, hardware, facility, construction, or other types of non-people interfaces inherent in the system being developed or constructed by the project.[3]

One might then describe the project manager's problem as that of "interface management."

THE PROJECT-FUNCTIONAL INTERFACE

In a matrix organization, the project organization and the functional organization are inseparable; one cannot survive without the other. Much of the confusion and concern about project management centers around a failure to

[1] *The American Heritage Dictionary of the English Language,* Houghton-Mifflin Company, Boston, 1976, p. 683.

[2] Linn C. Stuckenbruck, "Project Manager—The Systems Integrator," *Project Management Quarterly,* vol. 9, no. 3, September 1978.

[3] Ibid.

understand the complementary nature of the formal authority and responsibility relationships found in the project/functional, or matrix, organization.

Projects or task forces are essentially *horizontal* in nature; bureaucracy, the embodiment of the classical functional organization, as exemplified by the traditional organization chart, is *vertical* in nature. The basic dichotomy found in matrix organizations centers around a project-functional interface reflected in Table 13-1. The syntax of the statements in that table is somewhat peculiar, to provide a simple set of key words, as indicated by italics.

The "interface" described in Table 13-1 is a very broad way of portraying the authority and responsibility relationships in the matrix organization which can be used as a point of departure to develop an understanding of the *web of relationships* found in the matrix organization.

The interface clearly describes how project managers accomplish project ends by the managing of relationships within the total organization. There are few things which project managers can do alone. They must rely on the support and cooperation of other people within the organization. They must look to functional managers for specific support. Indeed, project managers *gets things done by working through others* in the classic sense of that phrase, which is often used as a definition of successful management.

This managing of organizational relationships is three-dimensional. Upward, project managers must relate to their boss, who is either a general manager or a *manager of projects*. Horizontally, they relate to members of their project team. Diagonally, they relate to functional managers and to representatives of other organizations—e.g., the customers.

Managing these sets of relationships is a most demanding task. It is nearly impossible, if care has not been taken to describe the formal authority and responsibility relationships that are expected within the organization. This means making explicit the network of relationships that project managers have in each of the three dimensions. To whom do they have to relate? What are the key relationships? What is the work breakdown structure around which action is expected? Who works for whom? In other words, a systems view of an organization is essential for them and for others to understand their role.

The authority and responsibility patterns can be effectively established along

TABLE 13-1
THE PROJECT-FUNCTIONAL INTERFACE

Project manager	Functional manager
What is to be done?	*How* will the task be done?
When will the task be done?	*Where* will the task be done?
Why will the task be done?	*Who* will do the task?
How much money is available to do the task?	*How well* has the functional input been integrated into the project?
How well has the total project been done?	

the lines portrayed in Figure 13-1. In this figure, the relative roles of the principal managers are indicated.

Figure 13-1 represents the central model governing the management of the project-functional interface. As such, it is one of the most significant pages in this book, and you should study it thoroughly.

Rather than spending many paragraphs discussing the figure, we shall discuss in detail the various organizational roles that are dictated and/or suggested by the depiction of the project-functional interface in Figure 13-1.

DEVELOPING ORGANIZATIONAL ROLES AROUND THE PROJECT-FUNCTIONAL INTERFACE

Once a need for a project model has been selected, the need is established for the development of the project organization. Selecting and defining a project organizational model requires an understanding of the work and the kinds of activities needed to reach the project's objective. The project manager must visualize the total project environment in selecting an organizational philosophy. This process of visualizing the total project environment will require an identification and alignment of the relative roles of all the project clientele. The project participants include any individuals or assembly of individuals in the project office, in the functional organizations, or in an outside organization—in short, clients who have an interest in the project affairs. Thus, this concept of the project environment is the same as the concept of the clientele of the overall organization.

In performing the function of planning, the project manager defines the tasks and suggests organizational alignments so that the members of the organization can build, develop, and maintain a structure and process of working relationships to accomplish the project objectives. The broad matrix organization becomes the structural and process framework through which all the project efforts are coordinated and integrated into the common objective. The project organization may therefore not be thought of as an independent entity operating in a vacuum; it is part of a larger organizational system.

The Project Manager

There is a great deal of similarity between a project manager and a general manager. Both deal with problems and opportunities that cut across organizational lines. A general manager's policies and practices are implemented by a project manager. It is helpful for general managers to realize that project managers are an extension of themselves and accordingly provide guidance to the project managers so that part of the project can be managed in accordance with organizational policies. Project managers should also have a general-manager outlook. They are concerned with solving problems through the making and execution of decisions in the accomplishment of a systems-oriented objective.

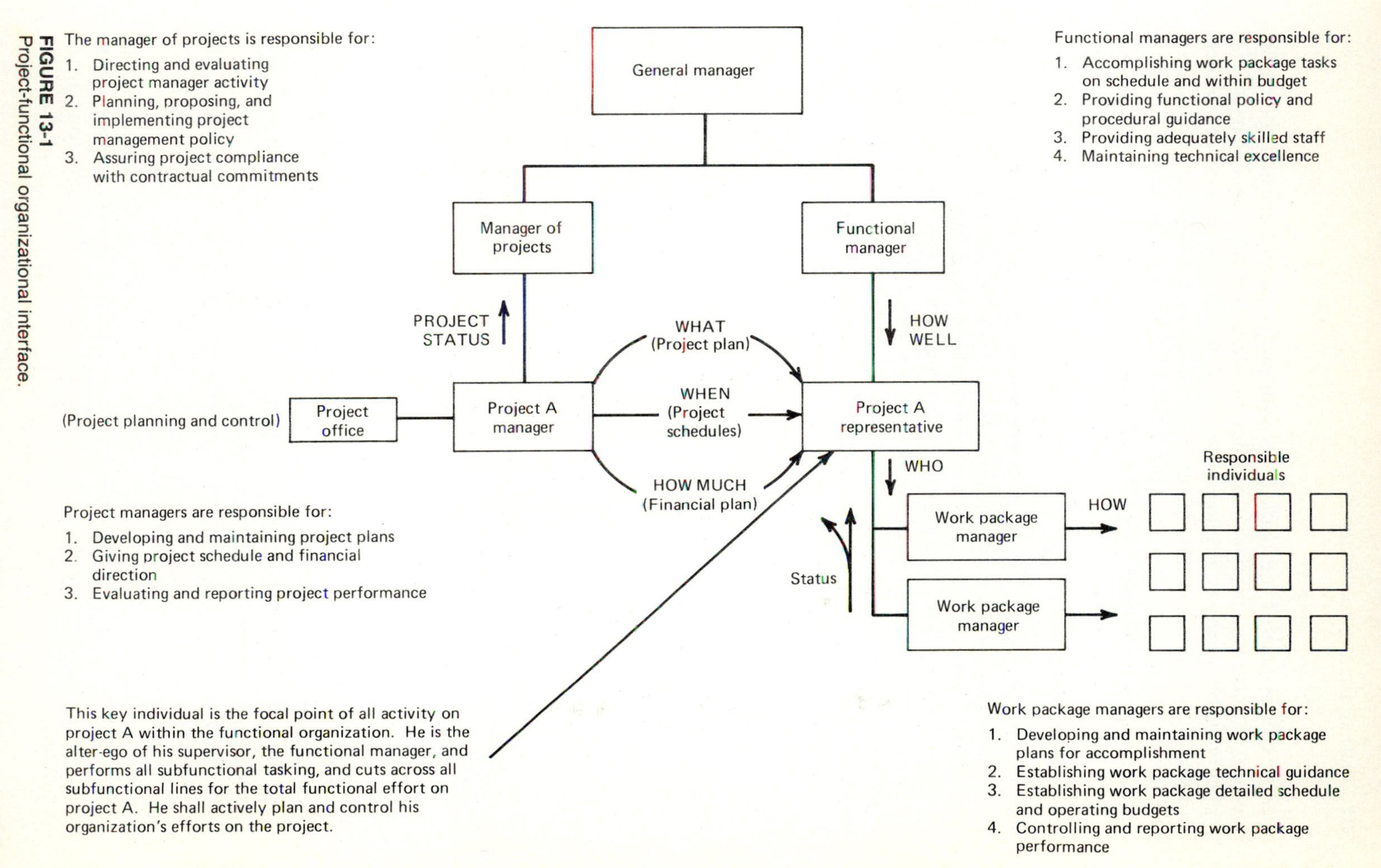

FIGURE 13-1
Project-functional organizational interface.

Most of this part of the book deals with project managers and their role. Because of this, and because the roles of other organizational participants implicitly involve project managers, we shall not explicitly treat the project manager's role in detail in this chapter. Instead, we shall focus primary attention here on the roles of the other "actors" which must be aligned and coordinated if a project-management model is to be effective.

The Project Team

The most effective project organization would be developed by assigning a project manager with perfect qualifications and giving this person clear-cut authority and responsibility. This situation would be ideal, but it is not realistic. The best manager available should be assigned, and the necessary human resources for support should be provided. To select the appropriate people requires that the overall project be divided into subtasks, and so on, until the project is represented by an alignment of rational, related, recognizable work units. This dividing should be accomplished by the project manager in collaboration with the functional managers who will be supporting the project. For example, in the development and production of a large project such as a ballistic missile, the makeup of the team to support the project activities would include functional groups with the responsibility for:

1 Developing the general requirements of the project, i.e., establishing the overall project strategic plan.

2 Preparing the *general* and *technical* specifications of the product or project. In this context, the general specifications express the customer needs and desires, while the technical specifications designate product parameters.

3 Fabricating, testing, and evaluating the prototype and production models.

4 Establishing the reliability, maintainability, and supportability requirements for the overall system and for individual equipments.

5 Identifying and selecting sources for the items of the project which are to be procured.

6 Negotiating and administering all contracts consummated with suppliers and participants in the project effort.

7 Developing, coordinating, and maintaining the master and supporting schedules required to produce the product within the time restraints.

8 Creating, publishing, and distributing technical manuals and reports required for the product in its intended operational environment.

9 Planning for, and subsequently installing, operating, and maintaining, the completed product.

10 Providing operational logistic concepts required to ensure the supportability of the product after it is produced. This group includes those individuals who are charged with the responsibility for providing a pool of spare parts, specialized supporting equipment, and trained personnel necessary to maintain the product.

11 Making a cost analysis to develop original costing and a method for continually tracking the project cost. This includes the financial management required to determine subsystem and system profitability and to develop a management information system for the project manager to use in maintaining surveillance of the schedule, cost, profitability, and technology of the project.

12 Establishing the product configuration and performance characteristics and tracking these characteristics throughout the project's life.

13 Providing technical leadership for the product line and maintaining the maximum state of the art in the project within the cost and schedule restraints. This is accomplished on the existing product generation, under the jurisdiction of the product manager, and is not to be confused with the overall R&D function to be carried out for the company as a whole.

14 Identifying and developing the personnel skills necessary to use the product.

For a project within an industrial firm, the representative elements of functional responsibility that might be integrated by the project manager include:

Engineering
Operations (Manufacturing)
Marketing
Financial management
Procurement
Product assurance
Contracts
Engineering operations

Because of the importance of the project-control function, project managers might have a staff of key individuals to assist them in this area, some assigned directly and some supporting them from the functional organizations. The remaining elements, however, are normally the responsibility of the appropriate functional managers. Each functional manager normally has assigned a key individual within the organization for liaison with the project manager and staff. This key individual, although reporting to the functional manager, represents the project manager within that functional organization, and is the focal point for the activity on that project within that organization.

One of this individual's main functions will be the *planning* of all the organization's responsibilities for the project, in consonance with the total project plan, and ensuring that all organizational effort satisfies the requirements of the project. He—or she—thus becomes the *alter ego* of the general manager for that project. His planning role is an important and continuous one, since he is required to maintain and update his portion of the total project plan as the project matures. His inputs and changes to the total project plan must be through the project manager, who ensures that all plans and changes thereto are in harmony and consistent with the total project objectives.

Depending upon the nature, size, complexity, schedule, phase, and potential of the project, certain specialized functional personnel may be organizationally assigned to the project manager for the duration of the project, or for as long as required to ensure successful passage through its most critical phases. The "borrowed" members of the organization who are assigned to the project represent their functional organization's responsibilities on the project, such as financial operations and contracts. These project-management organization personnel, who are provided by their functional home organizations on a loan basis, charge their time to the appropriate project, or overhead account, as specified by the project charter or as mutually agreed upon by the loaning (functional), borrowing (project), and financial organizations.

Assignment to a project team may be too parochial for people who wish to maintain their technical status and their identification with their technical reference group. These are important to the development of a technical capability in the organization, since the functional organization is where functional know-how is advanced, expertise is developed, standards of performance are established, and personnel and facilities assignments are made. The project staff can be drawn from the functional departments and are returned to them when the project or task is completed.

The Project Office

The size and composition of the project office depend on many factors. The size of the project, its importance to the organization, the degree of decentralization required, and the nature of the project (whether development or production) are some of the many factors that dictate the size and composition. The responsibilities of the project office include the direction, coordination, and control of all the project activities, wherever located in the supporting functional elements, as shown in Figure 13-1.

Criteria that give a hint as to the size and composition of the project office center around the responsibilities of the project manager in planning for and utilizing organizational resources to satisfy project objectives. Size and composition of the office should be such that the project manager is able to establish and maintain an effective line of communication between the customer-supporting organizations and the parent organization during the life cycle of the project.

In this connection, the collective responsibilities of the project office should make it possible to determine:

That the customer is advised of all tasks, schedules, and costs necessary to the success of the project.

That all tasks being performed on the project are authorized and funded by contracts or work-breakdown-structure documentation.

That deviations from the contract and the project plan and/or specifications are authorized by and communicated to all concerned parties.

That the customer and the supporting functional elements are continuously advised of project progress, as required by the contract and project documentation.

Given these collective responsibilities, the typical project-office size and composition will include the complement of skills necessary to be able to provide strategic project planning and control for the following representative categories of effort:

Business operations
Engineering
Manufacturing
Product reliability
Procurement
Integrated logistic support
Marketing and contracting

The point should be kept in mind that it is not necessary to bring into the project office the requisite skills for all this effort; these categories of skill are to be drawn from the supporting functional elements. On large projects, there may well be a requirement for people possessing all of these skills in the project office; however, they would report in a line capacity to the project manager and should have the necessary complement of skills to function as the *alter ego* of the project manager in providing strategic direction and evaluation of the project.

The Manager of Projects

The emergence of many projects in the organization and their importance to the long-term survival of the organization suggest the need for a separate *functional organizational entity* to facilitate the management of the projects. An individual heading such an organization may be called a *manager of projects* and is appointed at the same executive level as the major functional heads of the company. The organizational location of this position is illustrated in Figure 13-1. Projects that emerge in the R&D side of the organization will grow in maturity; at some point in maturity, these projects would come under the jurisdiction of the *manager of projects,* who is responsible for providing an environment whereby the stream of projects flowing through the company can be facilitated. The manager of projects is responsible for directing and evaluating all the individual project managers' activities, as well as planning, proposing, and facilitating the implementation of project-management policy. This manager is the focal point for ensuring project compliance with commitments and for advancing the state-of-the-art as far as project-management theory and techniques are concerned. Other specific activities of the manager of projects are as follows:

Provide an added "check and balance" to measure the consistency of emerging projects with the strategic objectives of the organization.

Work with other functional heads to allocate the projects and resources employed. Problems that arise concerning conflicting priorities among the projects can be resolved by the manager of projects and the appropriate functional head, thereby relieving the chief executive of this task and giving the chief executive time to think through the long-range strategy of the organization.
Resolve problems among the individual project managers concerning the allocation of resources.
Define and assist in the development and operation of project-management systems, such as information systems, technical-performance measurement techniques, and project-control measures.
Ensure management consistency among projects, and ensure that changes in one project are integrated with the cost, schedule, and technical performance objectives of all the other projects.

One real advantage of using a manager of projects is the opportunity it provides for the chief executive to delegate to the manager of projects much of the detail—resolution of conflict, routine allocation of resources, etc.—which exists in the project-oriented organization. If the chief executive becomes embroiled in too much of this activity, it can seriously detract from the time available for overall organizational strategic planning, an important element of responsibility. By placing a manager of projects in the organization, the chief executive simply recognizes that the project-management activity is a major thrust of the organization, just as finance, production, and marketing are also major thrusts of activity.

The Functional Organization

Functional organizations emerged out of worker specialization and the need to pool workers of common occupational activities into a cooperative effort. Thus, engineers are placed in an engineering organization, financial people in the finance department, etc. Functional organizations, according to Carlisle, reflect some weaknesses:

1 Functional organizations tend to emphasize the separate functional elements at the expense of the whole organization.
2 Under functional departmentation there is no group that effectively integrates the various functions of an organization and monitors them from the "big picture standpoint."
3 Functional organizations do not tend to develop "general managers."
4 Functional organizations emphasize functional relationships based on the vertical organizational hierarchy.
5 Functional organizations tend to fragment other management processes.
6 Functional organizations tend to be closed systems.
7 Functional organizations develop a strong resistance to change.
8 Functional segregation through the formal organization process encourages conflict among the various functions.

9 The emphasis on the various operating functions focuses attention on the internal aspects and relations of the company to the detriment of its external relations.[4]

While these weaknesses may be found in functional organizations, the primary purpose of the functional organization is to provide a pool of expertise that can be applied to the various projects in the organization. It should again be mentioned that the *project* and *functional* organizations are interdependent—one cannot survive without the other. One aerospace company highlighted this project and functional complementariness as follows:

> The project manager can be considered as the customer's representative as far as the acceptability of the finished product is concerned, and he is the general manager's representative as far as interdepartmental conflicts involving his project are concerned. Product performance and delivery must satisfy him. When either of these is jeopardized by an interdepartmental conflict, he provides the solution most acceptable to the customer. The project manager is the cornerstone of project finality.
>
> The functional manager is responsible for staffing and organizing a group which will have the technical competence to handle any project within its province. He will also be responsible for establishing the operating policies and procedures necessary for assuring the perpetuation of a healthy, efficient operation. The functional manager is the cornerstone of organizational continuity.[5]

There are difficulties in bringing together project team members drawn from different organizational functions and levels within the organization. Distance is often a complicating factor. The hostility of traditionally oriented functional managers who suspect an interference in their "territory" can be a problem. Functional managers often feel that the project could be handled better within functional units—preferably their own. Also, functional managers may feel that their authority is being undercut.

In the matrix organization, questions are often raised by functional managers as to why functional structure is retained at all if the project team approach is to be used. At times, this question is asked in a more subtle manner: "I find all I am doing as a functional manager is supporting the project manager; I don't have anything else to do." *This is most assuredly not true* and represents an excessively narrow view of the functional role in the matrix organization.

The responsibilities of functional managers for the overall maintenance of the base of technology in their disciplines still remains whether the organization is running on a matrix basis or not.

In addition, the use of a matrix organization and the formation of project teams can create an insecure feeling on the part of the team members who feel no security inasmuch as they are assigned to support the temporary project team. As one project reaches completion, the team members on that project may move on to some other project, often involving a different team. This can heighten the sense of insecurity.

[4]Howard M. Carlisle, "Are Functional Organizations Becoming Obsolete?," *Management Review,* January 1969, pp. 4–6.

[5]A. E. Roden, "A Case for Coexistence," *Machine Design,* Feb. 17, 1967, p. 149.

An individual in a functional element who serves on a project team may also be simultaneously serving in several different project teams. There can be a problem of role identity as far as this individual is concerned.

Of course, the ability to assign people as their talents are needed on the various teams that are working does provide flexibility for the organization, and it is a requirement for the teams to be successful. It provides for better use of personnel because then the project teams can be augmented with those people who have the skills that are needed at a particular point in time. Indeed, it may be wasteful for a team to be assigned to just one project, particularly if the members' time is not effectively used. At all times, and even when the project work is ebbing and flowing as it does, it is better to have the flexibility of moving people in and out of project teams as they are needed. It is during this period of flexibility and fluidity that a strong functional element as a home base for the team members is useful.

Functional managers should spend significant time in support of the personal and organizational well-being of their assigned people.

Another reason for retaining the functional structure is to maintain a sense of discipline within the "technology" that function provides, to keep track of the functional state-of-the-art, to track changes in the state-of-the-art, and to constantly provide the checks and balances to ensure that a wide base of functional expertise is maintained in the organization. In addition, some functions are best carried out on a centralized basis, such as general data processing and accounting.

Functional managers may themselves inordinately feel the impact of the matrix organization, inasmuch as their role appears to be dramatically changed. They may feel threatened by the matrix organization approach, sensing the risk of loss of status, authority, and control. Functional managers who have been very protective of their "territory" can become anxious when they see the project manager making demands on the functional community.

However, the role of the functional manager has, in actuality, changed very little. Functional managers continue to have a major responsibility in providing a strategic sense of direction *for their own disciplines* as practiced in the organization. This strategic direction has to be carried on regardless of how their people are assigned, whether on a matrix basis or on some other basis.

A permanent assignment base is needed for the matters of personnel administration, for such things as hiring and performance reviews, promotion, termination compensation, and any general need for "someone I can talk to" who is in the chain of command and with whom career opportunities and problems can be discussed. The functional element also provides a basis for professional association and career development, especially for those individuals who see themselves becoming an ongoing part of the professional, technical area on a long-term basis.

Thus, the role of the function is as strong in the matrix organization because functional managers have to maintain that broad technology base within the organization, provide a permanent home of assignment for the people who are working on the teams, maintain a sense of discipline in a way that technology is applied, and look out for the long range of their disciplines within the organization and in the organization's marketplace. These are all major responsibilities and are not changed at all by virtue of there being a matrix organization in existence.

The Project Engineer

The role of the project engineer is sometimes confused with that of the project manager. Typically, the project engineer exists to coordinate all the technologies and to ensure a satisfactory technical design of the system as an integral whole. The project manager embraces a much wider range of activities, being responsible for interpreting the strategic requirements and objectives of the project in such terms that it is compatible with the organizational mission. The project manager also controls expenditure of the funds and integrates reliability, maintainability, legal protection, safety, welfare, and other specialized aspects into the project. The project manager has a much broader direction and is concerned with the implementation of the organization's policy as applied to the project, which involves objectives, criteria, outlining and approving budgets, monitoring performance, making appointments, etc. The project engineer has a much narrower role.

The use of the two titles can, however, be ambiguous; in some organizations where projects are primarily technical in nature, an individual with the "Project Engineer" title functions as the project manager. In more complex situations, where both technical and nontechnical issues are involved, the two functions are quite separate.

Role of Top Management

A question that is often raised by the chief executive in an organization that is initiating a matrix organization is the matter of the chief executive's own role.

The general manager's basic role is to integrate the total organizational effort so that overall objectives are attained. Within a project-management context, the general manager delegates "project" authority to accomplish certain things in the organization to the project managers. The general manager has the responsibility for making it clear just what is expected from the project organizations. For instance, one general manager of an engineering organization published a policy document which contained a listing of "the general manager's expectations for project management in the organization" as follows:

Managing significant engineering projects, from initial scoping, through the design process, through successful introduction.

Developing engineering long-range strategies and plans that capitalize on opportunities presented by new product/process technologies to meet the defined needs of the business while achieving quality, safety, cost, standardization and product performance objectives.

Assuring that appropriate and uniform product specifications are established.
Acting as the focal point for the planning, scheduling, coordinating, and measuring performance against objectives for assigned projects.
Developing creative approaches which will optimize product benefits and resource expenditures in accomplishing project objectives.
Recognizing problems at an early date and establishing corrective plans.
Structuring project information for project reviews, priority determination, and resource allocation.
Acting as the direct extension of the department general manager managing daily operating issues to provide an increase in time available to the general manager for advanced planning, upward communication, and corporate issues.

The chief executive's broadest responsibility is to provide a cultural ambience for the matrix organization to flourish and produce objective related results. More specifically, the chief executive's role can be delineated thus:

1 The chief executive should provide for a rigorous definition of the relative authority and responsibility of the principal managerial positions in the matrix organization. Commensurate with this definition should be provided an executive development project to improve the knowledge, skills, and attitudes of these managers. This program could consist of a series of workshops on the theory and practice of the matrix organization and how it is intended to operate in the organization.

2 The chief executive should develop and promulgate a philosophy on how resource priorities will be determined in the matrix organization and how conflict over these resources will be resolved.

3 Since a project manager is inherently a decision maker, the chief executive should define the parameters and limitations of the project manager's decision power. This definition should include the legitimate right of the project manager *to see that a decision is made* if someone other than the project manager has the right to render a decision. This decision should include the right to force an appeal of a decision if a conflict, as described above, exists.

4 Since projects are related to organizational strategies, the chief executive, in approving strategic plans for the organization, should prescribe the objectives and goals against which a project manager's and other managers' performance is to be judged. If the project is a profit center, the performance objective can be easily delineated. If the project is not a profit center, then some other performance standard has to be developed against which to judge the project manager's performance.

5 An important part of the top manager's role in the matrix organization is to recognize and plan for the development of a *project-management system.* Supporting subsystems such as planning and control are required to support the project. An information system is required which provides information to the project manager so that time, cost, and technical parameters of the project can be determined. A policy on project review must be established.

Thus, there are numerous facilitating services which support projects—

budget, finance, accounting, public relations, to name a few. The chief executive should ensure that these services are made available on a timely basis for the support of projects.

6 As an organization gains experience with project management, it will become apparent that project managers are unique in terms of fitting into the standards of the prevailing wage and salary classification schema. A project manager may have direct line authority over only one or two people in the project office, yet that manager may be responsible for working productively with many other clientele. Standards for salary level typically depend on the number and classification level of the people supervised. In establishing a project-management system, top management should consider a new salary classification. The amount of resources the project manager controls, the organizational location, and the complexity of the project are key criteria to be used in determining the project manager's salary classification code. Other factors include:

Duration of the project
Importance of the project to the company
Importance of the customer
Annual project dollars
Payroll and level of people that report to the project manager through the matrix organization
Payroll and level of people on the project manager's staff that report directly to him
Material dollars (both direct and indirect)
Complexity of the interfaces with the project clientele
Percentage of research, design, and manufacturing dollars by year
Executive level of the individual to whom the project manager reports
Corporate investment involved to sustain the project
Potential payoff to the corporation if the project is successful
New business opportunities that the project might create
Business and commercial analysis and experience which the project manager must have
Complexity of the project planning and control
Importance of an adequate monitoring progress and problem-solving system
Technical know-how required of the project manager
Maintenance of balance of power between the project office and the functional elements of the organization (See Chap. 12.)

The matrix organization's success depends a great deal on the support that is given by the top executives in the organization. These executives must be convinced of the merits of the matrix organization and understand how the organization works, particularly in its interfacing with myriad client groups. Top managers must be vocal in developing and defending the use of matrix organizational techniques. The role of the top leader is indeed critical, for this leadership will ultimately be reflected in the success of the matrix organization.

Other Organizational Roles

The organizational roles that we have been describing are major roles within the matrix organization.

Table 13-2 shows how one company defined the relative roles of the project manager, design manager, and design engineer in a design engineering department. One of the greatest values of having a document such as this is the dialogue and meeting of the minds that comes about in developing the relative roles. In defining these roles, usually long, and at times emotional, meetings are held. In this case, each participant had the opportunity to express a viewpoint. Several issues of the "Who does what?" variety had to be referred to the common line supervisor for resolution. Eventually, every participant had a "day in court" and the document contained in Table 13-2 was published as part of project-management policy of the engineering department. (You should note the importance placed on communication in the design engineering department —making it effectively the responsibility of all managers and engineers.)

ADVANTAGES OF THE MATRIX ORGANIZATION

The matrix organization that has been described here has many advantages over other organizational forms (as described in Chapter 10):

1 The project is emphasized by designating one individual as the focal point for all matters pertaining to it.

2 Utilization of personnel can be flexible because a reservoir of specialists is maintained in functional organizations.

3 Specialized knowledge is available to all programs on an equal basis; knowledge and experience can be transferred from one project to another.

4 Project people have a functional home when they are no longer needed on a given project.

5 Responsiveness to project needs and customer desires is generally faster because lines of communication are established and decision points are *centralized.*

6 Management consistency between projects can be maintained through the deliberate conflict operating in the project-functional environment.

7 A better balance between time, cost, and performance can be obtained through the built-in checks and balances (the deliberate conflict) and the continuous negotiations carried on between the project and the functional organizations.

8 Interfunctional competition tends to be minimized by the intervention of the project manager.

Within the matrix organization, many advantages of the functional organization are retained:

> Each functional manager supervises a pool of functional talent to fill the shifting needs of the various projects; he directs assignment and reassignment and arbitrates

TABLE 13-2
PROJECT MANAGEMENT/FUNCTIONAL RESPONSIBILITIES (DESIGN ENGINEERING)

What, When, How Much	**Who, How**	
Project manager	**Design manager**	**Design engineer**
• Lead scoping activity for proposed projects Inside Outside • Represent project objectives to contributing functions Cost Schedule Performance • Lead project teams Project Definition Project Schedules Guidance & Direction • Ensure allocation of supporting resources • Integrate and communicate project information Corporate Product Board Outside • Track and assess progress against plan • Resolve conflicts • Communicate	• Participate in development of project plans • Participate in determination of project resource needs • Define design workload • Assign design personnel, consistent with project needs • Maintain technical excellence of resources • Recruit, train, manage the people in organization • Assess quality of design activities • Provide technical guidance and direction • Communicate	• Prepare detailed plans and schedules for design tasks consistent with overall project plan, including initial definition of support requirements • Execute design tasks • Represent design engineering on project teams, in project reviews, for cost estimates, etc. • Communicate

conflicting project demands; and he is in a position (because of the employee's long-term career dependence on his functional manager) to promote adherence to desired technical standards.[6]

Of course, there are some disadvantages to a matrix organization. The balance of power between the functional and the project organizations must be watched so that neither one erodes the other. The balance between time, cost, and performance must also be continually monitored so that neither group favors cost or time over technical performance.

Each of the forms of organizational structure has certain advantages, but none can be considered best for all applications. The form to be used depends on the environmental requirements, which change continually as the project goes through its life cycle and as the number of projects and the product mix of the

[6]J. Gordon Milliken and Edward J. Morrison, *Aerospace Management Techniques: Commercial and Governmental Applications,* Denver Research Institute, University of Denver, Denver, November 1971, p. 108.

company change. The organization must be changed as the environment changes. Rotating personnel among projects and functions can be a valuable technique in executive development. Individuals gain perspective in the project and functional ways of thinking; they develop an understanding of the other person's problems.

Changes in organizational structure are often necessary and can have a more far-reaching effect than merely rotating people from one position to another. Structural changes sometimes affect the human element adversely, however; they strike at the core of human motivation—status, security, acceptance. Informal working arrangements can be broken up and morale damaged, to the extent that the general efficiency of the organization declines. Regardless of these problems, change is necessary, and flexibility is desirable.

SUMMARY

Many managers and professionals work together at common boundaries of interest in the matrix organization. If their roles are properly delineated and effectively communicated throughout the organization, the likelihood of a favorable outcome of project management is enhanced. The chief executive has a major part in prescribing organizational roles and facilitating a cultural ambience to foster successful project management. The project-functional interface is the critical boundary of interest in the matrix organization. Properly defined and understood by the managers and professionals, this interface is the heart of the project-management system.

DISCUSSION QUESTIONS

1 There are different key roles to be carried out in the project-driven matrix organization. Identify and describe the characteristics of such key roles.
2 There are many different roles that a project manager can play in an organization and still have the focal responsibility for a project. What are some of these roles?
3 What are the differences between personal, organizational, and system interfaces?
4 Why might the project manager's role be described as interface management?
5 Interface can be described as the coming together at common boundaries of interest. Why is an understanding of Figure 13-1 important for the key managers of an organization engaged in project activities?
6 A project team can be compared to a football team. Do you agree? What are some of the common characteristics?
7 Most project managers would rather have the members of the project team assigned directly to them in an administrative reporting sense. Why do project managers tend to feel this way?
8 One of the authors has often said in his lectures, "The project manager's office ought to be small and uncomfortable." Why do you think he makes this point?
9 Under what circumstances might it be advisable to assign a manager of projects in the organization?
10 What are some of the organizational activities that a manager of projects can

discharge? A profit-center manager might view the role of manager of projects with some skepticism. Why?

11 What are some of the weaknesses of the functional organization? Some of the strengths?

12 What are some of the responsibilities of the functional manager, in addition to providing support to the project requirements?

13 Success or failure in project management often depends on the attitudes of senior executives of the organization. Why might the attitudes of these executives cause project management success? Failure?

14 What are some of the considerations in determining the salary classification of the project manager? Why are traditional salary classification techniques not always useful?

15 Take the interdependent roles of a project manager, a functional manager, and a professional in a matrix organization. Describe the nature of their interdependency in the context of authority, responsibility, and accountability.

RECOMMENDED READINGS

Carlisle, Howard M.: "Are Functional Organizations Becoming Obsolete?," *Management Review,* January 1969.

Davis, Stanley M., and Paul R. Lawrence: "Problems of Matrix Organizations," *Harvard Business Review,* May–June 1978.

Labovitz, George H.: "Managing Conflict," *Business Horizons,* June 1980.

Milliken, J. Gordon, and Edward Morrison: *Aerospace Management Techniques: Commercial and Governmental Applications,* Denver Research Institute, University of Denver, Denver, November 1971.

Roden, A. E.: "A Case for Coexistence," *Machine Design,* February 17, 1967.

Sears, Woodrow H.: "Conflict Management Strategies for Project Managers," *Project Management Quarterly,* June 1980.

Stuckenbruck, Linn C.: "Project Manager—The Systems Integrator," *Project Management Quarterly,* vol. 9, no. 3, September 1978.

Vickers, G.: "Management of Conflict," *Futures,* vol. 4, 1974.

Waterman, Robert H., Jr.; T. J. Peters, and J. R. Phillips: "Structure Is Not Organization," *Business Horizons,* June 1980.

Wilemon, David L., and John B. Cicero: "The Project Manager—Anomalies and Ambiguities," *Academy of Management Journal,* vol. 13, 1970.

Wright, Norman H. Jr.: "Matrix Management: A Primer for the Administration Manager," *Management Review,* June 1979.

"When Bosses Look Back to See Ahead," *Business Week,* Jan. 15, 1979.

Case 13-1: Project Manager Candidate Specifications

Develop a profile of a project manager's ideal *characteristics* and *modus operandi.*

Situation As a result of "the need for enhanced effectiveness and productivity," project management is being introduced in your organization and you have just been named manager of projects.

Your vice president has informed the entire organization that all managers should make their people available (to you) to staff the Project Management Section. The vice president also stated that, as manager of projects, you will determine and communicate *specifically* what you are looking for (background, experience, personal style, future potential, energy level, human resources capabilities, etc.). You agree with your employee relations manager's suggestion that candidate specifications should be documented and must be easily communicated and "defendable," meaning obviously related to the work which must be done.

In developing the list of specifications, it has occurred to you that since the work of project management is to complement the contribution of other functions, then specifications for these functions might also need to be reconsidered.

Your Task After considering work content, define a set of candidate specifications for project manager positions, to be used in staffing your new Project Management Section. Be prepared to present your viewpoint.

CHAPTER **14**

PROJECT CONTROL

The best laid schemes o' mice and men . . .[1]

In managing a project, the cost, scheduling, and technical performance must be controlled throughout the life cycle. Control of the project requires that adequate plans be formed, suitable standards developed, and an information system set up that will enable the project to be monitored through comparisons of planned performance with actual performance. When the inevitable deviations of actual performance from planned performance occur, corrective action can be taken to realign the project resources in order to "get back on track."

This view of establishing a project plan from which it is anticipated that deviations will occur is an integral part of systems-management philosophy. You establish the best possible plan and hope that it will be fulfilled to the letter. However, since you cannot control all of the forces that impinge on project performance, there will be unexpected deviations from even "the best of all plans." When these "anticipated but unpredictable" events occur, the key to project success is to detect them quickly and to take control action to get the project reoriented toward its goals.

This interdependence of planning and control is exemplified by the experiences of the Fluor Corporation, a construction firm with an annual sales volume of over $2 billion.

Within Fluor Utah, Inc., after hundreds of projects had been completed, it

[1]Robert Burns, "To a Mouse," stanza 7, 1785.

became apparent that many projects successfully achieved their objectives while others did not do so well. The history of all of these projects was carefully reviewed to identify conditions common to successful projects, vis-a-vis those conditions which occurred on less successful projects. *A common identifiable element on most successful projects was the quality and depth of early planning by the project-management group. Execution of the plan, bolstered by strong project-management control over the identifiable phases of the project, was another reason why the project was successful.*[2]

CONTROL SYSTEMS

The description of a control system calls to mind our earlier definition of a system as a collection of interrelated and interdependent parts, or subsystems, collectively seeking an objective or goal. In a *control system,* a number of elements are found:

1 An *objective* or function, which is the purpose of the system
2 *Inputs* consisting of information, material, or energy
3 *Outputs,* which are the end results of the conversion of inputs
4 A *sequence,* which is the precedence of actions for converting inputs into outputs
5 *Resources,* both human and material, which assist in the conversion of inputs into outputs
6 *Feedback* loops, which permit the adjustment of inputs and outputs to accomplish systems objectives
7 An *environmental setting* within which communication can be effected to responsible officials to reallocate resources

The term "control" carries with it negative connotations of external wills being imposed on a system. However, this is not typically the case. For instance, the control tower and the air controller serve to facilitate aircraft operations in ways which do not severely impinge on the authority of pilots to command their aircraft. Pilots accept the use of a control system because it avoids confusion, ensures safety, and makes their job easier. They accept such a control system, apart from the logical need for a system, because they understand it and are active in the operation of the system.

To extend this illustration, consider the plight of pilots in operating aircraft if they did not understand the air traffic control system and procedures. Yet it is not uncommon to find managers, project engineers, or workers, who really do not understand the control system that interfaces with them and do not see their

[2]Robert K. Duke, "Project Management at Fluor Utah, Inc.," *Project Management Quarterly,* vol. 8, no. 3, September 1977. See also Walter McQuade, "Bob Fluor, Global Superbuilder," *Fortune,* Feb. 26, 1979, pp. 54–61.

role in the larger system. Good communication and involvement are just as essential for the successful operation of a project-control system as they are for the operation of an air traffic control system.

Defining Control

The defining of "control" varies, depending on the specific function or area to which it is being applied. In describing the role of control in management, the emphasis may be on

1 The organic function of management, e.g., the control elements of supervision
2 The means for regulating an individual or organization, such as the specific tools a comptroller employs
3 The restraining function of a system (control is the objective of feedback and is defined as the monitored state of a system)

In complex organizations, there are at least two levels of control. Anthony, et al., refer to these as "management control" and "operational control."[3] *Management control* is the process by which it is ensured that activities and resources are directed toward the efficient and effective accomplishment of objectives as specified in the planning process. *Operational control,* on the other hand, is "the process of assuring that specific tasks are carried out effectively and efficiently."[4]

Thus, the "management control" level actually involves both planning and control, since it incorporates the allocation and reallocation of resources, if necessary, in order to maintain progress toward the goal. The "operational control" level is more concerned with the assessment of specific task accomplishment. *Both* play a role in project control.

The Need for Control Systems

Managerial controls, whatever the level, provide the project manager with the tools for determining whether the project team is proceeding toward the planned objective. *Control systems* advise the project manager of the extent of deviations and of the recommended corrective action or alternative course of action which will put things on course.

The need for the development of a *project-control system* arises out of the basic need for control of the activities of the various organizational units performing on the project. Projects emerging as technological ideas are the basic

[3]R. M. Anthony, J. Dearden, and R. F. Vancil, *Management Control Systems,* Richard D. Irwin, Inc., Homewood, Ill., 1965, chap. 1.
[4]Ibid., p. 7.

building blocks of organizational strategy, and they therefore provide a convenient and workable point from which to control the organization's application of resources. Circumstances which prompt the design of project control include such factors as cost overruns, the necessity of dealing with a large number of projects ranging widely in value and complexity, and the requirement for keeping top management apprised of the status of projects which are in, or face, difficulty. As a corollary, project managers will each tend to better assume responsibility for the outcome of the project if an organizational control system exists which provides organizational-wide visibility over the project, and if they feel that they have the critical information available which puts them in command of the project's situation.

Butler suggests that project-control systems need to be more sophisticated than those to which the functional executive has been accustomed.[5] Because of the dynamic nature of projects and their finite life-span, such systems require the functional executive to predict and report expected work difficulties in terms of milestones set by the project manager, rather than reporting difficulties selected from events which have already occurred. Functional executives also tend to be vulnerable to trade-off decisions by the program manager which may not accord their function the appropriate weight in terms of functional optimization.

Prerequisites of a Control System

The sophistication of the control system depends on the complexity of the project and the ability of the participants to administer it. A simple project may require only a few indicators to determine whether or not it is progressing on schedule and within the desired cost and performance constraints. On the other hand, a major project will require an extensive control system that will identify and report many conditions that reflect its progress. Regardless of the complexity of the project, however, certain basic conditions must be met in order to have a workable control system:

It must be understood by those who use it and obtain data from it.

It must relate to the project organization; since organization and control are interdependent, neither can function properly without the other.

It must *anticipate* and *report* deviations on a timely basis so that corrective action can be initiated before more serious deviations actually occur.

It must be sufficiently flexible to remain compatible with the changing organizational environment.

It must be economical, so as to be worth the additional maintenance expense.

It should indicate the nature of the corrective action required to bring the project back into consonance with the plan.

It should reduce to a language (words, pictures, graphs, or other models)

[5]Arthur G. Butler, Jr., "Project Management: A Study in Organizational Conflict," *Academy of Management Journal,* March 1973, p. 93.

which permits a visual display that is easy to read and comprehensive in its communication.

It should be developed through the active participation of all major executives involved in the project.

Even in the most complex projects, these requisites should be adhered to. The understanding of users and the usefulness of data to them cannot be sacrificed merely because the project is large and complex.

If the control systems have a logical fit into the project-management system, the interdependency of control to the rest of the project activities will be sensed by the project team. The control system for a project is often centered in a project control room. Sometimes such a room is called the "war room." This is a room assigned to the exclusive use of the project where all data concerning the project are displayed, and where meetings to review project status can be held.[6] During these project review sessions, the interdependency of the control subsystems with the project management subsystems should be obvious.

A control system, to be effective, must anticipate and report deviations on a timely basis so that adequate time is available to institute corrective action before serious deviations happen. Flexibility is important to the effective operation of a control system; the system should remain compatible with the changing organizational environment. Performance estimates laid down at the beginning of a project should not be so hidebound that comparing *actual* with *planned* performance becomes meaningless. In an R&D project, adjustments of objectives will occur as the knowledge base on a project is broadened.

The operation of a control system should be worth its cost. A large project that is being monitored will require sophisticated computer support. A small project may best be controlled by a simple set of records and frequent communication with the project clients. The essence of any control system is to know what is going on, to know what to do if something appears to be going wrong, and then to do it.

Cost is a consideration obviously related to size and complexity. Professionals who design and operate control systems, if left to their own pursuits, will develop a control system that satisfies their objectives but not necessarily the objectives of the manager who requires the system. We know of a highly specialized R&D organization where exotic sophisticated air/vehicle systems are developed for the U.S. Air Force. The manager of this organization, a technologist of international fame, eschews any complex formal control system. He prefers to have in-depth face-to-face reviews with his staff and the project managers who are heading up the various development efforts. He uses seventeen basic ratios as data base points on the utilization of organizational resources to keep abreast of the status of projects currently being developed. Maintenance of this control system is relatively simple, basically consisting of

[6]See Charles C. Martin, *Project Management: How to Make It Work,* AMACOM, New York, 1976, pp. 207–215, for a description of a typical project control room.

keeping key time, budget, and technical performance status charts posted. The most distinctive features of this control system are its simplicity and the personal engagement of the key executives in its operation.

Corrective action, such as reallocating resources, reassigning people, approving an engineering change, or whatever is required to get the project back on course, is usually systems-oriented. An engineering change to correct a technical performance deficiency has reverberations throughout the project. Costs and schedules require revalidation, technical documentation must be changed and field engineering support reprogrammed, etc. The control system, by its inherent nature, will implicitly suggest what corrective action should be undertaken to complement the project plan.

An Illustrative Project Control System

The nature and complexity of a project-control system can be illustrated by one developed for the Erie Nuclear Power Plant project.[7]

In this project-control system, several major modules were prepared. First was a Project Identification System, a structure for identifying hardware and software components of the projects. The structure for this Project Identification System consisted of five groups of numerical characters. Group 1 contained the unit number; group 2 contained the system number and was common for all items; groups 3, 4, and 5 varied by item being identified. This identification system was used for such things as correspondence, budgeting activities, budget items, calculations, drawings, specifications, inquiries, requisitions, purchase orders, and vendor documents. To apply the Project Identification System effectively, the elements were incorporated as part of the plant. The Erie plant contained approximately 60 major systems, 300 subsystems, and 120 general contracts. These general contracts (covering items from several systems) and systems were defined numerically as follows

010–099 General Contracts
100–199 Civil/Structural Systems
200–299 Steam Supply and Associated Systems
300–699 Mechanical Systems—Balance of Plant
700–799 Instrument and Control Systems
800–899 Electrical Systems

By initially defining and identifying each general contract, the control system broke down and eliminated many of the common barriers to good communications. Consequently, in this project the first element of the project-control system was a common communication system.

The project plan coordinated the efforts of the client, architects/engineer,

[7]Paraphrased from Barry M. Miller and Charles D. Williams, "Management Action through Effective Project Controls: A Case Study of a Nuclear Power Plant Project," *1978 Proceedings of the Project Management Institute,* Los Angeles, October 8–11, 1978.

general contractor, nuclear steam system supplier, and turbine generator supplier. The scheduling approach chosen for this nuclear project consisted of developing a critical path method (CPM)[8] which contained all essential project activities from the beginning of the design through plant startup.

The project plan schedule provided several reports which were used by the project manager and others to monitor and gauge the project's progress. Some of these reports included:

1 A milestone report/bar chart, which was a list of approximately 35 activities that marked the completion of major work efforts.

2 Monthly analysis report, which was a written report prepared by the project planner providing a discussion of the most pressing schedule problems.

3 A work action report, which was a listing of all activities that had to be started, maintained in progress, or completed within a preset interval.

4 A procurement status report, which was a listing of the detailed procurement scheduling activities associated with the procurement specifications.

5 The primary equipment report, which provided a listing of major equipment activities such as drawings and delivery dates related to the nuclear steam system supplier and the turbine generator supplier.

6 A project cost-control program, which covered the preparation and maintenance of current and detailed cost information for the project.

7 A project estimate, which was an essential element providing for budget monitoring and cash-flow projections. This estimate was based on plant layout drawings, flow diagrams, instrumentation drawings, electrical one-line drawings, and equipment sizing.

8 Also within the project cost control program was a cash-flow system, which provided total integration with the project schedule, project budget, and field cost-accounting system. The field cost-accounting system enabled the accounting staff to maintain current accounting records, generate prompt reports, and develop unitized plant property records on the project. Typical computer reports available from the field cost-accounting system included construction ledger; commitment and invoice record; commitment and expenditure record; property record; and budget monitoring and property record.

Another aspect of the project cost-control program was the Budget Monitoring System. This system provided for an orderly system of records and reports, measurement of work accomplishments, and reports on unit work-hours, unit costs, and other factors, in order to develop a base of total probable expenditures. The Budget Monitoring System was maintained to a current status by continuing reviews of the detailed design for the project.

Even a brief review of the control system for a large project such as the construction of a nuclear power plant keynotes the complexity such systems can entail.

It also illustrates the "prerequisites" in a variety of ways. For instance, the

[8]This will be discussed in Chap. 15.

project schedule and CPM network[9] were developed *by* the participating parties, rather than being developed for them by outside consultants. This ensured that they understood the system's capabilities as well as its limitations.

Configuration Management and Control

As a second example of the complexity of project-control systems, we introduce the notion of "configuration management." Most people recognize the need to plan and control budgets, schedules, and even performance specifications for a system that is under development. However, in complex projects, the elements that must be controlled are much more varied and detailed than is generally understood.

Configuration control represents one such level of specificity. In the development of complex weapons systems, changes in the configuration of hardware and software reverberate through the system to cause problems with budgets, schedules, etc. Thus, such changes must be *directly* addressed and controlled. It is inadequate to attempt to effect control solely at a higher level.

"Configuration management" is the discipline which provides for the integration of the technical and administrative actions of identifying the functional and physical characteristics of a system (or product) during its life cycle. Configuration management provides for controlling changes to these characteristics and provides information on the status of engineering change/contract change actions.

The essence of a discipline of configuration management in a system comprises three major areas of effort: (1) Configuration Identification; (2) Configuration Status Accounting; (3) Configuration Control. These areas are briefly discussed below.

Configuration Identification Configuration Identification is the process of establishing and describing an initial system "base line." This base line is described in technical documentation (proposal terms, specifications, drawings, etc.). The identification function provides for a systematic determination of all the technical documentation needed to describe the functional and physical characteristics of items designated for configuration management. Configuration identification also ensures that these documents are current, approved, and available for use by the time needed.

The concept of a base-line system requires that the total system requirements and the requirements for each item of the system be defined and documented at designated points in the evolution of the system. An evolutionary life cycle of the system from paper study to inventory items is prepared to plan for development and production status and to permit changes in the scope of the system.

There must be a recognized and documented initial statement of requirements. Once stated, any change in the system's requirements must be document-

[9]This will also be discussed in Chap. 15.

ed so that the current status may be fairly judged for performance requirements. A base line is established when it is necessary to define a formal departure point for control of future changes in performance and design. Configuration at any later time is defined by a base-line model, plus all subsequent changes that have been incorporated. This base-line model provides a point of departure to manage future engineering/contract changes.

Configuration Status Accounting Configuration Status Accounting is the process of recording and documenting changes to an approved base line for maintaining a continuous record of configuration status of individual items that make up the system. Configuration status accounting also provides management information that gives visibility to actions that are required and to completed engineering changes. Status accounting will identify all items of the initially approved configuration, then continually track authorized changes to the base line.

Configuration Control Configuration Control is the process of maintaining the base-line identification and regulating all changes to that base line. Configuration control prevents unnecessary or marginal changes while expediting the approval and implementation of those changes which are necessary or which offer significant benefits to the system. Necessary or beneficial changes are typically those that (a) correct inefficiencies; (b) satisfy a change in operation or logistics support requirements; (c) effect substantial life-cycle cost savings, or (d) prevent or eliminate slippage in approved scheduling.

A *Configuration Control Board* may be established which will be an official joint agency of the project clientele to act on all proposed changes. The Configuration Control Board would be a body for recommending final decisions on engineering changes and for installing good engineering-change discipline in the system. This board can provide a single-point authority for coordinating and approving engineering-change proposal requests.

THE PERSONAL NATURE OF CONTROL

The previous emphasis of control systems tends to imply that control is a rather mechanical process of tabulating data and using the information to identify appropriate corrective actions. In fact, as any project manager knows, these may be the least important aspects of the overall control system.

Control is a very *personalized* activity within the organization—all control systems operate through people—and people differ in personality, skill and ability level, organizational location, motivation, emotional response, work assignment, etc. Control becomes progressively more challenging as the work becomes more complex; and, to a greater degree than ever before, people expect control to be subtle and considerate of them as individuals, not mechanical and impersonalized.

Control is dependent on the knowledge, ability, and motivation of the people in the organization. People seem to dislike the idea of being "controlled," for it carries an implication of inadequacy and resentment, particularly if someone from the outside reviews, evaluates, and critiques performance. Scientists and engineers are often cited as being especially sensitive about their prerogatives, particularly if they sense that the control procedures might damage their professional images. Too often, control systems fail, not because of a failure to collect and process information but because people who are the object of the control systems do not understand the total organizational benefits that can be gained.

In the operation of any system, individuals' perception of how well they are doing in supporting organizational goals and in accomplishing their personal goals can be clouded by what they want to believe and what they want to reject. The degree of self-control, objectivity, and the individual's commitment to the organizational goals influences the information going into the feedback loop and ultimately influences the corrective actions taken in the control system. Control systems often operate differently than managers desire because of the opportunity for people to:[10]

1 Resist or fail to comply with the full requirements of the control system

2 Manifest antagonism (often subtle in nature) to the control system and to the people who are responsible for administering the system (the resistance to staff agencies performing a data collection process for control purposes is well known)

3 Report erroneous or misleading information; hide the true facts of the situation; report only what is literally required and fail to support the spirit of the control system

4 Neglect to build bonds of trust and interpersonal confidence with subordinates, superiors, peers, and associates

Resistance and noncompliance can often be attributed to communication failures. Douglas McGregor speaks of the unintended consequence of managerial control systems which yield compliance to some degree, but in addition may yield:

1 Widespread antagonism to the controls and to those who administer them.

2 Successful resistance and noncompliance. This occurs not with respect to a few people but with respect to many. It occurs not alone at the bottom of the organization, but at all levels up to the top (and sometimes there also).

3 Unreliable performance information because of (1) and (2) above.

4 The necessity for close surveillance. This results in a dilution of delegation that is expensive of managerial time as well as having other consequences.

5 High administrative costs.[11]

Control is indeed a subtle function in organizations when the design and

[10]Paraphrased from Douglas McGregor, *The Professional Manager,* McGraw-Hill Book Company, New York, 1967, p. 118.

[11]Ibid., pp. 117–118.

operation of control systems must take into consideration the behavioral aspect of people. It is not our purpose to delve into behavioral science and correlate the abundant information that can be found there with control theory. There are, however, a few pragmatic suggestions that can be put forth that should make it easier for project managers to "keep control of the human element."

First, project managers should use linear responsibility charting techniques to develop an understanding of the myriad of authority and responsibility relationships that are woven throughout the project team. Once these formal relationships have been designed, they should be continually reviewed to ensure currency and acceptance. (See Chap. 11.)

Second, project managers should be missionaries for the project and frequently visit the customer and functional people who support the project. Functional managers are concerned with supporting a "stream of projects," so project managers cannot expect these functional people to come to them.

Third, project managers should work at building a climate of trust and interpersonal confidence with the customer's people, as well as with the functional people supporting the project. They must not surprise the customer; they must show by example that they will not betray a confidence.

Fourth, project managers must take time to learn about the "informal" organization and the role that such organization can play to influence productivity, morale, loyalty, and communication. The role that the informal organization can play can be most significant. Jucius and Schlender have said:

> Though not stated in any manual of organization or of office procedures, informal controls are nonetheless very significant in their own way. A person who transgresses the informal "folkways" of an organization can be effectively stopped from further progress in the company. Moreover, he will fail to get his daily work done with success because cooperation of various parties will not be forthcoming. The importance of these informal controls is seen in the warning that any newcomer to a business situation should "learn the ropes" before taking important action. This is merely a slang expression stating that besides the formal organizational structure, procedures, and executive channels of command, there are informal relationships of significance in every enterprise.[12]

Fifth, project managers cannot depend on the formal control system to keep informed of what is going on; they must develop communication with different levels of people who can provide valuable insight into what can be expected in the project's future. They can do so, in part, by making a habit of talking frequently with the key project participants on the question "How are things going?" In large bureaucratic organizations there are many levels through which information must pass, which levels introduce filtering, distorting, and disappearance of information.

In one large project of national prominence, the project manager said that he never depended on the formal information system to keep him informed. The

[12]Michael J. Jucius and William E. Schlender, *Elements of Managerial Action,* Richard D. Irwin, Inc., Homewood, Ill., 1965, p. 114. (See Chap. 17 for more details.)

chain of command through which the information had to flow and the "fine tuning" that was inevitably done at each level distorted the key information so much that it was not always credible. Instead of depending on the formal system, this manager called or visited his key supporting people frequently, to be briefed on the project. These visitations and telephone conversations often brought to light potential problems that required immediate corrective action. If this project manager had depended on the formal reporting system, valuable time would have been lost while these problems filtered up the chain of command.

OVERALL PROJECT CONTROL

Project managers will want to know several basic things about their project during its life, such as the status of cost, schedule, and technical performance parameters of the project. If there are engineering changes on the project, the status of these changes and the effect on other elements of the project are essential knowledge. The interrelationships of costs, schedules, and technical performance are very important—particularly if trade-offs are to be considered in the development and production of the project. Finally, project managers should have an assessment of how the customer evaluates the progress on the project. Having the customer participate in the project reviews is one way to gain this assessment.

One of the more distressing trends in project management is the tendency to overstress control. Various types of management information systems have been developed recently which can provide an abundance of data concerning the project. Many of the information systems extend through many management levels and require the project participants to report a large amount of data. Also, the number of groups seeking data is often large; in the case of a major project, there can be several dozen. This proliferation of requirements for information can generate a need for augmenting personnel assigned to the performance and monitoring groups, since the more time they spend providing data, the less time they have to perform their other tasks.

Another distressing trend is the current tendency to rely heavily on complicated, sophisticated management systems. There is a very real danger that project managers can become so preoccupied with the system that they fail to exercise enough personal management of a project. Control is of a personal nature, so it is important that project managers use control techniques that reflect their personality and are consistent with the complexity of the project. The use of network-based systems such as PERT [13] may be justified for large projects; less sophisticated methods may be adequate on smaller projects, however, where the interaction between the manager and technicians concerned is adequate to control the project. Nevertheless, the greater insights into the structure of the project which may be gained through the use of a project network device such as PERT may well warrant its use. This is particularly true of the initial stages of

[13] These topics are discussed in Chap. 15.

planning a project, when such insights may be of great value. However, the value of PERT techniques in the subsequent control phases of a project is questioned by Baker, et al., who found that PERT techniques contribute little success compared with other factors. In fact, PERT techniques were concluded to be overused, overdetailed, creating excessive control, and detracting from project effectiveness.[14]

When to Initiate Control

Control begins when the first germ of an idea for a project appears in the organization—a discernible effort which will require an expenditure of organizational resources (human effort, time, money, physical resources) and lead to an organizational objective. After a project has been officially recognized, it should be reviewed, initially and during its life cycle, to determine if further development work should be undertaken—or the project terminated.

Chief executives can consider control of a project in two senses: *first,* to ascertain if the project is progressing as planned; and *second,* to check out the compatibility of the project with the strategic direction of the organization.

Control continues throughout the life cycle of the project, as one of the organic functions of the project manager. Project control is the nerve center of the project and provides a framework for decision making, both as to what decisions *should* be made and which *should not* be made.

Project control is facilitated by the creation of effective documentation for use by the project team. In the next section, we suggest some workaday documentation that is useful in managing and controlling a project.

Project Documentation and Review

There are many forms of documentation for the management of a project. A myriad of policy instruments can be created to establish the objectives, provide operating guidance, delegate authority, and compare performance with standards.

Our intent here is to identify and briefly discuss those forms of project documentation that may be used to control the project properly. We shall not be concerned with the detailed documentation, such as the many records required to track the cost of a project, but rather with the documentation that the project manager needs to provide overall control of the project, as opposed to control of a specific functional element.

Master Project Manual The basic project documentation is the master project manual. This manual provides an unambiguous definition of the policy

[14]Bruce N. Baker, Dalmar Fisher, and David C. Murphy, *Factors Affecting Success of Project Management,* paper presented to Annual Meeting of the Project Management Institute, Toronto, Canada, October 1973.

framework and many facts relating to the project. The master project manual should not be confused with the organizational policy and procedures manuals, although it may contain some of the same information. The master project manual expresses a policy framework for the project in terms of the interfaces between the project and the organizational operations. The specific content of the project manual will, of course, depend on the requirements of the particular project. Appendix 2 outlines what should typically be included in a project manual.

Because each project is unique, some of the sections of the project manual described in Appendix 2 may be applicable, and some not. The manual is most useful if prepared in loose-leaf form, since its value depends heavily on its being kept current. If maintained properly, such a manual can be the primary means by which the project manager keeps up to date on major aspects of the project. The time and money such a manual would save would be difficult to measure. What would be the value of having timely information on hand to give a customer, for example, or a top organization official calling about a particular problem?

Developing the Manual The manual should be developed as the project grows. The initial distribution of the manual should precede initiation of work on the project, since the manual will play an integral part in the project's management. If the project manager and the key project clientele work together in preparing the manual, important benefits can result. The project clientele will be more apt to support the manual if they have participated in its development and upkeep. This participation will help to ensure better knowledge by the clientele on project status. In addition, their participation should help to ensure their commitment to the project.

In a sense, the project manual is the plan of the project, since it summarizes all its major dimensions. This is the information project managers require to maintain activities in accordance with the plan. On small projects, they might be able to store this information in their heads or in several special documents, but on a project of any complexity, they will need more formalized documentation to keep order in the project.

Project Status Reports The heart of any project-control system is timely and relevant information—concerning the cost, schedule, and technical status of the project—which enables managers to make an assessment of the project's status and of the corrective action needed. After the basic information for control has been selected and reporting instructions have been formulated, a *project status report* format can be developed to present control-oriented information in summary form. In addition to basic information describing the project, customer, nature of the contract, contract value, key personnel, etc., such a report should contain:

1 Overall qualitative and quantitative information on project cost, schedule, and technical performance, e.g., "on time," "over cost"

2 Indication of whether notification of exceptions has been conveyed to top management, to the customer, and to other relevant clientele

3 Graphs showing expended and committed funds, from the time period when commitments began to the present

4 Charts showing major project milestones, in terms of scheduled completion dates and actual project status

5 Description of project status and highlights; areas of concern; status of notification of any clients regarding concerns.[15]

Project Review Periodic formal reviews are effective in comparing progress with the progress plan.

The project review should be accomplished by the project manager in collaboration with key clientele. The project manager may designate one individual from the staff to act as a central point of contact during the review. The areas to be examined during the project review should generally include the following:

Authority and responsibility of the project manager
Project charter
Project identification
Project priority
Dollar priority
Dollar size and complexity
Applicability of project-management techniques
Project history
Project visibility
Project staffing
Communications channels
Reporting procedures
Project status reviews and evaluations
Management information system
Financial management
Planning
Technical direction

Each of the foregoing major areas should be subdivided into specific questions, the answers to which would provide salient information on the project. If inconsistencies appear during the review, the specific area involved should be investigated further.

[15]See Robert A. Howell, "Multiproject Control," *Harvard Business Review,* March–April 1968, pp. 63–70, for a detailed example of such a report.

An agenda is essential for each review session. If a problem (or opportunity) is not resolved during the review session, responsibility for its resolution can be pinpointed. The problem (or opportunity) should remain on the agenda until successful resolution is effected.

Project-Review Team Project review should be conducted with as little disruption of the project effort as possible. The continuity of the project effort can be preserved by appointing an ad hoc team of qualified executives to head up the review. The skill mix of this team should be compatible with the organic activities of the project; specialists in cost, schedule, and technology (or performance) should be included. At the conclusion of the review, the team should present its *independent* findings, conclusions, and recommendations to the project manager, who, at his or her discretion, communicates the report with appropriate notations to the top-level managers.

The team's report should be given wide distribution in the project, particularly among the major contributors. This will provide cross-fertilization of ideas, which will contribute substantially to the overall understanding of the project problems and work.

Project-Management-Review Checklist The project-review team needs some structure to follow in appraising the project. The questionnaire outlined in Appendix 3 can provide a framework against which to measure the status of the project. Of course, the checklist will vary depending on the project.

The checklist technique is used extensively to ensure the proper sequencing in complex jobs; it is used, for example, in the operation of a highly automated manufacturing shop. The best example of the value of a checklist is in the field of aviation where it is used to ensure that the system is prepared for operation; the checklist with the proper sequencing is required for the human subsystem, as well as for the physical system. "Fasten seat belts" and "Start engines" are two separate actions in getting an airliner into operation, and they should be done in a predetermined order. In a sense, the checklist for the project is a summary of the documentation: it embodies the main features of the project manager's plans, policies, and management philosophy. At the same time, the checklist can serve as a technique for identifying areas where all may not be well with the project and for causing them to be investigated.

Although the checklist implies a mechanistic approach, project clientele should not be offended by it, since the list is not sacred. It serves the same purpose in project review as a financial ratio does, i.e., to signal where further study and investigation are required. A deviation signaled by the checklist could have significant meaning to the company, or it could mean nothing depending upon what the investigation discloses.

We feel that the project manager will need to use a checklist or other similar device to periodically review the status of the project. Therefore, we recommend that the technique be formalized to the degree required to facilitate proper control.

SUMMARY

In this chapter, we have examined the main tenets of control as well as some of the more subtle factors affecting it. We have reviewed the philosophy of control, some prerequisites for control, and some of the documentation that facilitates control of the project. We have reviewed the requirements of a project information system. Our view has been that the project information system is that complex of communications which provides intelligence about the project.

In the next chapter, we shall discuss some of the techniques for portraying information which is used to evaluate project status.

DISCUSSION QUESTIONS

1 What is meant by control of the project? Who has basic responsibility for such control? Why?
2 Control and planning are interrelated in the management of a project. Explain the nature of this interrelationship
3 What are the elements to be found in a control system?
4 Define control. Relate the nature of control to the other management functions.
5 If a project is "out of control," serious questions can be raised concerning the effectiveness of the project planning. Would you agree? Why or why not?
6 What are the prerequisites of a project-control system? How does the control system fit into the project-management system?
7 A control system must have predictive characteristics. What is meant by this?
8 What role does project review play in the control of the project?
9 Take a project for the construction of a plant. What are some of the key factors that should be evaluated to determine how well the construction of the plant is being carried out?
10 What are some of the key reports that could come out of an effective control system for the construction of the plant mentioned in the previous question?
11 What is configuration management and control? Can the concept be applied to small as well as to large projects? How might the concept be applied to a project in which there is a lot of software?
12 In the final analysis, control is a very personal factor. What does this mean?
13 Control and information systems are highly interdependent in the control of a project. Explain the rationale for this statement.
14 Select a small project and describe the key features of a control system for this project.
15 What is some of the essential documentation involved in the control of a project? Who has the responsibility to see that such documentation is developed and used?

RECOMMENDED READINGS

Anthony, R. M., J. Dearden, and R. F. Vancil: *Management Control Systems,* Richard D. Irwin, Inc., Homewood, Ill., 1965

Baker, Bruce N., Dalmar Fisher, and David C. Murphy: *Factors Affecting Success of Project Management,* paper presented to Annual Meeting of Project Management Institute, Toronto, Canada, October 1973.

Butler, Arthur G., Jr.: "Project Management: A Study in Organizational Conflict," *Academy of Management Journal,* March 1973.
Chapman, Chris: "Large Engineering Project Risk Analysis," *IEEE Transactions on Engineering Management,* vol. 1, EM-26, no. 3, August 1979.
Duke, Robert K.: "Project Management at Fluor Utah, Inc.," *Project Management Quarterly,* vol. 8, no. 3, September 1977.
Gunz, Hugh P., and Alan Pearson: "How to Manage Control Conflicts in Project Based Organizations," *Research Management,* vol. 22, March 1979.
Howell, Robert A.: "Multiproject Control," *Harvard Business Review,* March–April, 1968.
Jucius, Michael J., and William E. Schlender: *Elements of Managerial Action,* Richard D. Irwin, Inc., Homewood, Ill., 1965.
McGregor, Douglas: *The Professional Manager,* McGraw-Hill Book Company, New York, 1967.
McQuade, Walter: "Bob Fluor, Global Superbuilder," *Fortune,* February 26, 1979.
Miller, Barry M., and Charles D. Williams: "Management Action through Effective Project Controls: A Case Study of a Nuclear Power Plant Project," *1978 Proceedings of the Project Management Institute,* Los Angeles, Oct. 8–11 1978.
Van Steelandt, Frank V., and Ludo F. Gelders: "Financial Control in Project Management: A Case Study," *IEEE Transactions on Engineering Management,* vol. 1, EM-26, no. 3, August 1979.

CASE 14-1: Project Review

Situation Any ongoing project must be reviewed on a periodic basis.

Your Task Develop a strategy for a project manager in an industrial firm to follow for the review of a project. In doing this, consider such factors as:

Frequency of review sessions
Who attends?
What is reviewed?
Where is review held?
Who manages review sessions?
How to conduct review meetings

What do you think is important in developing a strategy for the review of a project? Select a team leader to present your findings.

CHAPTER **15**

PROJECT PLANNING AND CONTROL TECHNIQUES[1]

And yet, the PM (Project Manager) must know cost progress and cost performance status not only for the project as a whole, but also for tasks . . . in order to control the project. But this is only part of the problem confronting him; he should also be able to forecast where likely trouble spots may develop in each of these levels of effort.[2]

The techniques outlined in this chapter concern the planning and control of projects having the following general characteristics:

An objective that is known and can be specified
Actions and activities to accomplish the objectives that can be determined in advance
A desired or required sequence for performing the activities

Thus, the chapter is concerned with *project* planning and *project* control.

It is important to distinguish the planning activities of this chapter from those discussed earlier. In Chapter 2 and in the chapters in Part Three of the book,

[1]Some of the material in an earlier version of this chapter is based, with permission, on a report by Thomas L. Healy, The National Cash Register Company, Dayton Ohio, Apr. 1, 1963.
[2]John Stanley Baumgartner, *Project Management,* Richard D. Irwin, Inc., Homewood, Ill., 1963, p. 44.

attention was on overall organizational *strategic* planning and decision making. Here, the emphasis is on planning and decision making *within the context of a specific project.* Thus, these planning activities are more tactical in nature, in that they are concerned with the *efficient achievement of a previously made strategic decision,* such as development of a new product or installation of a new computer system.

There is considerable overlap between the techniques of systems analysis and those of project planning and control. This is natural because both areas focus on decision making. There is also considerable overlap between project planning and control techniques and modern concepts of management information systems, since comprehensive MISs have a built-in capability for some decision analysis, and since these techniques are readily programmed for computers. We shall deal more fully with the MIS topic and its relationship to these other elements in Chapter 16.

PROJECT PLANNING

Project control starts with project planning, since the project plan is the key to the development of adequate control procedures and mechanisms. Starting with a statement of work, performance standard specifications, and associated documentation, the objective of a project planning phase is achieved by:[3]

Determining what is to be done and translating it into a *work breakdown structure* (WBS). The WBS, as the name implies, is a technique for breaking down a total job into its component elements which can be displayed to show the relationship of the elements to each other and to the whole.

Establishing a project team based on the major tasks. This project team, working with the existing functional organizations, determines the *who* of the project-functional effort.

Coupling the tasks and resources, which determine the *who does what* through the development of a linear responsibility chart.

Creating the key planning and control documentation, for subsequent derivation of schedule and cost criteria against a single framework.

Developing an event-logic network and, perhaps, associated CPM or PERT analysis. This network is then translated into a schedule, taking into consideration the resource constraints and other projects' priorities. This then defines the *when* of the project-functional effort.

Establishing manpower, facilities, and subcontractor requirements by task. From these factors, task and subtask cost estimates are developed for project control.

[3]Adapted from "Project Management and Control," Fecker Corporation, Space-Defense Division Owens-Illinois, Pittsburgh, Pennsylvania, unpublished, undated documentation.

Finally, after integration of the component elements, and after resolution of problems, the overall plan is submitted for general-manager approval.

PLANNING AND CONTROL TECHNIQUES

The techniques to be treated in this chapter are:

- Work breakdown structures (WBS)
- Project planning (Gantt) charts
- Network plans
  - Precedence diagrams
  - CPM and PERT network plans
- Critical path method (CPM)
- Program evaluation and review technique (PERT)
- Network analysis, using historical estimating behavior
- Network simulation
- Simulations using historical estimating behavior
- Graphical evaluation and review technique (GERT)
- Line of balance (LOB)

Some of these techniques are extremely useful and very widely used. Others have been found to be less useful, because they are complicated or require the manipulation of great volumes of data. We shall try, in this chapter, to distinguish among these various techniques in terms of their usefulness.[4]

The Work Breakdown Structure

The work breakdown structure (WBS) is the "organization chart" which schematically portrays the products (hardware, software, services, and other work tasks) that completely define the projects. The work breakdown structure describes the project tasks and provides a relationship between tasks and objectives, providing a basis for the planning and control of the project. Once completed, a work breakdown structure tends to relate to organizational structure and proceeds from a mission or objective.[5] Each successively smaller subsystem is identified into lower-level subsystems, for planning, implementation, and control; ultimately the smallest system is identified as a *work package*. A *work package* is a specific job which contributes to a clearly defined specific task of accomplishment toward the system objective. A work package may be a design, document, hardware item, or service. Figure 15-1 shows an example of a hardware-oriented work breakdown structure.

[4]C. W. Dane, C. F. Gray, and B. M. Woodworth, "Factors Affecting the Successful Application of PERT/CPM in a Government Organization," *Interfaces*, November 1979.

[5]We described the concept of a work breakdown structure and its associated work packages in an organizational context in Chap. 7.

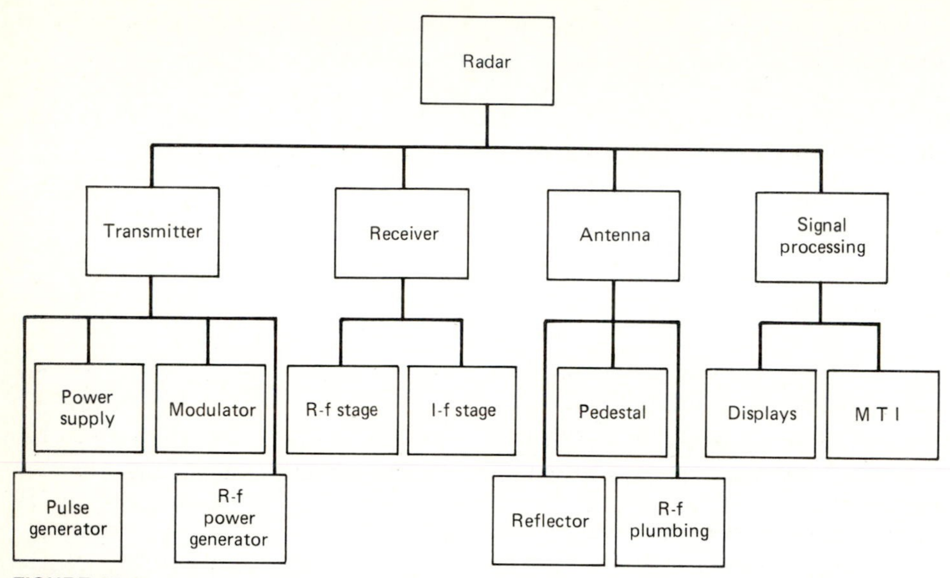

FIGURE 15-1
Product-oriented work breakdown structure. (From Robert W. Miller, *Schedule, Cost, and Profit Control with PERT,* McGraw-Hill Book Company, New York, 1963. Used by permission.)

Project Planning (Gantt) Charts

The most widely used technique for simple project planning and scheduling is based on the Gantt chart. This chart, which may be directly based on a WBS, consists of a scale divided into units of time (e.g., days, weeks, or months) across the top and a listing of the project elements down the left-hand side. Bars, lines, or other symbols are used to indicate the schedule and status of each element, in relation to the time scale.

Such charts usually entail a sequence of steps necessary to complete the project, together with the time available for completion and a summary reporting technique in terms of the total job. In the original Gantt charts, the governing factor in accomplishing activities and tasks was the capacity of the workers and machines. The progress chart, the worker and machine record chart, and the load chart are examples of these early charts. A division on the chart represented both an amount of time and the amount of work to be done in that time. Horizontal lines across the time divisions on the chart showed the relationship between the amount of work accomplished in a specific time span and the amount scheduled.

In extending the applications of these techniques, Gantt's associates recognized that, for some project-type efforts, time, rather than resources, was the governing factor. In such cases, resources would be made available within practical limitations to complete the project in the shortest possible time. The chart developed for these projects was called the *project planning chart* and was based on the progress chart developed by Gantt. Figure 15-2 is an example of a project planning chart for the development of an electronic device. The elements of the project (in this case, functional units) are listed on the left-hand side, and the units of time in workdays are shown at the top. The light horizontal lines indicate the schedule for the project elements, with the specific tasks or operations written above the schedule line. The starting and completion times are indicated by opening and closing angles, ⌈ and ⌉. Work accomplished is indicated by a heavy line below the schedule line.[6] The large V on the time scale at the top of the chart marks the time to which progress has been posted. Progress is posted at regular intervals.

The system line in Figure 15-2 indicates that the project as a whole is six days behind schedule. At the last posting, the receiver video amplifier unit was the furthest behind schedule. The display and antenna units were ahead of schedule. If the work on the video amplifier and available personnel skills permit, therefore, the chart might suggest that the project manager should consider transferring personnel temporarily from the display and antenna units to the video amplifier unit. Other elements are behind schedule, too, but not as much as the video amplifier. Thus, bringing the video amplifier up to schedule might have priority.

A project planning chart is usually prepared as follows:

Analyze the project to determine the method and approach to be used.

Break the project down into elements to be scheduled.

Estimate the time required to perform each element. (Time estimates should be made by the persons who will accomplish the work, or in conference with them.)

List the elements down the left-hand side in sequence of time, considering those which must be performed sequentially as well as those which can be performed simultaneously. (If the completion date has been specified, the elements can be sequenced by working backward from the completion date.)

To post progress on the project, the amount of time that was estimated to accomplish that portion of work completed is determined, and a heavy solid line is extended from the left margin to represent that time increment. The span of time between that work-accomplished line and the charted completion date for that element, therefore, represents the amount of time required to complete that element.

The primary advantage of the bar chart is that the plan, schedule, and progress of the project can all be portrayed graphically together. It is particularly effective in showing the status of the project elements and identifying the elements that are behind or ahead of schedule. The time the project is behind the schedule is usually determined by the maximum delay of any element from the schedule.

[6]Project planning charts often use open bars (hence the name "bar charts"), with the ends indicating the start and completion times. Accomplished work is indicated by filling in the bar.

Description	May			V	June					July
	4	11	18	25	1	8	15	22	29	6
System		Design	and Fabrication					Ass'y	System	Test
Receiver										
Mixer and oscillator	Design	Fab								
TR amplifier and detector	Design	Fab								
Video amplifier	Design		Fab. &	Test						
Unit						Ass'y	Test			
Transmitter										
Magnetron		Procurement								
TR switch			Design	Fab.						
Modulator	Design		Fab.							
Unit					Ass'y	Test				
Power supply	Design	Fab.								
Display	Design		Fab.	and	Test					
Antenna										
Dish	Design	Fab.								
Support	Design	Fab.								
Drive	Design		Fab &	Test						
Unit					Ass'y					

FIGURE 15-2
Project planning chart.

The bar chart has some disadvantages when applied to projects:

Planning and scheduling must be considered simultaneously. The time dimension associated with the chart requires alternative plans to be evaluated in terms of the schedule established when the plan is originally chartered. Thus, the course of action must be selected almost entirely on the basis of the adopted schedule; little or no opportunity is provided for considering alternative plans with different schedules.

It provides no means for assessing the impact of an element's being behind or ahead of schedule. Simply because a project element is behind schedule does not necessarily mean that the project is behind schedule by that amount. For most projects, only a few dates are critical in the sense that any delay in them will delay the project by a corresponding amount. The impact of slips in schedule dates depends upon the interrelationships between elements, which are not easily portrayed.

It does not present sufficient detail to enable the *timely* detection of schedule slippages.

Since it is usually maintained manually, the chart may tend to become outdated.

The chart's simplicity precludes the portrayal of schedule-progress information for large and complex projects; its greatest value is in depicting gross progress (or lack of it) in the major elements of a project and in communicating the overall status of the project to top management.

Network Plans

Network plans represent projects in terms of the interrelationships among the critical project elements. These pictorial displays are constructed around the technological and time requirements of a project in a way which separates the planning and scheduling functions. This permits the consideration of alternative plans. When scheduling is subsequently performed, it may be done on the basis of the availability of resources and the demands of other projects.

With project planning charts, a linear calendar format is used. This forces the schedule to be prepared simultaneously with the project plan. The technological and time-requirement aspects of project planning become intermingled with the resource-allocation problems of scheduling. As a result, alternative plans are usually evaluated on the basis of their schedules. Furthermore, because planning and scheduling proceed in a step-by-step fashion, trade-offs between planning and scheduling cannot be determined so as to arrive at a preferred course of action.

Only a small percentage of the tasks and jobs are critical to the overall time requirement for completing most projects. Furthermore, tasks which are critical in one plan may not be critical in another, and noncritical tasks may become critical because of the way they are scheduled. Knowing which tasks are critical to a project plan facilitates scheduling the project and allocating the resources necessary to accomplish it.

In network planning and scheduling techniques, the plan is prepared in the form of a network or flow diagram. Using a network rather than a bar chart alleviates many of the problems associated with planning and scheduling, since the analysis of the network enables the criticality of each task to be determined in a quantitative and objective manner.

Precedence Diagrams The simplest form of network plan is the precedence diagram, which portrays major project elements in terms of their logical precedence relationships. Figure 15-3 shows a precedence diagram for a project involving the development of a processing plant.

CPM and PERT Network Plans The network plans utilized with the widely known techniques CPM and PERT are somewhat more complex than precedence diagrams. Such a network is developed by studying the project to determine the approach, methods, and technology to be used and then breaking it down into elements for planning and scheduling purposes. The elements of a project can be classified as follows:

Project objectives. These are the goals to be accomplished during the course of the project. In most cases, the project objectives are specified before the plan is prepared; the plan merely prescribes the course to be followed in achieving the objectives.

Activities, tasks, jobs, or work phases. These elements identify and describe the work to be performed in accomplishing the project objectives. They normally utilize time and other resources.

Events or milestones. These are points of significant accomplishment—the start or completion of tasks and jobs, the attainment of objectives, the completion of management reviews and approvals, etc. They are convenient points at which to report status or measure and evaluate progress.

After the elements of the project have been determined, they are arranged in the sequence preferred for their accomplishment. This is a synthesis process that must consider the technological aspects of the activities and tasks, their relationships to one another and to the objectives, and the environment in which they will be performed. A network is used to reflect these factors as it portrays the sequence in which the project elements will be accomplished.

Networks are composed of events which are represented by points interconnected by directed lines (lines with arrows) which represent activities. Constraints are also represented as directed lines. Elements of the network correspond to elements of the project as follows: points in the network represent project objectives, events, and milestones; the lines between the points repre-

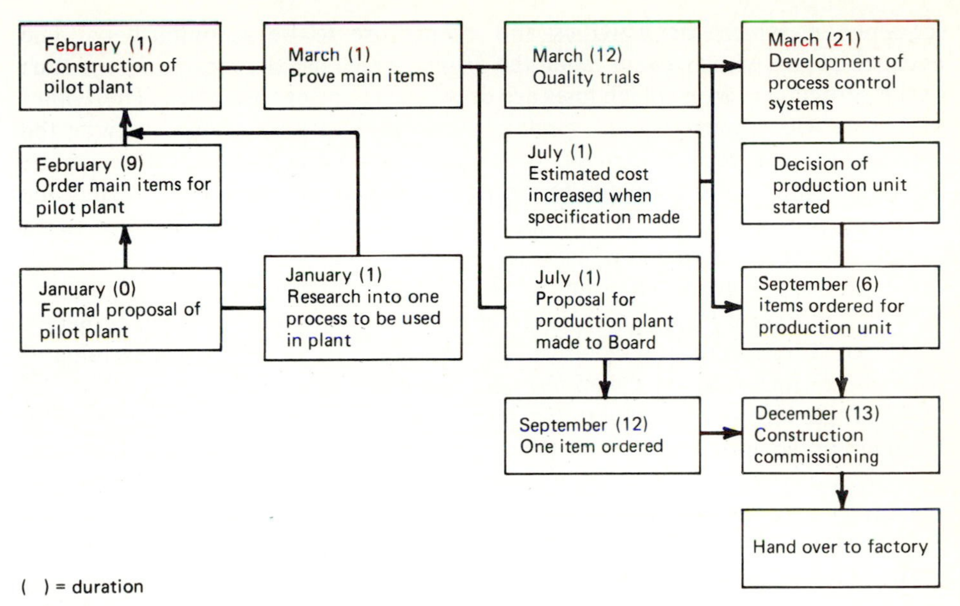

FIGURE 15-3
Precedence diagram. (From A. Wilkes and A. W. Pearson, "Project Management in Research and Development," *British Chemical Engineering and Process Techniques,* November 1971, pp. 1009–1011. Used with permission.)

sent project activities, with the direction of the line indicating a precedence or sequential relationship; and directed solid or dashed lines indicate constraints.

Activities are the jobs and tasks, including administrative tasks, that must be performed to accomplish the project objectives; activities require time and utilize resources. The length of the line representing an activity has no significance (in contrast to Gantt charts, where it is the significant factor). The direction of the line, however, indicates the flow of time in performing the activity.

Events are usually represented by circles or squares. Numbers are inserted in these circles and squares, which are used to identify the events and the activity that connects two events. Events represent particular points or instances in time, so they do not consume resources; the resources to accomplish an event are used by the activities leading up to it.

Constraints in network plans represent precedence relationships resulting from natural or physical restrictions, administrative policies and procedures, or management prerogatives. Constraints, like activities, are represented in a network plan by directed lines. However, constraints indicate precedence only; they do not require resources and normally do not require time. Those constraints which require neither time nor resources are represented by broken directed lines, which are often referred to as "dummy" activities.

The network plan is constructed by drawing directed lines and circles in the sequence in which the activities and events are to be accomplished.[7] The network begins with an event called the *origin,* which usually represents the start of the project and from which lines are drawn to represent activities. These lines terminate with an arrow and a circle representing an event, which may be the completion of a project element or an activity. All activities that are to be performed next are then added to the network plan by drawing a directed line from the previous event. For example, suppose activities B and C are to be simultaneously performed upon completion of activity A. These three activities and their precedence relationship would be represented in the network plan as indicated in Figure 15-4. Activities and events are then added until the project is complete. Constraints are added where required. The network plan terminates with one or more events, called *terminal* events.

To progress from one event to the next requires that an activity be performed. Each activity begins and ends with an event. The event at the start of an activity is called a *predecessor* event; and that at the conclusion, a *successor* event. Time flows from a predecessor event to successor event, as indicated by the arrow, and is normally from left to right throughout the network. As each activity is added to the network, its relationship to other activities is determined by answers to the following questions:

What activities must be completed before this activity can start? Activities that must be completed first are predecessor activities.

What activities can start after this activity is completed? Activities that can start after are successor activities.

What activities can be transformed at the same time as this activity? Those activities are *concurrent,* or parallel, activities.

In preparing the network plan, administrative activities must be included, such as the preparation of contracts, the procurement of parts, and the preparation of test procedures, specifications, and drawings. Technical work often cannot begin until a contract has been awarded or long-lead-time subsystems have been procured. A test cannot be started until test procedures have been written, and tooling cannot commence until specifications and drawings have been prepared and approved.

Two activities with a predecessor-successor relationship are called *sequential* activities. Performing activities in sequence requires that the start of the successor activity depend upon completion of the predecessor activity. Activities performed concurrently must be independent of one another. However, activities independent of one another cannot always be performed concurrently; for

[7]There are two general methods which are used in actual construction of a network plan. This section describes the *forward* method, where construction begins with the start event and activities and events are added in sequential fashion to reach the end event. In the *backward* method, construction begins with the end event and proceeds backward to the start event. The backward method of network construction is often preferred to the forward method because attention is directed to the project objectives. With the objectives firmly in mind, the activities and events required to accomplish those objectives are often more easily determined.

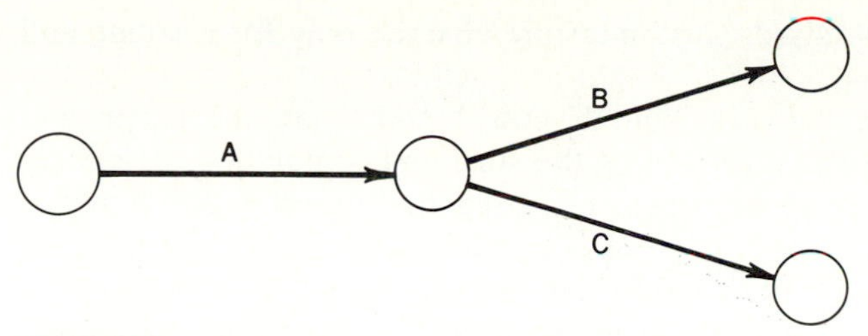

FIGURE 15-4
Simple network plan.

example, if one activity creates a safety hazard for the other, they must be performed in sequence. Independent activities may have a common predecessor event or a common successor event, but not both.

Suppose, for example, that activities B and C can be performed concurrently but that both are dependent upon the completion of activity A; activity D can be started after both B and C are completed. The relationships would then be represented as illustrated by Figure 15-5. The constraint, or dummy activity, is needed between activities B and D so as to identify activities B and C uniquely by their predecessor and successor events. The constraint could just as well have been added between activities C and D.

To illustrate the preparation of a network plan, let us consider as a project the servicing of an automobile at a service station. This example will be slightly exaggerated in order to emphasize the interrelationships between project activities that must be considered. It is also somewhat out of date for people in areas that are not now familiar with "full-service" stations!

The project situation is described as follows: Automobiles arrive at a service station for gasoline. Services provided by the station include cleaning the windshield and checking the tires, battery, oil, and radiator. Sufficient personnel are available to perform all services simultaneously. The windshield cannot be

FIGURE 15-5
Network plan: correct predecessor-successor relationship.

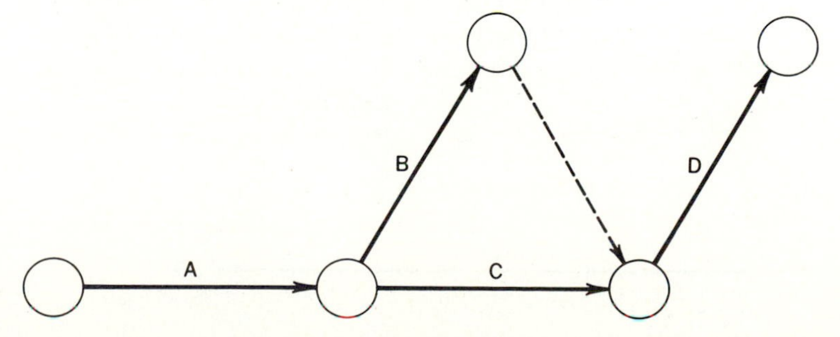

cleaned while the hood is raised. Customers are charged only for gasoline and oil.

Figure 15-6 shows the network plan. Events 1 and 9 are the origin and terminal events, respectively, representing the start and completion of service. Three constraints, or dummy activities, are used to sequence the activities properly.

The constraint between events 3 and 5, denoted as activity 3–5, is used so that the activities "check radiator" and "check battery" will not have common predecessor and successor events. The dummy activity 4–5 is used for the same reason. The constraint 4–6 is used to indicate that the activity of computing the bill cannot start until the activities "check oil" and "add gas" have been completed. Suppose that the dummy activities had been sequenced as shown in Figure 15-7*a*. This implies that computing the bill also depends upon completing the check of the battery and radiator, which is not true because there is no charge for servicing the battery or the radiator. On the other hand, suppose that the dummy activities had been sequenced as shown in Figure 15-7*b*. This implies that the hood cannot be lowered until the gas has been added, which obviously is an improper relationship unless the gas-tank cap is under the hood. This simple case illustrates the care that must be exercised in sequencing and constraining activities.

Only a few practical guidelines can be given for general use in preparing network plans, and most of these apply to the construction of the network only after the project activities and events have been determined. One question that

FIGURE 15-6
Network plan for servicing an automobile.

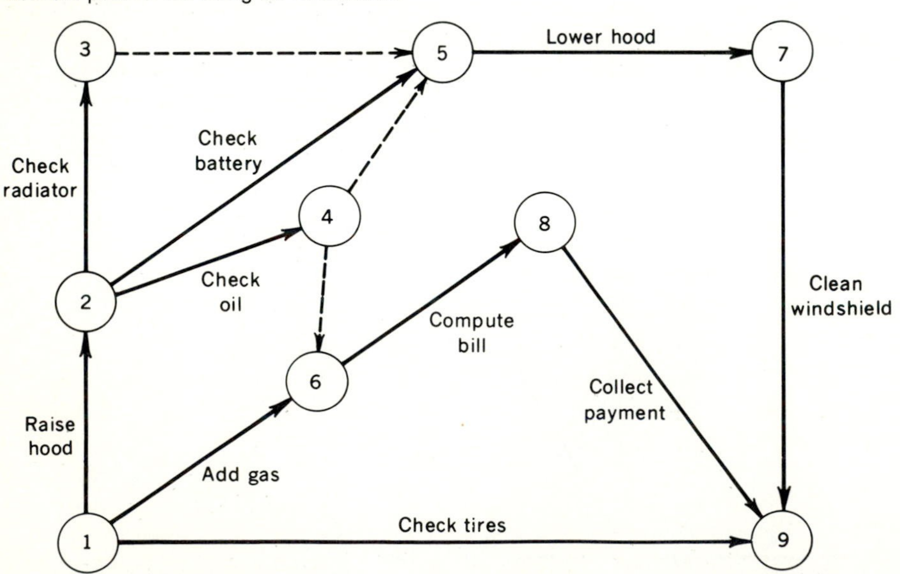

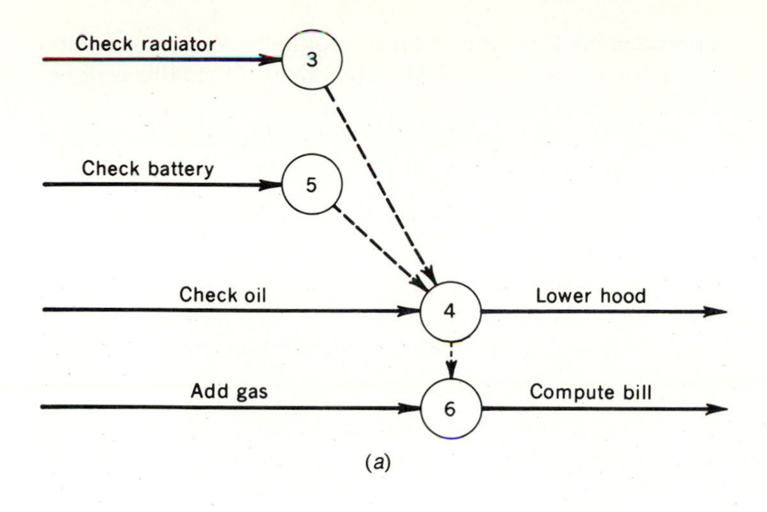

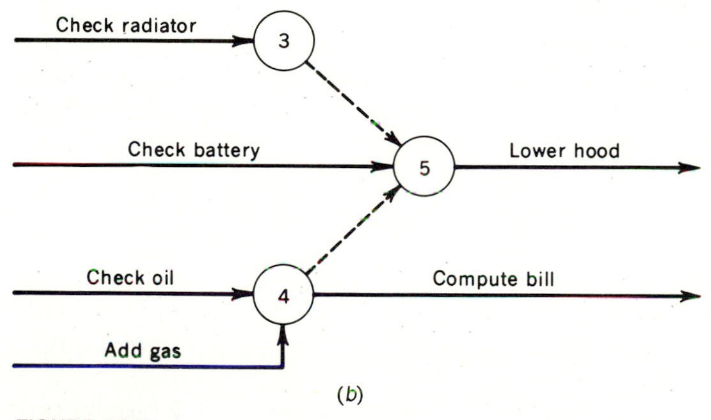

FIGURE 15-7
Two cases of improper sequencing of activities in Figure 15-6.

invariably arises in preparing a network plan is: "What level of project detail should be contained in the network?" By this is meant the magnitude and scope of the events and activities to be shown. Obviously, if the network is to serve as an adequate plan, it must contain sufficient detail for scheduling the project and measuring progress against the plan.

The quality of a plan is often evaluated in terms of the amount of detail it contains. Network planners sometimes become obsessed with this precept and go to great effort to incorporate the smallest details. Each activity, regardless of its nature, can be divided into smaller activities. The correct procedure is to consider carefully which activities are required, which are not clearly defined, which are not essential, and how they are interrelated. The primary emphasis during the preparation of a network should be on planning the project, that is, developing a sound approach to how the project should be carried out and determining what must be done to accomplish the objectives. Once this is done, the network becomes a means for depicting the project plan.

Some of the guidelines suggested for establishing the correct level of detail for projects of substantial size and duration are:

Activities should represent efforts requiring four to six weeks to complete. The level of detail should be *one level below the level of responsibility;* for example, the network for the design of a system should contain activities and events pertaining to the design of each subsystem.

Standardized activities and events should be included in network plans prepared for similar projects.

These guidelines cannot be universally applied effectively. Determining the correct level of detail for a network plan is a matter of experience and judgment. The logical manner in which a network plan is prepared leads to incorporating more detail than is normally provided in other planning techniques. This happens frequently in networks for large projects where many people from different organizations participate in the preparation. In such cases, it is best to set an arbitrary level to use in preparing the initial network. After this initial network has been reviewed, a second networking session will usually produce a network with the appropriate amount of detail.

Analysis of Network Plans

The project network plan displays the activities, events, and constraints, together with their interrelationships. For the network to be useful in planning and controlling the project, time estimates must be made for the various activities which constitute the project.

In order to facilitate understanding of the ideas of network analysis, we shall for the moment assume that single time estimates have been obtained for each of the activities in a project network. Later, we shall discuss some of the ways which have been devised for obtaining these estimates. Here, our emphasis is on the *use* to which time estimates are put. The time estimates for each activity are typically expressed in terms of workdays or workweeks.

A *network path* is a sequence of activities and events traced out by starting with the origin event and proceeding to its successor event, then to another successor event, and so on, until the terminal event is reached. The *length* of a network path is the sum of the time estimates for all those activities on the path.

After activity time estimates have been made, an earliest and latest time for each event may be calculated. The *earliest time* for an event is the length of the longest path from the origin to the event. Thus, it indeed represents the earliest time at which the event can occur (relative to the timing of the origin event). The earliest time for the terminal event is the length of the longest network path. It therefore represents the shortest time required to complete the entire project.

The *latest time* for an event is the latest time at which the event can occur relative to the timing of the terminal event. If one imagines that the direction of each activity is reversed, the latest time for an event is determined by the length of the longest path from the terminal event to the event in question.

In calculating earliest event times, the general practice is to consider that the origin event occurs at time zero. The earliest time for each event is the sum of the earliest time for the predecessor event and the time for the predecessor activity. If an event has more than one predecessor event, this calculation is made for each of them, and the largest sum is selected as the earliest time for the event. This is because the earliest time is the length of the longest path from the origin to the event.

To calculate the latest time for an event, the latest time for the terminal event is usually initially set equal to the previously computed earliest time for the terminal event. Then, for each event, the time for its successor activity is substracted from the latest time for its successor event. The result is the latest time for that event. If an event has more than one successor event, this calculation is made for each, and the smaller result is used as the latest time for the event. This is compatible with the view of the latest time for an event as the longest path from the terminal event backward to the event in question.

Using these basic activity, event, and path measures, a number of network measures may be developed to aid in network analysis.

Event slack is the difference between the latest time and the earliest time for an event. The slack for an event is the difference between the length of the longest network path and the length of the longest network path through the event. Hence, event slack is a property of a particular network path. Consider Figure 15-8, which shows the last three events of a network plan, time estimates for the activities (the numbers on the lines representing each activity), and

FIGURE 15-8
Portion of network illustrating computation of event slack.

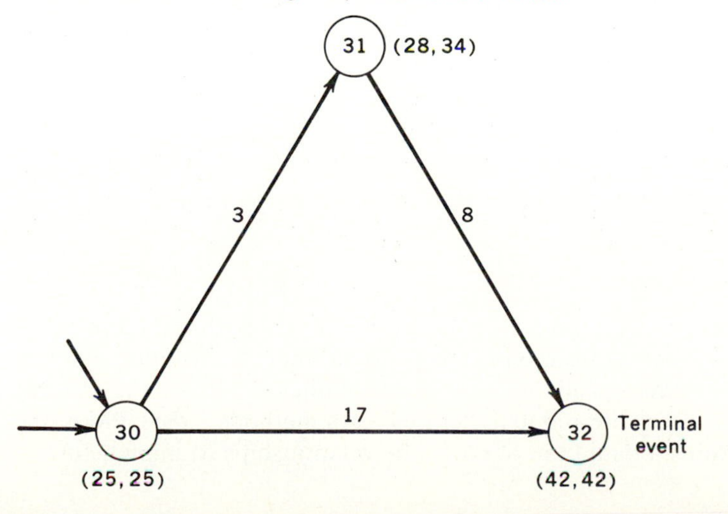

earliest and latest event times (the pair of numbers in parentheses at each event). The slack for event 31 is 6 (34 − 28). This can also be determined from the lengths of the network paths. The length of the longest network path is the earliest time of terminal event 32, which is 42. The length of the longest network path through event 31 is 36 (25 + 3 + 8). The difference between the length of the longest network path and the length of the longest network path through activity 31 is therefore 6 (42 − 36), which is the event slack.

The most important use of event slack is in identifying the critical path. The *critical path* is the longest network path. Thus, its length determines the minimum time required for completion of the entire project. *Critical events* are those events on the critical path. To identify critical events, we need only determine those events with the smallest amounts of event slack. Their identification is usually sufficient to identify the critical path; however, it need not uniquely identify it.

The operational significance of the critical events is that they are the pacing elements of the project. If the project is to be expedited, the accomplishment of at least one of the critical events must be expedited. If there is a delay in the actual accomplishment of *any* critical event, the completion of the project will be delayed.

Whether one is planning, scheduling, or controlling a project, the central idea involved in using network plans is the principle of *management by exception.* Stated simply, this means that it is the exceptions which require the primary attention of management. In the case of a project, the exceptions are the activities on the critical path, for it is they which pace the completion of the project.

If a project is to be expedited, some way must be found to hasten the accomplishment of critical events. Moreover, if the project is under way and the events on the critical path are not being accomplished according to plan, the project will be delayed if no way is found to hasten the completion of other critical events.

The application of the principle of management by exception in such projects usually takes the form of reallocating resources from noncritical activities to critical ones. This may be accomplished in either the planning or the control phase of the project, i.e., it may be done so that an earlier project completion date can be set up, or it may be done because the project is falling behind schedule. Presumably, such reallocations will permit faster accomplishment of critical activities and, hence, faster completion of the project itself.

Critical Path Method (CPM)

The critical path method recognizes the concept of a critical path and emphasizes the reallocation of resources from one activity to another, to facilitate efficient project completion. Although all the network techniques which use the concept of a critical path are often referred to as "critical path methods," *the* CPM is one method which requires a data base of cost-time relationships to implement.

Basically, this approach utilizes cost-time relationships for individual activities as a basis for reallocating resources among activities. However, the use of the approach requires that these cost-time relationships be known. This is a somewhat unrealistic requirement in the case of most complex projects, since no such data base usually exists. The alternative which is generally taken is to assume simple linear cost-time relationships. For instance, Figure 15-9 shows a hypothetical linear cost-time relationship involving "normal" activity time and cost versus "crash" activity time and cost. This indicates that under normal conditions the activity can be completed in six weeks at a cost of $20,000. If the activity is "crashed," through the use of additional resources, it can be completed faster (four weeks) at a $40,000 cost. The straight line is the linear relationship which is used to approximate the true, but unknown, cost-time relationship. It implies that the per-day cost of shortening the time required for the activity is uniformly $2,000 ($20,000 divided by 10 working days).

If such cost-time relationships were available for each activity, the project planner could make reallocations of resources from noncritical activities to critical ones, in order to shorten the critical path. The planner would be able to calculate the net cost associated with each such reallocation directly from such relationships and would thereby be in a position to "optimize" the project plan.

Of course, such linear relationships may be very poor approximations of the true underlying cost-time relationships. Even if they are reasonable, the

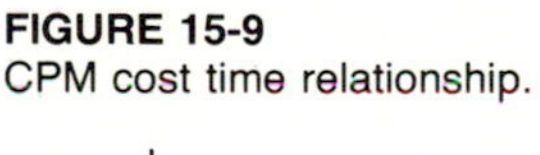

FIGURE 15-9
CPM cost time relationship.

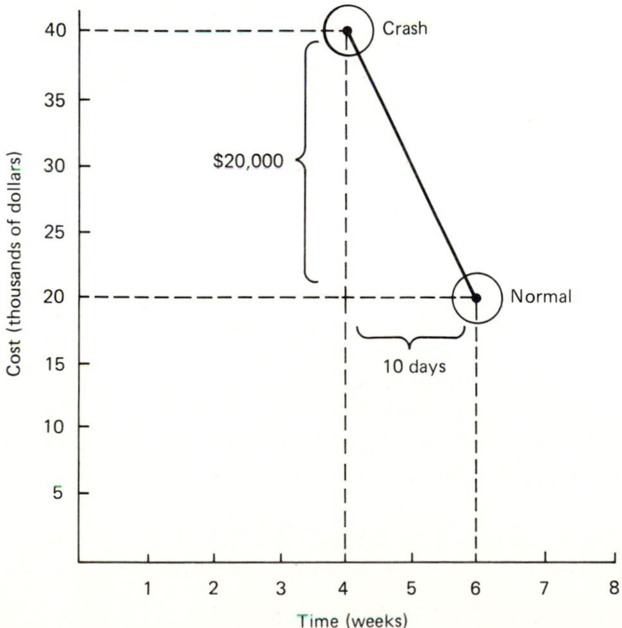

difficulty involved in "optimizing" large networks is great, so that the formal CPM technique is not itself widely used. However, it should be emphasized that the same statement is *not* applicable to the more generic use of the term; critical path *methods,* such as the less formal application of CPM and the use of other formal methods that are based on critical path ideas, such as PERT, are useful and much used.

Program Evaluation and Review Technique (PERT)

In developing the Polaris missile, the U.S. Navy became concerned with techniques for evaluating its progress. A schedule had been established for its development, and a system was set up for reporting status, progress, and problem areas in terms of accomplishment or slippage (actual or forecasted) of important program milestones. Major components were also evaluated and their status indicated by one of the following terms: "In good shape," "minor weakness," "major weakness," or "critical." These evaluations provided no measure of the impact on the overall program made by attaining a milestone or changing the forecast for its accomplishment. Tight schedules had been established for the program, so it was necessary to know the significance of a slip in schedule date, its impact on future scheduled dates, and the prospect for future slippages so that corrective action could be taken. As the slips in schedules and the prospects for future slips were studied, ". . . it appeared that the capacity to predict future progress was more limited than desired."[8]

An operations research team was formed of representatives from the special projects office; Booz, Allen, and Hamilton, Inc., a management consulting firm; and Lockheed Missile Systems division, the prime contractor for the missile. This team was to study the application of statistical and mathematical methods to the planning, evaluation, and control of the Polaris program. The following objectives were established:

To develop methodology for providing the integrated evaluation of progress to date and the progress outlook, changes in the validity of the established plans for accomplishing the program objectives, and effects of changes proposed for established plans

To establish procedures for applying the methodology, as designed and tested, to the overall FBM (Fleet Ballistic Missile) program

The team felt that the two major requirements for a program-evaluation methodology were (1) detailed, well-considered time estimates for future activities and (2) precise knowledge of the required or planned sequence in which the activities were to be performed. Since the time required to perform

[8]D. G. Malcolm, J. H. Rosebloom, C. E. Clark, and W. Frazer, "Applications of a Technique for Research and Development Program Evaluation," *Operations Research,* vol. 7, no. 5, pp. 646–669, September–October 1959.

development activities is often uncertain, a procedure for quantitatively expressing this uncertainty was desired; this led to the statistical estimation technique, which is a primary feature of PERT. The sequence requirement was fulfilled by use of network plans.

PERT, therefore, was originally developed as a technique for evaluating established plans and schedules, but its utility is not limited to this. PERT can also be used as a planning and scheduling technique. The PERT technique for estimating elapsed times provides a way of handling some of the uncertainties in estimating the time required to perform many types of activities.

After the network is prepared, the PERT planners obtain three elapsed time estimates for each activity: an optimistic one, a pessimistic one, and the most probable. These three estimates are used to compute the expected times required to perform each activity and a measure of the probability of accomplishing the activity in that time. The expected time estimate for each activity is used in analyzing the network. Variabilities in activity times are accumulated along the network paths in the same manner as activity times are accumulated, and they provide a measure of variability for each event. The variability associated with an event can be used to make statistical inferences about the occurrence of the event at a particular time, such as: "The likelihood that the project will be completed by its scheduled completion date is 34 percent."

The PERT approach usually involves obtaining the activity time estimates from the people who are responsible for performing or for supervising the performance of each of the activities. The person directly responsible for the activities should be asked to make the estimate because they are most knowledgeable concerning the inherent difficulties and the potential variability in accomplishment. Scheduled times cannot be used because they are not adequately responsive to changing conditions, contain no information on variability, and are often made under conditions and in an environment that do not reflect the technical aspects of the activity. A single elapsed time estimate would not, by itself, provide a measure of the variability in the time; this requires a range of estimated elapsed times. Estimates of the extreme times, reflecting the optimistic and pessimistic times, can usually be given with some degree of reliability, however, and the most likely time estimate lies somewhere within this range.

The three elapsed time estimates, referred to as the *optimistic,* the *most likely,* and the *pessimistic* times, are defined below.

Optimistic Time This is the shortest time in which the activity can be accomplished. There should be little hope of completing the activity in less time than this, but if everything goes exceptionally well, it should be possible to accomplish it in approximately this time.

Most Likely Time This is the normal, or most realistic, time required to accomplish the activity. If the activity were to be repeated numerous times under the same conditions and without any "learning curve" effects, it would be accomplishd most frequently in this time. (The most likely time is not the expected time, but an estimate based on experienced judgment; the expected time is a mathematically computed value.)

Pessimistic Time This is the longest time required to accomplish the activity, assuming unusually bad luck (e.g., major redesign or major reshuffling of planned action). The pessimistic time estimate should include such possibilities as initial failure and a second start, but not major catastrophic events such as strikes, fires, or tornadoes.

The range between the optimistic and the pessimistic time estimates is used in PERT as a measure of the variability, or uncertainty, in accomplishing an activity. If there is no uncertainty, all the time estimates will be the same, and the range will be zero. If there is considerable uncertainty, the range will be large. The time estimates must necessarily be based on planned assumed resources. The most likely time estimate must be based on the same level of resources that is used for estimating the optimistic and pessimistic times. For example, the optimistic time estimate must *not* be based on an extra shift or additional personnel, while the most likely time estimate is based on a normal shift and fewer personnel.

The most likely time estimate should be made first so that the estimate considers the available or planned level of resources and appraises the technical aspects of the activity realistically. The optimistic estimate can then be made, based on the same resources but with the assumption that everything goes exceedingly well. The pessimistic time estimate is made last, assuming that problems arise. The time estimates for each activity must be made independently and should not include a pad to cover possible delays. The possibility of padding is reduced by skipping around through the network when developing the time estimates.

The first step in analyzing the network is to convert the three elapsed time estimates to a single expected time estimate. This estimate, called the *expected activity time* (not to be confused with the most likely time estimate), is designated by t_e and calculated by the following formula:[9]

$$t_e = \frac{\text{optimistic time} + (4 \times \text{most likely time}) + \text{pessimistic time}}{6}$$

To illustrate, suppose the optimistic, most likely, and pessimistic time estimates for accomplishing an activity are 3, 5, and 9 weeks, respectively. The expected time for accomplishing the activity would be

[9]The deviation of this formula is based on the following assumption: The model for the probability distribution of an activity time is a beta distribution whose standard deviation is one sixth the range between the pessimistic and optimistic time estimates and whose mode is equal to the most likely time estimate. K. R. MacCrimmon and C. A. Ryavec discuss the implications of all of the basic PERT assumptions in "An Analytic Study of the PERT Assumptions," *Operations Research,* vol. 12, no. 1, 1964, pp. 16–37.

$$t_e = \frac{3 + (4 \times 5) + 9}{6} = \frac{32}{6} = 5.3$$

With PERT, the measure of uncertainty or variability is the standard deviation σ, and is estimated by the following formula:

$$\sigma = \frac{(\text{pessimistic time}) - (\text{optimistic time})}{6}$$

For optimistic, most likely, and pessimistic time estimates of 3, 5, and 9 weeks, respectively, the standard deviation associated with the expected time would be

$$\sigma = \frac{9 - 3}{6} = 1$$

The calculated expected times for each activity in a project network may be added to determine the expected duration of various paths through the network. Thus, these calculated expected activity durations are used in much the same fashion as the deterministic estimates that were discussed earlier for simple critical path analysis. Once the expected durations of each activity have been calculated from the pessimistic, optimistic, and most likely estimates, there is no mechanical difference between PERT and single-time-estimate critical path analysis in this regard.

However, there is a difference in the meaning of the expected times that are calculated from three time estimates and single estimates, the most obvious difference being that single estimates are analogous to the most likely estimate used in PERT and are not treated as though they were subject to uncertainty, as in PERT.

The uncertainty associated with the earliest time estimate for an event is determined by combining the uncertainties associated with the expected times for all the activities on the longest network path leading to the event. The network path that determines the earliest time for an event, therefore, also determines the uncertainty associated with that time.

The probability of accomplishing an event on, before, or after its scheduled date can be computed by using the earliest event time, the scheduled time, and the uncertainty associated with the earliest event time. If the scheduled time is prior to the earliest event time, the probability of accomplishing the event on or before its scheduled date is less than 50 percent. If the scheduled time is later than the earliest event time, the probability of accomplishing the event on or before its scheduled date is greater than 50 percent. As the scheduled date moves further away from the earliest event time, the probability of accomplishing the event on or before the scheduled date approaches an upper value of 100 percent. The following criteria are suggested for using the computed probabilities in program evaluation.

If the probability of meeting a scheduled date is less than 25 percent, the amount of risk associated with it makes it infeasible. In this case, the allocation of resources, the performance requirements, or the planned sequence of activities should be revised to obtain a probability greater than 25 percent, or the event should be rescheduled. With probabilities between 25 and 60 percent, the risk is normal and the allocation of resources is reasonable. With a probability greater than 60 percent, the activities should be examined for excessive resource allocations.

An important property of the computed expected times is that they are added to calculate an earliest time, and this earliest time is also an expected event time and has a probability of 50 percent. This probability would not hold if most likely time estimates were summed in a similar fashion.

PERT has attracted considerable attention, perhaps more than is warranted by its range of useful applications. Many feel that because the three time estimates are subjective, the estimator's personal bias will be introduced. A fundamental principle of PERT is that the three estimates are to be made by persons who are most familiar with the technical aspects of the activities and therefore are best qualified to make the time estimates reflecting uncertainties involved in technical activities. Asking for three time estimates tends to remove the psychological barrier often encountered when only a single estimate is given, since a time range does not imply a commitment such as a single estimate does, and allowing estimators to make a pessimistic time estimate permits them to provide for unforeseen contingencies that would probably be included as a "pad" in a single estimate. The effects of personal biases may be canceled in the analysis of the network, since estimates of optimists may be offset by estimates of pessimists.

Another controversial aspect of PERT pertains to use of computed expected times for scheduling. It can be shown that PERT assumptions provide optimistic expected times. Therefore, many feel that scheduled times should be later than computed expected times. But some argue that automatically setting schedules later than expected times may increase the likelihood of schedule slippages and that expected times should be automatically used for establishing schedules. The basis for this argument is that the computed expected times provide for slippage, and since roughly half the activities will be completed in less than their expected times and half will require more than their expected times, one will balance out the other. In actuality, however, R&D activities usually take as long as their schedules permit and are seldom completed ahead of schedule. Thus, schedule slippages occur in R&D activities which were not contemplated when schedules were prepared.

The validity of PERT expected time is another controversial matter. Where PERT is applied to the early stages of weapons-system development programs, the critical path is frequently 1⅓ to 2 times as long as the originally planned program. No doubt the greater attention to detail that is necessary in applying PERT accounts for part of the additional time. A study of completed Air Force weapons-system development programs conducted independently of any PERT considerations, however, indicated that extensions of development time by one-third to one-half over the originally planned program were the rule rather than the exception.

The deficiencies of PERT are really of two different varieties—theoretical and behavioral. Some of the assumptions on which techniques such as PERT are based have been criticized as deviating too far from project realities.[10] Perhaps more important, the degree of credibility of the techniques in the minds of managers is affected by their practical failures and the apparently ethereal nature of the time estimates which form a basic input. Anything so nebulous as a subjective time estimate warrants suspicion, and when such subjective estimates have been used in applications where networks have not proved to be good project control devices, that verdict is verified in the minds of some managers.

Network Analysis Using Historical Estimating Behavior

A number of variations of traditional critical path approaches have been proposed and tested. These approaches generally are oriented toward the avoidance of some of the problems which are inherent in PERT.

One of these approaches focuses on the use of the historical estimating behavior of the estimator as a means of obtaining better results. The use of historical estimating behavior in project planning and control assumes that there is available a set of completed activities for which the actual activity durations are known. The completed activities for which a single estimating unit prepared time estimates constitutes the historical estimating data base for that unit. This data base may be used as a basis for adjusting estimates made by that estimating unit for ongoing or future activities.

The data base of completed activities may represent past projects in which the unit has participated. For instance, a government agency might treat a corporate bidder as an estimating unit and adjust the bidder's estimates for a project which is in the proposal stage by using information on the bidder's historical performance; or an individual engineering manager's estimating behavior may be viewed in a similar way by divisional or corporate analysts.

Alternatively, if information on past projects is unavailable, or if these past projects are deemed to be structurally different from current projects, historical data may be compiled on the early activities in a project to allow for adjustment of time estimates on subsequent activities. In either case, a set of completed activities, each with known estimates and known actual duration, forms the historical data base for each estimating unit.

The object of using historical behavior to adjust current estimates is to obtain better estimates. Some studies which have dealt with the question of activity-time accuracy have indicated that this is feasible.[11]

[10]For example, see K. R. MacCrimmon and C. A. Ryavec, op. cit.

[11]See William R. King and T. A. Wilson, "Subjective Time Estimate in Critical Path Planning: A Preliminary Analysis," *Management Science,* vol. 13, no. 5, pp. 307-320, January 1967, and William R. King et al., "On the Analysis of Critical Path Time Estimating Behavior," *Management Science,* vol. 14, no. 1, pp. 79-84, September 1967.

The most natural basis to use for adjusting activity-time estimates is one involving the multiplication of the single current time estimate for a future activity by the average ratio of actual activity duration (A) to estimated activity duration (E) for all activities in the estimating unit's data base. To do this, one would simply calculate the ratio A / E for each past activity, take the average, and apply it to current estimates. Such a procedure would represent a simple bias correction.

If b symbolizes the historical average A / E, the bias correction for the current time estimate for the jth activity (E_j) is:

$$E_j' = bE_j$$

Thus E_j' is the adjusted estimate for the jth activity. If the historical data base is composed of activities of highly variable absolute durations, it may be desirable to use a weighted average for determining the adjusting coefficient b to account for the fact that large absolute estimating errors on activities of long duration may be small, on a percentage basis, while small errors on an activity of short duration may be large, on a percentage basis.

In any case, the adjusted value may then be used in a critical path analysis in the same way that the unadjusted estimate might have been. There is some evidence to suggest that this will lead to better results than the more traditional method of analysis.[12]

Network Simulation

Another extension of PERT which is directed toward avoiding some of the PERT-induced problems involves the *simulation* of PERT networks. The simulations are based, as is PERT, on assumed probability distributions of the time required for each activity. However, simulations allow for the use of a wide variety of possible distributions, whereas PERT assumes a specific distribution.

Most network simulation models assume the form of the activity-time distributions to be known and fixed in advance. Their principal output is a "project duration" distribution, which is generated by simulating many possible situations and combinations of circumstances which *could* result from the assumed input distributions. The approach permits these many circumstances to be simulated and considered in their proper proportion, so that the output distribution which is generated is representative of the various project durations which might be expected to occur.[13]

[12]See William R. King and P. A. Lukas, "An Experimental Analysis of Network Planning," *Management Science,* vol. 19, pp. 1423–1432, August 1973.

[13]See R. M. Van Slyke, "Monte Carlo Methods and the PERT Problem," *Operations Research,* vol. 11, no. 5, 1963, pp. 839–860; and A. R. Klingel, Jr., "Bias in PERT Project Completion Time Calculations for Real Network," *Management Science,* vol. 13, pp. 194–201, 1967.

The simulation aproach obviates some of the theoretical difficulties with PERT analyses, since it simply replicates many realizations of the same network and bases its calculations on the empirical distribution of results which are generated. Thus, most of the often-criticized PERT assumptions, such as those of activity independence and the applicability of the Central Limit Theorem, are avoided.[14] However, the arbitrarily assumed input activity-duration distribution remains a weak element of most simulation analyses.

Simulations Using Historical Estimating Behavior

The simulation approach to network analysis has also been combined with the utilization of historical estimating behavior.[15] This combined approach utilizes the conditional probability distributions of the *A*'s (historical actual realized activity times), given the *E*'s (estimated times) as a basis for the simulation. Thus, when an estimate is made for an ongoing project, the distribution of all actual times which resulted for activities on which that same estimate was made in the past is used to generate simulated occurrences in their proper proportion. When this is done for all activities, it can lead to significantly better results than the more traditional approaches.[16]

Graphical Evaluation and Review Technique (GERT)

GERT is a technique for the analysis of networks in which the realization of events is probabilistic. Even though PERT is often thought of as permitting the consideration of uncertainty, it is, in fact, deterministic in that the critical path is assessed without regard to uncertainty. GERT addresses itself directly to the uncertainty which is inherent in real projects.

Unfortunately, the GERT approach is sophisticated and demands significant computer resources for the analysis of projects having only modest levels of complexity.[17]

PERT-Cost

PERT-Cost is a PERT-like technique which grew as a natural extension of the basic PERT (time) approach.[18] It is logical to assume that whatever could be

[14]See MacCrimmon and Ryavec, op. cit.

[15]William R. King, "Network Simulation Using Historical Estimating Behavior," *AIIE Transactions,* June 1971.

[16]William R. King, and P. A. Lukas, op. cit.

[17]The interested reader may explore further in A. A. Pritsker and G. E. Whitehouse, "GERT: Graphical Evaluation and Review Technique," parts 1 and 2, *The Journal of Industrial Engineering,* no. 5 and no. 6 (1966), pp. 267–274 and 293–301.

[18]H. W. Paige, "How PERT-Cost Helps the General Manager," *Harvard Business Review,* pp. 87–95, November–December 1963.

accomplished with PERT on the basis of time could as well be accomplished on a cost basis. However, for many of the same reasons (e.g., lack of data and difficulty in obtaining managerial understanding) that CPM (the technique, rather than the general methodology) has never made much impact, PERT-Cost has not been widely accepted and implemented.

Line of Balance (LOB)

The line of balance technique is oriented toward the control of production activities. Although it can be used in such developmental projects as those involving the production of prototypes, it emphasizes the extent to which the planned production of a quantity of items is actually being realized. Thus, its utility in project management is limited.

However, it is interesting to note that the weaknesses of LOB are generically similar to those of the other project planning and control techniques. It requires the estimation of the "percent completion" of component parts, and it is more difficult to understand than comparable techniques such as the Gantt chart.[19]

SUMMARY

This chapter deals with the process of planning and controlling a project and focuses on those techniques which have been widely applied and/or discussed in the literature of project management. The simpler techniques, such as WBS, Gantt charts, and simple network plans, are in widespread use in a wide variety of environments from the construction industry to highly sophisticated systems-development contexts. Many of the more complex tools, such as PERT, PERT-Cost, network simulation, and GERT have proved to be of use in specific situations, but they are not in widespread use. In part, this is due to their complexity and the requirements for the manipulation of great volumes of data.

As well, PERT was, at one time, *required* of all U.S. federal government contractors. Any such requirement for use of a management tool is certain to reflect unfavorably on that tool, whatever its intrinsic merits. As a result, the PERT requirement was subsequently dropped and PERT has had an unfavorable connotation (probably beyond that which it merits) ever since.[20]

Various computer programs for CPM, PERT, and straightforward network planning are available in time-sharing systems.[21] This makes it easy to gain access to network planning without a great deal of labor. The payoffs can be great.

[19]See J. N. Holtz, "An Analysis of Major Scheduling Techniques in the Defense Systems Environment," RM-4697-PR, The RAND Corp., October 1966. Reprinted in D. I. Cleland and W. R. King (eds.), *Systems, Organizations, Analysis, Management: A Book of Readings,* McGraw-Hill Book Company, 1969, pp. 317–355.

[20]W. G. Ryan, "Management Practice and Research—Poles Apart," *Business Horizons,* June 1977.

[21]J. R. Martin, "Computer Time-Sharing Applications in Management Accounting," *Management Accounting,* July 1978.

There is no question that simple tools such as WBS, Gantt charts, and basic critical path networks (although not perhaps the critical path *method,* per se) are extremely useful at many levels of project sophistication. Despite the negative connotation of such techniques as PERT, the use of the simpler tools is being extended into new areas such as auditing.[22]

The lessons of these techniques are clear. There is great value in using a formal model for the planning and control of complex projects. However, if the models are so complex that their data requirements become costly, they will not be used. So, too, with their level of complexity as it relates to managerial comprehension. If the model is not susceptible of being communicated to managers in a manner which permits them to understand its underlying logic, if not its every mathematical detail, it will not be used, regardless of its worth.[23] Finally, if a model does not appear to be realistic and to have realistic data requirements, even though it may be superficially accepted, it will probably not be used in making critical project decisions.

DISCUSSION QUESTIONS

1 What is the role of standards in planning and control? If a project manager is appointed to undertake a research project in an R&D laboratory, what kinds of standards might be developed to use in the planning and control of the research project?
2 Relate the concept of feedback to standards in the control function.
3 How can the control function be conceptualized to form the central core of the manager's job? In this conceptualization, relate other management functions to the control function.
4 Project planning involves "thinking about the future" of the project. Some projects have a life cycle of ten or more years. Considering this, how are project planning and strategic planning related in the context of the overall organizational mission?
5 Why can it be said that the project plan is the least important part of the planning process?
6 Projects have their genesis as ideas evolving out of the mainstream of organizational activities. What relationship does project planning have to the planning for a functional organization that would be supporting a variety of projects?
7 How is the objective of a project-planning phase achieved?
8 Identify and discuss some of the interrelated planning and control techniques that are in use today. Under what circumstances should these techniques be used?
9 For small projects, the Gantt chart may be all that is needed to track schedules. Defend or refute this statement.
10 How can precedence diagrams be used as a planning and control tool?
11 What are some of the key essentials to be concerned with in the analysis of network plans?

[22] J. L. Krogstad, G. Grudnitski, and D. W. Bryant, "PERT and PERT/Cost for Audit Planning and Control," *The Internal Auditor,* August 1979.
[23] C. W. Dane, C. F. Gray, and B. M. Woodworth, op. cit.

12 What role is played by "dummy" activities in a network plan? How do they differ from "real" activities? What are "critical" activities? In what sense are they critical?
13 What is "slack"? What is its importance to the project manager?
14 Summarize the management uses to which a project network can be put.
15 What are some of the values in using a formal model for the planning and control of complex projects? When might a formal model become too complex to use? Too costly to use?

RECOMMENDED READINGS

Albert, K. J.: *How to Be Your Own Management Consultant,* McGraw-Hill Book Company, New York, 1978.
Archibald, R. D., and R. L. Villoria: *Network-based Management Systems (PERT/CPM),* John Wiley & Sons, Inc., New York, 1967.
Carruthers, J. A., and A. Battersby: "Advances in Critical Path Methods," *Operational Research Quarterly,* vol. 17, no. 4, December 1966, pp. 359–380.
Childs, Marshall R.: "Does PERT Work for Small Projects?" *Data Processing,* vol. 4, no. 12, pp. 32–35, December 1962.
Clark, Charles E.: "The Greatest Finite Set of Random Variables," *Operations Research,* vol. 9, pp. 460–470, 1961.
———: "The PERT Model for the Distribution of an Activity Time," *Operations Research,* vol. 10, no. 3, pp. 405–406, May–June 1962.
Clark, Wallace: *The Gantt Chart,* 3d ed., Sir Isaac Pitman & Sons, Ltd., London, 1952.
Clark, Mrs. Wallace: "The Gantt Chart," in Harold B. Maynard (ed.), *Industrial Engineering Handbook,* McGraw-Hill Book Company, New York, 1956.
Clingen, C. T.: "Modification of Fulkerson's PERT Algorithm," *Operations Research,* vol. 13, pp. 629–632, 1964.
Dane, C. W.; C. F. Gray, and B. M. Woodworth: "Factors Affecting the Successful Application of PERT/CPM in a Government Organization," *Interfaces,* November 1979.
DeCostes, D. T.: "PERT-COST—The Challenge," *Management Services,* July–August 1965, pp. 13–18.
Davis, E. W.: "Resource Allocation in Project Network Models: A Survey," *Journal of Industrial Engineering,* vol. 17, no. 4, pp. 177–188, April 1966.
Donaldson, W. A.: "The Estimation of the Mean and Variance of a PERT Activity Time," *Operations Research,* vol. 13, pp. 382–385, 1965.
Dooley, A. R.: "Interpretations of PERT," *Harvard Business Review,* March–April 1964, pp. 161–172.
Dunne, E. J., Jr., R. F. Ewart, and D. M. Nanney: "What Happened to PERT?" *Defense Systems Management Review,* Winter 1976.
Fulkerson, D. R.: "Expected Critical Path Length in PERT Networks," *Operations Research,* vol. 10, pp. 167–178, 1962.
Grubbs, F. E.: "Attempts to Validate Certain PERT Statistics or 'Picking on PERT,'" *Operations Research,* vol. 10, pp. 912–915, 1962.
Hajek, V. G.: *Management of Engineering Projects,* McGraw-Hill Book Company, New York, 1977.
Hansen, B. J.: *Practical PERT,* American Aviation Publishing, Washington, D. C., 1964.
Hogarth, R. M.: "Cognitive Processes and the Assessment of Subjective Probability Distributions," *Journal of the American Statistical Association,* vol. 70, 1975.

Jodka, John: "PERT (Program Evaluation and Review Technique): A Control Concept Using Computers," *Computers and Automation,* vol. 11, no. 3, pp. 16–18, March 1962.

Kahalas, H.: "A Look at Major Planning Methods: Development, Implementation, Strengths and Limitations," *Long Range Planning,* August 1978.

Kelley, James E., Jr.: "Critical-path Planning and Scheduling: Mathematical Basis," *Operations Research,* vol. 9, no. 3, pp. 296–320, May–June 1961.

King, L. T.: *Problem Solving in a Project Environment,* John Wiley & Sons, Inc., New York, 1981.

King, W. R.: "Project Planning Using Network Simulation," *Pittsburgh Business Review,* vol. 38, pp. 1–8, 1968.

———: "Network Simulation Using Historical Estimating Behavior," *AIIE Transactions,* June 1971.

———, and P. A. Lukas: "An Experimental Analysis of Network Planning," *Management Science,* vol. 19, August 1973.

———, and T. A. Wilson: "Subjective Time Estimates in Critical Path Planning: A Preliminary Analysis," *Management Science,* vol. 13, no. 5, January 1967.

———, D. M. Wittevrongel, and K. D. Hezel: "On the Analysis of Critical Path Time Estimating Behavior," *Management Science,* vol. 14, no. 1, Sept. 1967.

Klingel, A. R., Jr.: "Bias in PERT Project Completion Time Calculations for Real Network," *Management Science,* vol. 13, pp. 194–201, 1967.

Krogstad, J. L.; G. Grudnitski, and D. W. Bryant: "PERT and PERT/Cost for Audit Planning and Control," *The Internal Auditor,* August 1979.

Kushnerick, J. P.: "How Dynasoar Managers Used PERT," *Aerospace Management,* vol. 7, no. 1, pp. 20–23, January 1964.

Lukas, P. A.: "A Theory of Estimating the Project Duration in Critical Path Planning," unpublished Ph.D dissertation, Graduate School of Business, University of Pittsburgh, 1968.

Lukaszewicz, J.: "On the Estimation of Errors Introduced by Standard Assumptions Concerning the Distribution of Activity Duration in PERT Calculations," *Operations Research,* vol. 15, pp. 326–327, 1965.

MacCrimmon, K. R., and C. A. Ryavec: "An Analytical Study of PERT Assumptions," *Operations Research,* vol. 12, no. 1, 1964.

Malcom, D. G., J. H. Rosebloom; C. E. Clark, and W. Frazer: "Application of a Technique for Research and Development Program Evaluation," *Operations Research,* vol. 7, no. 5, September–October 1959.

Martin, J. R.: "Computer Time-Sharing Applications in Management Accounting," *Management Accounting,* July 1978.

Moder J. J., and C. R. Philips: *Project Management with CPM and PERT,* Reinhold Publishing Corporation, New York, 1964.

Moskowitz, H., and W. I. Bullers: "Modified PERT Versus Practice Assessments of Subjective Probability Distributions," *Organizational Behavior and Human Performance,* vol. 24, 1979.

Naylor, T. H.; J. L. Balintfy; D. S. Burdick, and K. Chu: *Computer Simulation Techniques,* John Wiley & Sons, Inc., New York, 1966.

O'Brien, James J.: *CPM in Construction Management,* McGraw-Hill Book Company, New York, 1965.

Parks, William H., and Kenneth D. Ramsing: "The Use of the Compound Poisson in PERT," *Management Science,* vol. 15, no. 8, April 1969.

Paige, H.W.: "How PERT-Cost Helps the General Manager," *Harvard Business Review,* pp. 87–95, November–December 1963.

Pocock, J. W.: "PERT as an Analytical Aid for Programming Planning: Its Payoff and Problems," *Operations Research,* vol. 10, no. 6, pp. 893–903, November–December 1962.

Ryan, W. G.: "Management Practice and Research—Poles Apart," *Business Horizons,* June 1977.

Sadow, R. W.: "How PERT Was Used in Managing the X-20 (Dyna-Soar) Program," *IEEE Transactions on Engineering Management,* vol. FM-11, no. 4, pp. 138–154, December 1964.

Schoderbeck, P. O.: "PERT/Cost: Its Values and Limitations," *Management Services,* January–February 1966, pp. 29–34.

Van Slyke, R. M.: "Monte Carlo Methods and the PERT Problem," *Operations Research,* vol. 11, no. 5, 1963.

Vazsonyi, A.: "L'Histoire de grandeur et de la décadence de la méthode PERT," *Management Science,* vol. 16, pp. B449–B455, 1970.

DOD and NASA Guide: PERT Cost Systems Design, Office of the Secretary of Defense and National Aeronautics and Space Administration, Washington, D. C., June 1962.

"Making Project Management Easy," *Datamation,* April 1978.

CASE 15-1: A Project Plan

Situation No matter how small a project may be, it needs a plan that defines *what* is to be done, *by whom, when,* and for *how much.*

Your Task Develop a brief model of a "plan for a project plan" that can be used within an industrial organization as a standard model for project planning. In doing this task, consider what the basic contents of a project plan should be.

What do you think should go into a project plan? Have the team leader present your findings.

CASE 15-2: Building a Plant in "X" Country—Who Is Involved?

Situation Your team has just been selected by your division manager as a preliminary task force for the conceptualization, design, development, and construction of a new plant to be operational in country "X" within three years. In three weeks, the team is expected to present an *organizational plan* for getting the task force underway. One of your first priorities is to come up with a summary "work breakdown structure" for the plant and an identification of the key individuals you would want on the task force.

Your Task Be prepared to present a *summary outline* of the briefing you plan to give to your division manager. This briefing should be sufficient to convince your manager that your task force is off to a good start. In carrying out this assignment, you may make any assumptions you wish, such as:

Product(s) to be manufactured in the plant
Plant configuration
Market proximity
Political/social stability of the country

ORGANIZATIONAL SUPPORT FOR SYSTEMS-ORIENTED MANAGEMENT

CHAPTER **16**

INFORMATION SUPPORT FOR SYSTEMS PLANNING AND IMPLEMENTATION

Knowledge is of two kinds: we know a subject ourselves, or we know where we can find information upon it.[1]

All of the planning, analysis, and implementation activities that have been discussed in prior chapters depend intimately on a single common commodity: information. While some think that information is in so plentiful a supply that they are deluged with it, the fact is that management today is characterized by an *overabundance of data* and a *dearth of relevant information.*[2]

"Information," in this context, means data that have been evaluated to be useful for some purpose—to support the making of a decision, to make possible the detection of a problem in an ongoing project, etc. All too frequently, managers have stacks of computer printouts containing much *data* but too little of the critical *information* that they need.

In this chapter, we separately treat the role of information in the planning and implementation phases of management and discuss some *information systems* concepts that serve to integrate ideas that have been brought forth in the preceding chapters.

[1]James Boswell, *Life of Dr. Johnson.*

[2]See R. L. Ackoff, "Management Misinformation Systems," *Management Science,* vol. 14, no. 4, December 1967.

INFORMATION SYSTEMS AND DECISION SUPPORT SYSTEMS

The information-systems area is filled with acronyms—EDP, MIS, DSS, etc. These reflect a continuous concern, throughout the computer era, of providing information that is *managerially relevant*.

In the early stages of the computer era, electronic data processing (EDP) systems were developed to automate many of the organization's routine procedures, such as billing and ordering. The management information systems (MIS) era followed in which higher-level systems were given attention. The specific goal of these systems was to go beyond the *operational control* level at which EDP focused to the level of *management control.*[3] At the management control level, decision-relevant information is provided, in the form of fixed reports, by the MIS.

At the strategic planning level, the latest computer application is that of the *decision support system* (DSS).

In some sense, "DSS" is merely the latest term used to describe the application of computer-based systems to the support of management decision making. Much that is now written and said about the goals of the DSS is similar to those things that were espoused as the purposes of "management information systems" a decade or so ago.[4]

However, the DSS is a variety of information system that takes advantage of the latest technology to go beyond the providing of fixed reports to management. The most useful definition of a DSS is that it is an integrated system of subsystems that have the purpose of providing information to aid a decision maker in making better choices than would otherwise be possible.

The subsystems that are commonly associated with DSSs are:

a Decision-relevant models
b Interactive computer hardware and software
c A data base
d A data management system
e Graphical and other sophisticated displays
f A modeling language that is "user friendly"

There is no minimum set of such subsystems that must be incorporated into a system to make it a DSS. However, most such systems involve many of these subsystems.[5]

Thus, a DSS is an interactive computer-based system that utilizes decision

[3]See Chap. 14.

[4]G. B. Davis, *Management Information Systems: Conceptual Foundations, Structure, and Development,* McGraw-Hill Book Company, New York, 1974.

[5]See S. Alter, *Decision Support Systems: Current Practice and Continuing Challenges,* Addison-Wesley Publishing Company, Inc., Reading, Mass., 1980; E. Carlson, "Proceedings of A Conference on Decision Support Systems," *Data Base,* vol. 8, no. 3, 1977; P. G. W. Keen, and M. S. Scott Morton, *Decision Support Systems: An Organizational Perspective,* Addison-Wesley Publishing Company, Inc., Reading, Mass., 1978.

models, gives users easy and efficient access to significant data bases, and provides various display possibilities. As well, it usually incorporates a "user friendly" modeling language, such as IFPS or SIMPLAN,[6] to give users the opportunity to go beyond preprogrammed models to construct and use their own decision-aiding constructs. (The "user friendly" designation means that a nonspecialist can formulate models while directly interacting with the system.)

Such systems have obvious potential for application in the strategic planning area, where problems are not initially well defined and perhaps not even well recognized or understood. The flexible capabilities of a DSS give users (the managers or planners) the opportunity to ask for information, to test out alternative ways of viewing the problem, to subsequently ask for different information, to use preprogrammed models, to construct their own decision-aiding models, etc. Such a flexible iterative process is much like the way in which real-world strategic decision making is conducted. Many of the computerized models that had been proposed for use in planning before the DSS era[7] were not realistic in that they presumed "the problem" to be well understood and well formulated—a characteristic seldom present in strategic decision problems.

DSSs use modern technology to allow users to interact with the system and to obtain information from it that is relevant to their problem. For instance, Equitable Life Assurance Society has developed CAUSE, a DSS that provides computer assistance in the making of insurance underwriting decisions.[8] IBM's research division has developed GADS (Geodata Analysis and Display System) that permits great flexibility to users in viewing spatial arrangements of data in a wide variety of situations. This system has proved to be useful in analyzing various configurations of census tracts for political redistricting, police beat design, and a variety of other strategic choices that involve spatial configurations as strategic alternatives. Others, such as RCA, Citibank, Louisiana National Bank, American Airlines, and the First National Bank of Chicago have reported the successful implementation of DSSs.[9]

Rodriguez and King[10] and Dutta and King[11] have developed DSSs appropriate for the analysis of competitive strategy. One, called SICIS (Strategic Issue Competitive Information System), uses the concept of a "strategic issue" to help a user to interactively gain access to a complex data base consisting of

[6]T. H. Braun, "The History, Evolution and Future of Financial Planning Languages," *ICP Interface,* Spring 1980.

[7]W. F. Hamilton, and M. A. Moses, "An Optimization Model for Corporate Financial Planning," *Operations Research,* vol. 21, 1973.

[8]Alter, op. cit.

[9]Alter, op. cit., R. H. Sprague and R. L. Olson, "The Financial Planning System at Louisiana National Bank," *MIS Quarterly,* 1979, vol. 3, no. 3, pp. 35–46; G. M. Welsch, "Successful Implementation of Decision Support Systems' Pre-installation Factors, Service Characteristics, and the Role of the Information Transfer Specialist," Ph.D. dissertation, Northwestern University, Evanston, Ill., 1980.

[10]J. I. Rodriguez and W. R. King, "Competitive Information Systems," *Long Range Planning,* December 1977.

[11]B. K. Dutta and W. R. King, "A Competitive Scenario Modeling System," *Management Science,* vol. 26, no. 3, 1980.

information about competitors. The other, called COSMOS (Competitive Scenario Modeling System), enables users to test out proposed strategies by trying to "outthink" their competitors, i.e., by anticipating their likely responses. The system permits many combinations of factors to be considered after simple input information and judgmental data have been provided by the user.

INFORMATION SYSTEMS IN PLANNING

In assessing the role of information in the planning function, we may address information in terms of uses that are unique to the strategic level of the organization as well as in terms of the direct role of information in systems analysis.

Information Systems for Planning[12]

An organization's strategic planning effort is aimed at providing a sense of direction when approaching an uncertain future, the nature of which will only in part reflect the organization's own goals and choices. Forces in the environment—everything outside the organization itself—will also play an important role in determining the organization's future, so that effective strategic planning must operate to permit the organization to assess the environment, to forecast it, to develop strategies for taking advantage of it, and, to the degree possible, to alter it.

Environmental information is, therefore, critical to effective strategic planning. However, many organizations base their strategic planning more on judgment, intuition, partial data, and ad hoc studies than on objective, systematic information that is routinely collected and analyzed for strategic purposes. This is the case, in part, because they have justified information systems largely on the basis of cost efficiencies rather than on increased organizational effectiveness and, in part, because of the conceptual and practical difficulties inherent in the definition of systems designed to support strategic decision making.

Indeed, it is possible to argue that truly strategic decisions are of such a unique and unstructured nature that it is not cost-effective to develop an information system to support them. While this may be true to some degree, it is feasible, and even cost-effective, to develop information systems to support strategic planning *processes* (as opposed to individual strategic decisions).

Table 16-1 depicts a strategic planning *process* in terms of a number of key phases, identifies a number of key *sources* of environmental information, and gives descriptive names to a number of environmental *information subsystems*.

Several points must be made in explaining Table 16-1. First, there is no unique

[12]Portions of this section are adapted from W. R. King and D. I. Cleland, "Environmental Information Systems for Strategic Marketing Planning," *Journal of Marketing*, vol. 38, October 1974.

TABLE 16-1
INFORMATION SUBSYSTEMS RELATING INFORMATION SOURCES TO THE PLANNING PROCESS

Strategic planning process	Environmental information subsystems	Strategic information sources
Situation assessment (What is our current situation?)	Image subsystem Customer subsystem	Customers
Goal development (What do we want our future situation to be?)	Potential customer subsystem	Potential customers
Constraint identification (What constraints might inhibit us?)	Competitive subsystem Regulatory subsystem	Competitors Government
Selection of strategies (What actions should we take to achieve our goals?)	Intelligence subsystem	

correspondence between the elements in each row; the various subsystems and information sources are interrelated, and each impinges on more than one phase of the planning process. However, there is a general primary relationship that is identified by the elements in each row.

Second, the various subsystems need not be developed as computerized information systems. The term "system" is used here to describe a systematic, continuous, and formal set of activities that provide decision-related information.

Third, no inference should be drawn that it is necessary, or even feasible, for a single organization to develop the total system described here. The framework of Table 16-1 will be explained in subsequent sections in terms of specific systems in whose development we have participated. However, no single organization has, in fact, implemented all of these subsystems, and it may well not be cost-effective for any single firm to do so. The systems described here are illustrations of the kinds of environmental information systems that may prove to be useful and cost-effective to any given organization.

Illustrative Environmental Information Subsystems

The most basic assessment made by the managers of an organization is summed up in the "What is our current situation?" or "Where are we now?" questions, as indicated in the top left of Table 16-1. To function effectively, every organization must continually assess its status relative to both its history and its environment.

Such an assessment requires objective and subjective measurements. At the objective level, the necessary information is that which is readily obtainable from internal sources—data on profits, costs, the organization's financial status, and, in general, historical performance data that are produced by the internal accounting and financial information systems.

Image Information Subsystem These objective data can be readily complemented with subjective judgmental data from internal sources. Whether this is done formally[13] or informally, internal judgments are often overly biased by the influence of the readily available objective data. More important, internally generated judgmental data do not provide critical information concerning the firm's external *image* as it is projected to and perceived by the customers and potential customers on whom the firm depends for its success.

Our experiences suggest that there are great discrepancies between a firm's image of itself and the image held by its customers. Often these discrepancies are less significant in their impact on the firm's current operations than in terms of their potential impact in the future. For instance, the firm that sees itself as technically superior in an era when cost is becoming more significant may find that its image of being "high-priced" is more important to its future success.

A firm's image may be assessed in two general areas: product image (price, quality, reliability, etc.) and organizational image (quality of personnel, responsiveness, integrity, etc.). The basic techniques to use in the formal image survey are *structured and unstructured personal interviews* of key customer personnel. A questionnaire to serve as a guide for the conduct of these interviews can be developed and tested within the seller's organization. This in-house testing can be used as a basis to define and operationally describe the important dimensions of the product and organizational characteristics that are deemed to be important to the seller's image. For example, in one such survey that we conducted, the customer interviews centered around an evaluation of the following product and organization characteristics areas:

General characteristics
Personnel image
Ability to communicate with customers
Project-management skills and capabilities
Ability to meet normal customer requirements
Responsiveness to customer's special requirements
Negotiating skills
Special capabilities
Product characteristics

The overall image that emerged in this case was surprising to the executives of the sponsoring organization. It depicted an honest and technically competent organization that lacked marketing aggressiveness. This lack of agressiveness

[13]See William R. King, "Human Judgment and Management Decision Analysis," *Journal of Industrial Engineering,* vol. 18, pp. 17–20, December 1967, for an assessment of using formal assessments of human judgment in management; and William R. King, "Intelligent Management Information Systems," *Business Horizons,* vol. 16, pp. 5–12, October 1973, for a description of the incorporation of human judgmental data into information systems.

was reflected in the customers' perceptions of virtually all aspects of customer contact, from the bureaucratic lack of responsiveness to customer inquiries to the lack of contact of top management with customers. Such specifics as the failure to communicate to customers about key personnel changes in the organization and deficiencies in the technical proposals presented to customers were also pointed out. The seller's products were rated high in terms of operating characteristics—performance, reliability, and ease of maintenance—but customers raised serious questions about the seller's overall capability to manage a technical product-development effort and still maintain cost and schedule credibility.

The image survey was also conducted internally by querying personnel within the sponsoring company. The contrast between customer perceptions and internal personnel perceptions led management to take a number of specific actions designed to have a short-run impact on the image as well as to formalize the incorporation of image considerations into the strategic planning activities of the firm. This led, for the first time, to specific concern with the image that the company wished to project and the actions that it could take to reach this image goal.

Such incorporation of image information as an integral and continuing part of the strategic planning process requires that some type of formal information subsystem be established. In the case in point, the economic impracticality of continuing large-scale surveys led the firm to integrate the continuing image-monitoring activities into other information subsystems where image-related surrogate measures were monitored and assessed. In this firm, the overall image assessment is to be updated at two-year intervals.

Customer Information Subsystems In most firms, the area of customer information is the best developed of all of the environmental information subsystems. However, much of the existing customer information is not systematically used for any decision purposes, much less for strategic planning.

Two types of customer information are most useful for strategic planning: aggregate information and trend information. Thus, while data on a specific customer may be useful in the short-term decisions of the sales manager, long-range decision making requires that sets of customers who form important market segments be identified and analyzed. Such segments are made up of customers who are homogeneous in some sense that is relevant to strategic planning—for example, a common industry, common behavior, common responses to changes in the business cycle, and the like.

Trend information, both in terms of individual customers and for market segments, is also important to strategic planning. For instance, is a given market segment likely to increase or decrease in importance in the future? Will a given segment be changing so that a different strategy will be required to retain them as customers? These are questions related to strategic planning that can only be answered through analyses of aggregates and trends.

Thus, the keys to creating customer information subsystems that are supportive of strategic planning are two-fold: first, new varieties of information in the form of aggregates and trends; and second, a built-in analytic capability that permits the objective analysis of the strategic customer information.[14] While many customer information systems are in existence, few have significant capabilities in these areas.

Potential Customer Information Subsystems While most organizations have some form of organized information about current customers, few have similar information on potential customers. Yet such information is of equal importance for the development of strategic goals, since potential customers represent the opportunities that will ultimately determine the organization's future. Information on potential customers permits the organization to make rational choices concerning its future products, services, and markets.

The development of a potential customer information subsystem is not a straightforward task for most organizations. The list of potential customers is infinite, so some rational culling of this list must be performed. This may be begun by using a criterion that reflects the potential of a particular segment of the overall market. For instance, one commercial bank determined that many small manufacturing firms in the local area could avail themselves of a variety of bank services. They began to construct a data base using commercially available services such as Dun & Bradstreet's State Sales Guides[15] and those provided by various manufacturers' associations. They then assessed the potential of various segments of the market through personal contacts made on a test basis.

Another firm, after having built the data base and having identified high-potential firms, developed a "clipping service" for collecting and assembling published references to these potential customers. In this way, a great deal of intelligence information concerning the performance, plans, new products, finances, and the like, of other organizations was obtained. Although this approach may seem naïve to the uninitiated, it is the essence of any good intelligence function, and firms that have tried it have often found it to be of surprising significance.

Competitive Information Subsystem Many organizations do not have a great deal of nonhearsay information about competitors. Often the limited hearsay information that is available is misleading and, in any case, such unsystematic competitor information usually does not provide a sound basis for strategic planning.

One of the most useful tools in developing a competitive information system is a profile of each competitor. Such a profile should delineate the business "character" of the competitor. One company constructed profiles of all competitors to focus on such factors as:

[14]The various levels of analysis that may form an intrinsic part of an information system are treated in William R. King and David I. Cleland, "Manager-Analyst Teamwork in MIS Design," *Business Horizons*, vol. 14, pp. 59-68, April 1971.

[15]Published for various states by Dun and Bradstreet, Inc., New York, New York 10008.

a Background of key competitor personnel
b Characteristics of projects on which competitive proposals were made
c Characteristics of projects on which competitive proposals were not made
d Mix of competitor's in-house business
e Assessment of competitor's marketing strategy
f Assessment of relative value placed by competition on various performance measures—for example, product quality, service capability, and the like

From a compilation of basic public information and informed inferences about competitors emerged clear pictures that had not previously been perceived by the firm. For instance, one competitor clearly bid only on projects having a key common characteristic. Another was seen to be solidly in the control of managers with engineering backgrounds. The recent behavior of a third competitor was explained by the backgrounds of a number of nontechnical people who had recently moved into key executive positions. When these profiles were reported and discussed, some critical decisions were made concerning the company's future marketing strategy. The key to the strategy was the company's ability to identify a place for itself in the market—one that provided it with a comparative advantage over the competition.

The profile concept can be instituted as a regular part of the information system. It should be linked to a clipping service and updated on a continuing basis. In more advanced applications, it can be supplemented with competitive image assessments made parallel with the firm's own image assessment. Such information can form a data base when key questions or issues are being dealt with, as well as a source of valuable information that can be summarized for use in the ongoing strategic planning process.

Regulatory Information Subsystem Every organization operates in an environment that imposes formal constraints on it and its activities. The most obvious such constraints are government regulations. Moreover, every organization has individuals who are knowledgeable about the existing regulatory environment. However, their knowledge is often used only in an informal way, and usually after commitments have already been made in ignorance of the constraints.

The basic nature of strategic planning—which involves *new* and unfamiliar areas for an organization—normally mitigates against such regulatory information being readily available to those who are doing the planning. Managers may know the regulatory environment for the products and markets that they are used to dealing with, but they cannot be expected to be familiar with the regulations surrounding *new* areas. Thus, the strategic planning environment is fraught with the danger of expending planning and development resources in ignorance of crucial regulatory constraints. Such a situation cries out for a formalized data base with easy access by the many managers who participate in strategic planning.

The basic characteristic of a regulatory information subsystem is the same as that of any information retrieval system. The development of such a system requires that a taxonomy of the regulatory environment be developed. Then key descriptors can be used by managers to access specific domains of the taxonomy. In this way, the regulations that are relevant to a particular product, industry, or political subdivision can be furnished to planners who have need for comprehensive regulatory information as it applies to a specific area for which planning is being accomplished.

Intelligence Information Subsystem As used here, the term "intelligence" refers to specific facts or the answers to specific questions concerning happenings in the environment. For instance, the answer to a question concerning a competitor's intentions to bid or not bid on a project is an intelligence item, as is an assessment that a potential customer will soon be changing suppliers.

The usual definition of the term is broader than that used here. For instance, the competitor profiles and other aspects of the various information systems previously discussed also qualify under the more widely accepted military use of the term. Here, however, they are incorporated into other subsystems and specifically excluded from the intelligence subsystem.

The critical aspects of intelligence gathering are *organization* and *systemization.* It is not the purpose here to enumerate the myriad data sources and data collection requirements for a good business intelligence system but, rather, to establish the desirability of having a formalized intelligence system, and the authority and responsibility patterns that are appropriate for effective intelligence activities.

The critical point in the intelligence subsystem, as in the potential customer and competitive subsystems, is to gather intelligence systematically, to have it evaluated, aggregated, and analyzed by trained people, and *to ensure that it is distributed to those decision makers who can make use of it.* If this can be done in a parsimonious fashion to ensure that the great amount of redundant and irrelevant information already flowing around in the organization is not merely made larger, the benefits can far outweigh the costs of such an operation.

In the development of an intelligence system, the most important element is the people who will develop and implement it. Moreover, the most important factor in determining its effectiveness is the recognition that everyone in an organization is involved both in the marketing function and in the process of intelligence gathering. The engineer who discusses specifications with the customer is both a marketer and intelligence agent, as is the field marketing representative. Indeed, technical people can often have marketing impact of a far different and more significant variety than can the professional marketer or undercover agent. So, too, is top management involved both in marketing and in the collection of market intelligence. One of the most significant results of the image survey example described earlier was the recognition by the company that top managers who had preached a customer-oriented marketing approach for years, were not themselves personally customer-oriented.

An effective approach to ensure that all play their marketing intelligence roles is to specifically integrate them into the intelligence-gathering network. When engineers are to have customer contact, they must be made aware of the critical information that is needed and who in the customer's organization is likely to have it. Top management should be similarly briefed before their visits to customers and debriefed on return. In this way, a great deal of relevant information can be garnered and provided to those decision makers who are in need of it.

Of course, all of this presumes that an office in the organization has been set up for the collection and analysis of intelligence information. The analysis of intelligence involves a determination of the relevance, credibility, value, and appropriate dissemination of intelligence data. This central office can also perform the function of gathering together the key questions and identifying the voids in the knowledge necessary for effective strategic marketing planning. These questions can be asked in a routine fashion of field personnel and others who might be expected to have relevant information. Often, these people have the desired information in one form or another, but without a formalized intelligence system they have no way of getting it to the right people or of having it integrated with other information to form useful information aggregates.

This same intelligence organization—with its focus on analysis, eliminating redundancies, posing questions, and disseminating information to those who are in need of it—can also function as a part of the competitive and potential customer information subsystems. For example, data provided by clipping services require much the same analysis whether they relate to competitors or customers.[16]

The Nature of Planning Information Systems

To emphasize that all of these varieties of planning subsystems need not be of the fullest computerized variety, Table 16-2 is presented to illustrate three "environmental scanning" system models that two studies have found to exist in many firms.[17]

Table 16-2 characterizes the environmental scanning and forecasting activities so crucial to planning in terms of three models, termed "Irregular," "Periodic," and "Continuous," in increasing order of sophistication and complexity. Some of the selected dimensions which are operationally useful in distinguishing among the three categories are defined in the row labels in the table.

Irregular systems are characterized by the reactive nature of planning as well as environmental scanning. These systems respond to environmentally generated crises. Such systems are not really systematic. Their focus is on specific problems which tend to be short-term in nature. Methodologically, these systems rely on simplistic tools which primarily utilize information from the past. At the extreme, they may focus on the near-term budgetary impact of the events

[16]For more details, see D. I. Cleland and W. R. King, "Competitive Business Intelligence Systems," *Business Horizons,* vol. 18, no. 6, December 1975.

[17]L. Fahey and W. R. King, "Environmental Scanning for Corporate Planning," *Business Horizons,* August 1977; and L. Fahey, W. R. King, and V. K. Narayanan, "Environmental Scanning and Forecasting in Strategic Planning: The State of the Art," *Long Range Planning,* February 1981.

TABLE 16-2
A TYPOLOGY OF ENVIRONMENTAL SCANNING AND FORECASTING SYSTEMS

	Irregular	Periodic	Continuous
Impetus for scanning	Crisis-initiated	Problem-solving decision/issue oriented	Opportunity finding and problem avoidance
Scope of scanning	Specific events	Selected events	Broad range of environmental systems
Temporal nature	Reactive	Proactive	Proactive
(a) Time frame for data	Retrospective	Current and retrospective	Current and prospective
(b) Time frame for decision impact	Current and near-term future	Near-term	Long-term
Types of forecasts	Budget-oriented	Economic- and sales-oriented	Marketing, social, legal, regulatory, culture, etc.
Media for scanning and forecasting	Ad hoc studies	Periodically updated studies	Structured data collection and processing systems
Organization structure	(1) Ad hoc teams (2) Focus on reduction of perceived certainty	Various staff agencies	Scanning unit, focus on enhancing uncertainty handling capability
Resource allocation to activity	Not specific; (perhaps periodic as "fads" arise)	Specific and continuous but relatively low	Specific continuous and relatively substantial
Methodological sophistication	Simplistic data analyses and budgetary projections	Statistical forecasting oriented	Many "futuristic" forecasting methodologies
"Cultural" orientation	Not integrated into mainstream of activity	Partially integrated as a "stepchild"	Fully integrated as crucial for long-range growth

over budget
over time

which they monitor. The organizations which use these systems generally have not created a "strategic planning culture." More important, however, these systems attempt to reduce uncertainty in the current and near-term future environment and, in doing so, generally fail to detect opportunities to facilitate the creation of radically new solutions to problems.

Periodic systems, on the other hand, are more sophisticated and complex. While the focus of these systems is still problem solving, they exhibit greater proactive characteristics. These systems look more toward the future, but they emphasize near-term environmental changes. As a result, while they are forecasting-oriented, the forecasts that they produce are limited in their scope and methodologies: they emphasize economic and sales projections using simple statistical methodologies. There is often a partial integration of the activity into the mainstream of the organization, as evidenced by a specific and continuous resource allocation to the activity. In general, organizations which have systems of this genre tend to treat them as a necessary evil.

Continuous systems, as delineated in Table 16-2, are the ideal portrayed in planning literature. Here the focus shifts from more problem solving to opportunity finding and the realization that planning systems contribute to the growth and survival of the organizations in a proactive way. As a result, these systems draw on expertise varying from marketing to cultural analysis, and information gathering becomes a structured activity. In other words, *these systems attempt to enhance the organization's capability to handle environmental uncertainty rather than to reduce perceived uncertainty.* The time horizons which are treated are considerably longer—varying from "long" to "futuristic"—and there is a substantial continuing resource allocation to these activities in the organization.

Information in Systems Analysis

The conceptual role of information in the systems analyses that support the planning function is important to the overall role of information in planning.

The *models* that are such an important element of systems analyses, as discussed in Chapters 4 to 6, may be viewed as devices for *transforming data into more valuable information.* This is shown in Figure 16-1 in terms of a flow of data into a decision model and the output of decision-relevant information.

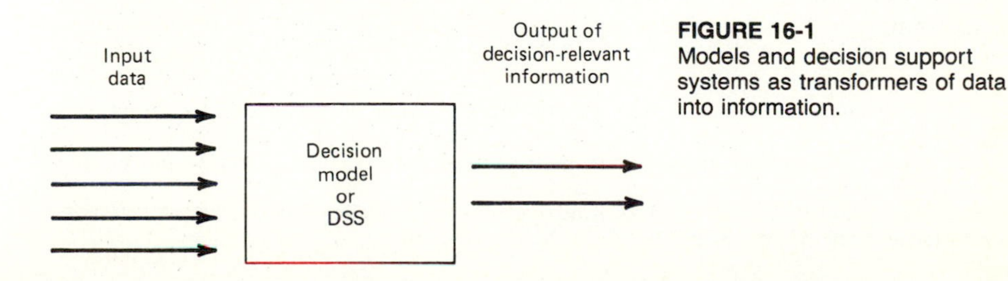

FIGURE 16-1
Models and decision support systems as transformers of data into information.

Thus, a decision support system (DSS), which inherently involves both models and data bases, provides access to information that is of one of two varieties:

a Model output information
b "Evaluated" data

Both represent ways of *transforming data.* These transformations are performed by the system to benefit the analyst who is the user of the system.

All varieties of decision models serve this information-transforming purpose. For instance, a descriptive model, such as one of the business position matrices often used in planning,[18] summarizes a great deal of diverse data concerning the potential of a product or business on a firm's competitive position in a pictorial form that serves to greatly enhance the value of the information.

Similarly, a predictive model, such as an econometric model of the economy,[19] transforms data on the current and past state of the economy and on "leading indicators" into predictions of future economic situations. An optimization model carries this information transformation process a step further through the evaluation of the alternative actions that are available, and the prescription of an alternative that is best according to some criterion such as profit maximization.

"Evaluated data" are data that have been evaluated to be relevant and useful for the making of particular decisions or classes of decisions. Unlike a management information system, which may facilitate access to much of the basic operational data of the organization, the data bases that the DSS user can call on are generally more focused, and directed toward the specific uses for which the DSS is designed. For instance, a strategic DSS would not normally access *operating* data. Thus, one of the varieties of "transformed data" that the DSS entails is sets of data that have been collected; assessed for relevance to a particular set of decisions, quality, and validity; and stored in a data base that is accessible to the DSS user. In effect, this process transforms "data" into "information," and despite the fact that the user of the system may be given direct access to a data base, the "data" therein have already been substantially evaluated and transformed.

The data bases which most corporate models[20] access are of this variety in that they entail financial, market, and other data at a fairly high level of aggregation. Their use does not normally require, and the system may not readily permit, access to detailed operational-level data such as those which concern business transactions.

[18]D. F. Abell and J. S. Hammond, *Strategic Marketing Planning,* Prentice-Hall, Inc., Englewood Cliffs, N.J., 1979.

[19] O. Eckstein, "Decision Support Systems for Corporate Planning," *Data Resources, U.S. Review,* February 1981.

[20]R. H. Hayes and R. L. Nolan, "What Kind of Corporate Modeling Functions Best?," *Harvard Business Review,* vol. 52, no. 3, 1974.

INFORMATION SYSTEMS IN IMPLEMENTATION

The variety of information systems that can be developed to aid in the implementation phase of management is more control-oriented than decision-oriented (as are planning-oriented systems).

Project Management Information Systems (PMIS)

An information system is as essential to the effective control of projects as it is to meeting external reporting requirements, to reporting project progress to top management, and to performing effective project planning.

Often, project managers develop ad hoc project information systems which suit their individual needs. Some project managers are able to develop information systems which permit them to foresee potential problem areas and to act in time to alleviate them. Other project managers may have systems which allow them to monitor the progress of their projects but which are not truly effective in indicating areas of future trouble within the project.

The development of a project information system involves the gathering and coalescing of data from the functional units supporting the project. An information system should be designed to serve two purposes: first, visibility for the functional manager, in terms of his input to the PMIS, and second, visibility for the project manager, in terms of cost, time, and performance. Many project managers rely on historical information, which depicts the past and present rather clearly but which does not allow them to foresee the future in a manner which would permit control actions to be taken before minor project problems become major ones. In any case, whether the various information systems developed by individual project managers are good or bad, the various systems are often not well integrated. Thus, while each information system may provide the individual project manager with the information desired, the output of the various systems may not provide information adequate for decision making and planning at the overall organizational level.

Basic Objectives of a PMIS A project management information system (PMIS) may be developed within an organization to serve as a model information system for all projects. The term "model" is used here in the sense of a basic information system structure which will provide essential information to the project manager and to top management, and, at the same time, will be:

1 Sufficiently flexible so that it can be modified to suit the unique needs of the individual project manager
2 Adaptable to many different projects
3 Adaptable to differing customer information requirements

Such a model PMIS will provide a basic information source to meet the requirements of functional manager, project manager, top management, and customer information. It can serve as a point of departure for project managers who desire to implement their own systems. By virtue of this flexibility, it should provide all levels of management with "personalized" information; and, at the same time, it should provide standardized information which is integrated into an overall planning and control system.

PMIS Criteria In addition to the requirements for flexibility and adaptability which are inherent in the concept of a "model" PMIS, a number of criteria for the design of such a system can be readily stated. It must:

1 Provide essential information on the cost-time-performance parameters of a project and on the interrelationships of these parameters.

2 Provide information in standardized form to enhance its usefulness to top management for multiproject control and long-range planning.

3 Be decision-oriented, in that information reported should be focused toward the decisions required of the project manager and top management.

4 Provide for customer-reporting requirements such as those outlined in current government specifications.

5 Be exception-oriented, in that it will focus the manager's attention on those critical areas requiring attention rather than simply reporting on all areas and requiring the manager to devote attention to each.

6 Fit into the overall organization information system and strategic planning system.

7 Be prospective in nature rather than retrospective, in that it should give special attention to potential problem areas within the project; it should, in effect, be an "early warning" system for the project manager.

8 Incorporate both external and internal data to provide a capability for keeping track of evolving projects in the customer's organization. In this fashion, the PMIS can interface with an organizational strategic information system.

9 Be consistent with existing project management guides and procedures.

10 Be consistent with policy documentation developed earlier by the organization.

11 Provide a capability for routine reporting, exception reporting, and special analyses (such as statistical analyses) which may be desired by the project manager or top management.

12 Provide for measurement at the critical project-functional interface, so that the project manager and functional manager will have data on which to base those decisions for which authority and responsibility are shared.

13 Provide a basic data requirement for functional managers to furnish to the project office and to facilitate in-house functional visibility.

14 Provide for project visibility during the various phases of a project life cycle.

Of course, the best of information systems will not keep project managers out of trouble, but it will keep them from being surprised when trouble comes. The information system is only as good as its inputs. A standardized reporting system can be set up that will give "visibility" to the project. Such a reporting system will make the delegation of authority easier, because everyone can tell quickly who is in charge of each part of the project and where to reach this person. Without an adequate information system, it can be difficult to tell who is responsible for a particular aspect of the project and *how that individual is doing in his or her work.*

Reports should be specified for each project and reviewed frequently for applicability and need. The use of existing information should be encouraged; only where a new problem exists, or a new approach is being tried, should new techniques be employed. The tendency to apply *all* currently popular administrative and control techniques indiscriminately should be resisted.

SUMMARY

Effective systems-oriented management requires high-quality, timely information that can serve as a basis for decision making. The computer era has often provided systems managers with a plethora of data and a lack of crucial relevant information.

Information systems and decision support systems, if properly designed, can be effective aids to managers in both the planning and implementation phases. Such systems must provide environmental information as well as the internally generated information that has been supplied in the past.

DISCUSSION QUESTIONS

1 The single greatest commodity in the management of a project is information. What kinds of general information should the project manager be concerned about?
2 What is meant by relevant information? What impact has the computer had in the development and production of relevant information?
3 What is the relationship between information systems and decision support systems?
4 What is a decision support system? To what use can such a system be put by the project manager? What are some of the subsystems commonly associated with a decision support system?
5 What is a competitive business information system? Why should a project manager be concerned about such a system? What responsibility does a project manager have for the analysis of the competition?
6 What are some of the differences between an information system for operational and strategic decision making?
7 Information subsystems relating information sources to the planning process are described in Table 16-1. Take this table and relate it to an actual situation of strategic planning in a corporation.
8 How might a project manager go about determining what image the project has with the project claimants? To what use could such information be put? Should the project manager be concerned about the project's image?
9 Every project has a competitive project. Defend or refute this statement.
10 There are certain environmental scanning and forecasting activities that are crucial to strategic planning. What are these activities?

11 What models can be used to represent the transforming of data into more valuable information?

12 What are the key objectives of a project management information system?

13 A project management information system is the foundation of information systems for a corporation engaged in project management. What is the rationale for this statement?

14 What are some of the criteria that should be developed for a project management information system?

15 The cooperation of the functional managers in the development of a project management information system is critical. Why is this so?

RECOMMENDED READINGS

Abell, D. F., and J. S. Hammond: *Strategic Marketing Planning,* Prentice-Hall, Inc., Englewood Cliffs, N.J., 1979.

Ackoff, R. L.: "Management Misinformation Systems," *Management Science,* vol. 14, no. 4, December 1967.

Alter, S.: *Decision Support Systems: Current Practice and Continuing Challenges,* Addison-Wesley Publishing Company, Inc., Reading, Mass., 1980.

———, R. H. Sprague, and R. L. Olson: "The Financial Planning System at Louisiana National Bank," *MIS Quarterly,* vol. 3, no. 3, 1979.

Braun, T. H.: "The History, Evolution and Future of Financial Planning Languages," *ICP Interface,* Spring 1980.

Carlson, E.: "Proceedings of a Conference on Decision Support Systems," *Data Base,* vol. 8, no. 3, 1977.

Cleland, D. I., and W. R. King: "Competitive Business Intelligence Systems," *Business Horizons,* vol. 18, no. 6, December 1975.

Davis, G. B.: *Management Information Systems: Conceptual Foundations, Structure, and Development,* McGraw-Hill Book Company, New York, 1974.

Dutta, B. K., and W. R. King: "A Competitive Scenario Modeling System," *Management Science,* vol. 27, no. 3, 1980.

Eckstein, O.: "Decision Support Systems for Corporate Planning," *Data Resources, U.S. Review,* February 1981.

Fahey, L., and W. R. King: "Environmental Scanning for Corporate Planning," *Business Horizons,* August 1977.

——— and V. K. Narayanan: "Environmental Scanning and Forecasting in Strategic Planning: The State of the Art," *Long Range Planning,* February 1981.

Hamilton, W. F., and M. A. Moses: "An Optimization Model for Corporate Financial Planning," *Operations Research,* vol. 21, 1973.

Hayes, R. H., and R. L. Nolan: "What Kind of Corporate Modeling Functions Best?," *Harvard Business Review,* vol. 52, no. 3, 1974.

Keen, P. J. W., and M. S. Scott Morton: *Decision Support Systems: An Organizational Perspective,* Addison-Wesley Publishing Company, Inc., Reading, Mass., 1978.

Kingdon, Donald R.: *Matrix Organization: Managing Information Technologies,* Harper & Row Publishers, Inc., New York, 1973.

King, William R.: "Human Judgment and Management Decision Analysis," *Journal of Industrial Engineering,* vol. 18, December 1967.

———: "Intelligent Management Information Systems," *Business Horizons,* vol. 16, October 1973.

——— and David I. Cleland: "Manager-Analyst Teamwork in MIS Design," *Business Horizons,* vol. 14, April 1971.

———: "Environmental Information Systems for Strategic Marketing Planning," *Journal of Marketing,* vol. 38, October 1974.

Radford, K. J.: *Information Systems for Strategic Decisions,* Reston Publishing Company, Reston, Va., 1978.

Rodriguez, J. I., and W. R. King: "Competitive Information Systems," *Long Range Planning,* December 1977.

Terry, P. T.: "Mechanisms for Environmental Scanning," *Long Range Planning,* vol. 10, June 1977.

Welsch, G. M.: *Successful Implementation of Decision Support Systems' Pre-installation Factors, Service Characteristics, and the Role of the Information Transfer Specialist,* Ph.D. dissertation, Northwestern University, Evanston, Ill., 1980.

CASE 16-1: Supporting Systems for Project Management (Matrix Management)

When matrix management is introduced into an organization, a series of cultural changes is set in motion. One of the first visible changes is the change in organizational structure. As a result of the structural change, managerial and professional roles change as well. Indeed, the realignment into a matrix structure sets in motion a "system of effects" which are often not considered when matrix management is initially introduced. The ultimate success of matrix management often depends on how supportive these accommodating systems have been.

For example, the existing information systems have to be modified to provide information to matrix managers and matrix teams as well as to the functional and general managers.

Your Task Develop a summary briefing of how you would expect the "system of effects" of matrix management to impact on the culture of an organization of your choosing. You might want to think about how an organization has been, or will be, changed by the introduction of some form of matrix management.

Prepare a summary of your findings for presentation to the other teams.

CHAPTER 17

THE CULTURAL AMBIENCE OF SUCCESSFUL PLANNING AND IMPLEMENTATION[1]

A corporation's culture can be its major strength when it is consistent with its strategies.[2]

In other chapters of this book, we have presented the concepts of systems analysis and project management. This chapter develops the concept of an organizational *culture* as the ambience within which these processes exist.

Culture is a set of refined behaviors that people have and strive toward in their society. "Culture," according to anthropologist E. B. Taylor,

> . . . is that complex whole which includes knowledge, belief, art, morals, law, custom, and any other capabilities and habits acquired by man as a member of society.[3]

Anthropologists have long used the concept of culture in describing primitive societies. Modern sociologists have borrowed this anthropological concept of culture and used it to describe a way of life of a people. Here, we use the term "culture" to describe the synergistic set of shared ideas and beliefs that is associated with a way of life *in an organization.*

[1]Portions of this chapter are adapted from David I. Cleland and William R. King, "Developing a Planning Culture for More Effective Strategic Planning," *Long Range Planning,* September 1974; and David I. Cleland and William R. King, "Organization for Long-Range Planning," *Business Horizons,* June 1974.

[2]"Corporate Culture," *Business Week,* Oct. 27, 1980, p. 148.

[3]E. B. Taylor, *The Origins of Culture,* Harper & Row, New York, 1958. (First edition, published in 1871.)

THE NATURE OF AN ORGANIZATIONAL CULTURE

The culture associated with an organization has distinctive characteristics that differentiate it from others. In the IBM Corporation, the precept "IBM means service" sets the tone for the entire organization, infusing all aspects of its environment and generating its distinctive culture. At 3M, the motto "Never kill a new product idea" reflects an organizational atmosphere of inventiveness and creativity. In some large corporations such as Hewlett-Packard, General Electric, and Johnson & Johnson, the crucial parts of the organization are kept small to stimulate a local culture, which encourages a personal touch in the context of a motivated, entrepreneurial spirit of teamwork.

An organization's culture reflects the composite management style of its executives, a style that has much to do with its ability to adapt to change. For instance, one company selected an acquisition opportunity which had passed the test of potential profitability; analysis showed it to be a potential winner through environmental and competitive assessment. After several years of trying to make the acquisition work, the company sold it off at a loss. The business opportunity had failed because it was not compatible with the prevailing corporate culture. There was not a "cultural fit."

A successful business culture, on the other hand, can lead to productivity and satisfaction. For instance, at Delta Air Lines a key element of the cultural ambience is a philosophical attitude called "the Delta family feeling," which was nurtured by the airline's founder, the late C. E. Woolman. Delta's other leaders have institutionalized this feeling of caring within the company. This attitude is found in the sense of cooperation and teamwork exhibited in the company. During periods of peak travel, it is not uncommon to find executives and flight crews pitching in and helping with baggage. Cross-training is used to help employees understand how their jobs fit in with overall company goals. Delta has had the fewest complaints per passenger boarded of any major airline. Only pilots and flight dispatchers—about 1 percent of the work force—are unionized.[4]

Thus, a business organization's culture consists of shared agreements among the members of the organization as to what is desirable or undesirable by way of values, beliefs, standards, and social and management practices. A company formed by the association of its people in a free market environment takes on characteristics influenced by that environment, both within the company and in its environmental and competitive system.

THE IMPORTANCE OF THE ORGANIZATION'S CULTURE

The American economy of the 1980s has been afflicted with stagnating productivity, high unemployment, surging inflation, and high interest rates. The average age of plant and equipment is about 20 years; the rate of investment in R&D and capital equipment has been sliding. All of this has created tangible

[4] *The Wall Street Journal,* July 7, 1980, p. 16.

consequences for the United States: a declining standard of living, inflation, and too few jobs. Tax laws, government regulations, pollution-control expenses, unsuitable economic policies, labor costs, and the high price of imported oil are a few of the factors that have played a part.

There is a growing consensus that the performance of American management is lacking in something. The management policies that led the United States to develop a productive economy that the rest of the world held in awe have lost their effectiveness.

If this is so, it undoubtedly reflects both organizational and national cultural phenomena. For instance, Akio Morita, the cofounder and chairman of the Sony Corporation, manages a large and highly successful multinational corporation. In his view, two key attitudes account for the business problems of the United States: American managers have been too concerned about short-term profits and too little concerned about their workers.

This criticism can be illustrated by drawing on an example cited in the *Harvard Business Review*. A manager with cost pressure facing him elected to cut the work force when a downturn in business occurred. The cutback in the work force had extremely poor timing, coming just before the Christmas holidays. This cutback was done in spite of a crucial strategy of the company in remaining nonunion, an advantage that it held over its competition. Angry workers held a representation election and in a few months had the entire company unionized. The manager met his short-range goal of keeping costs down and improving his profit picture, but in so doing traded away one of the company's important strategic competitive advantages.[5]

The short-term profit motive reduces the opportunity to replace aging equipment and make strategic investments. A lack of concern for the worker creates an ambience of antagonism between management and labor. This, says Morita of the Sony Corporation, "is most counterproductive."[6] Morita argues that the Japanese approach to labor management—lifetime employment, corporate paternalism, worker participation, and consensus decision making—can be transplanted into the United States. As evidence, at Sony's plants at San Diego, productivity has risen steadily and is now close to the company's factories in Japan.[7]

Whether this theory is true or not, it clearly shows how different cultures can be important in determining performance. The idea of a culture applied to a business organization can be illustrated by examining the characteristics of well-run companies. McKinsey & Company, a management consultant company, has studied the management practices of 37 U.S. companies that are well-run organizations. Eight common characteristics were found in these companies:

[5]Robert L. Banks and Steven C. Wheelwright, "Operations vs. Strategy: Trading Tomorrow for Today," *Harvard Business Review,* May–June 1979, pp. 112–120.

[6]Quoted in *The New York Times Magazine,* Jan. 4, 1981, p. 17.

[7]Ibid.

- A bias toward action
- Simple form and lean staff
- Continued contact with customers
- Productivity improvement via people
- Operational autonomy to encourage entrepreneurship
- Stress on one key business value
- Emphasis on doing what they know best
- Simultaneous loose-tight control[8]

In effect, these are some cultural elements that *may* be thought to be associated with business success. Many are also characteristics associated with the systems tenets of this book. In the remainder of this chapter, we shall explicitly deal with those cultural elements associated with successful application of these systems ideas in the planning and implementation aspects of management.

THE CULTURAL AMBIENCE FOR EFFECTIVE PLANNING

To aid in defining a "planning culture" in something other than a pontifical fashion, it is useful to consider the status of long-range and strategic planning in organizations and the needs for improved planning which are thereby implied.

The Need for a Planning Culture

While most large organizations formally recognize the need for systematic forward thinking and analysis and most have established organizational entities to perform planning, by whatever name they may be known, few have developed the variety of strategic systems planning capability discussed in this book. It is much more typical to find planning activities which in reality involve only financial extrapolations or which focus solely on extensions of the present products and services of the organization. When these activities are creative, they are often unstructured as well, so that they have little real impact.

One of the most important reasons for the restricted view of strategic planning which is adopted by even the most progressive organizations is the difficult question of integrating an inherently vague, future-oriented, nebulous activity into the practical day-to-day activities of the organization. Good managers, typically, are far too concerned with solving important immediate problems to spend valuable time in addressing future issues which are uncertain, relatively undefined, and, in any one instance, of uncertain potential.

There is a good deal of empirical evidence which suggests that high-level managers recognize the great significance of strategic planning, but that, paradoxically, they devote only a small proportion of their time to it. As a consequence of this incongruity, planning is often done either by staff people, with top managers playing merely a reviewing role, or by line managers who devote only modest attention to parochially defined areas of concern.

[8]"Putting Excellence into Management," *Business Week*, July 21, 1980, p. 196. Clearly, the study was exploratory since no "control" group was used. The results are only suggestive.

Various organizational modes and planning procedures have been developed to address these difficulties. The effectiveness of most of these organizational approaches is thwarted by a number of characteristics which are inherent in most *people* and in most *organizations.*

Disciplinary and Organizational Parochialism The most significant personal characteristics of managers which inhibit effective planning have to do with their tendency to view all the world and everything in it from the standpoint of their own experiential and educational backgrounds.

In *disciplinary parochialism,* managers still think (even unknowingly) in a narrow specialized discipline in which they were educated and in which they won their first kudos. Having won their credentials and a degree of success in a specialty (e.g., engineering), they never totally recognize the narrowness of their education and view that discipline as the most important one in the organization. Such an individual is inclined to spend excess time supervising the engineering aspects of a strategy and to neglect an evaluation of the interdisciplinary elements of it.

Since their success is reinforced by other specialists who have assisted them in moving up the ladder, disciplinary parochialists often surround themselves with those who are known and have been trusted within their field of specialty. Thus, even as they move up the ladder into positions requiring a broader perspective, their view is not broadened by their access to advisors.[9]

Organizational parochialism is similar; it is the tendency of managers to view their organization as the center of affairs, such as by focusing attention on a coveted product line which has already been successful—another form of myopia. This pervasive tendency of some managers to spend product development funds in areas which are known and familiar—thus extending existing product lines—even in the face of overwhelming evidence of a changing market, is indicative of this sort of narrowness and reliance on the choice of "comfortable" alternatives.

Organizational Characteristics Inhibiting Planning In addition to these pervasive personal traits which can inhibit planning, there are a number of organizational characteristics which have a similar effect:

A bureaucratic organizational structure designed more for maintaining *efficiency* and *control* in current operations than in fostering long-range *innovation*

The lack of an "organization" designed specifically for bringing out the *innovation* necessary to develop new products and services

An assumption that the chief executive or, alternatively, a professional planning staff should *do* the long-range planning

An incentive system wherein performance oriented toward an ability to

[9]Warren Bennis has described this condition as "The Doppelganger Effect." According to Bennis: "By and large, people at the top of massive organizations tend to select as key assistants people who resemble them." See "The Doppelganger Effect," *Newsweek,* Sept. 17, 1973, p. 13.

produce short-range results is rewarded more highly than that oriented toward long-range opportunities

The introduction of radically new planning systems into organizations without proper concern for their effect on the motivations and behavior of those who will implement them

The natural consequence of these characteristics is an organizational structure which is too tied to tradition and to today's problems to give due concern to the future's opportunities and one which articulates long-range strategic goals while allocating the bulk of its resources to short-range problems.

Basic Premises of a Planning Culture

The resolution of these inhibiting personal and organizational characteristics in a fashion which permits effective strategic systems planning is not simple. However, over the last 20 or so years, a variety of organizational approaches for the development of strategic plans have been tried. From this experience has come an empirically tested viewpoint that the success of strategic planning in an organization tends to be less sensitive to the parameters of the planning techniques than it is to the overall *culture* within which the planning is accomplished. Such a culture is based on a set of premises which form its underlying rationale.

The basic premise of the cultural approach to strategic planning described in this chapter is that meaningful strategic planning must be done by those individuals and organizational units who will ultimately be responsible for executing the plans. If this were not true, the question of organizing for planning would have a deceptively simple answer: organize a "planning department." However, although professional planners clearly have an important role in planning, there are few illustrations of "delegated planning" which are truly successful. Those few instances where success with such an approach is claimed tend to measure success in terms of planning-related performance measures rather than overall organizational performance measures. If consciously effected positive change is the measure to be applied to the success of planning, the "planning department" approach seems to be a failure.

An alternative to the simplest "one-dimensional" approaches to planning, which are exemplified by the vesting of planning responsibility in a planning department, is the conscious creation of a *planning culture*. This multidimensional concept of a "culture" is interpreted here to mean *an accepted and demonstrated belief* on the part of organizational participants as to what is important and valued within the organization. This culture reflects the accepted knowledge, attitudes, and philosophies of both those who guide the organization and those who operate it.

In the strategic systems planning context, a planning culture is a clearly established bias concerning the importance and effectiveness of strategic planning which is accepted and demonstrated in daily actions and attitudes by both planning participants and organizational operatives. This means that strategic systems planning is an accepted and continuing way of life in the organization—not just a necessary evil which is carried on cyclically.

The development of a planning culture is obviously not a simple task, since it requires that planning be made endemic to the value system of managers; fully five to eight years may be required under optimum conditions. Yet to ignore strategic planning is to make oneself the victim of the planning of others, and to "go through the motions" of planning without inducing a planning culture is to simply make the same error in a more resource-consuming way.

The need for a strategic planning culture within an organization can be deduced from a number of premises concerning the processes that are involved in developing a future sense of purpose for an organization. These premises are:

1 Professional planners can facilitate a planning process, but they cannot themselves do the planning.

2 Planning activities should be performed by the managers who will ultimately be responsible for the implementation of the plans.

3 Creative planning is inherently a group activity, since it must involve many different subunits of the organization and many different varieties of expertise.

4 Strategic planning involves much more than numerical extrapolations of trends; it involves the final selection of strategic alternatives as well.

5 The strategic planning involvement of managers must be broader than that usually achieved in a "management by objectives" framework. (Chap. 7 discusses MBO.)

6 Managers must be motivated to spend time on strategic planning through a formalized system which permits their contribution to be assessed.

7 The planning process must provide for the development of relevant data bases—qualitative as well as quantitative—which facilitate the definition and evaluation of strategic alternatives.

The emphasis on *managers,* rather than planners, in these premises is obvious. It emanates from the recognition that *if managers play an essentially passive role in the planning process, the output will reflect the values of the planning technicians rather than those of managers.* Thus, when the crucial definitional tasks of problem formulation, specification of alternatives, and so forth, are left to technicians, the managers who play only a reactive reviewing role lose a degree of control over the firm's destiny.

THE CULTURAL AMBIENCE FOR EFFECTIVE IMPLEMENTATION[10]

The cultural ambience that is necessary for, and reflected in, project-oriented organizations may be described in terms of a set of cultural characteristics of the matrix organization.

[10]Some of the ensuing material in this chapter has been adapted from David I. Cleland, "The Cultural Ambience of the Matrix Organization," *Management Review,* November 1981, with permission.

Organizational Openness

A propensity toward organizational openness is one of the most characteristic attributes of the matrix design. This openness is demonstrated through a receptiveness to new ideas, a sharing of information and problems by the peer group. Newcomers to a matrix organization are typically accepted without any concern. There is a willingness to share organizational challenges and frustrations with the newcomer. This openness characteristic of project team management is described in one company as "no place to hide in the organization."[11]

Participation

Participation in the project-driven matrix organization calls for new behavior, attitudes, skills, and knowledge. The demands of working successfully in the matrix design create opportunities for the people as well as for the organization. For the people, there are more opportunities to attract attention and to try their mettle as potential future managers. Because matrix management increases the amount and the pattern of recurring contacts between individuals, communication is more intense. The resolution of conflicts is also of a more intense nature than in the traditional organization, where conflict can be resolved by talking to the functional boss. In a matrix design, at least two bosses have to become involved—the manager providing the resources and the manager held accountable for results. These two managers, locked in a conflict, may appeal as a last resort to the common line supervisor for resolution. Matrix management demands higher levels of collaboration. But in order to have collaboration, trust and commitment are needed on the part of the individuals. In order to be committed and to maintain trust, the individuals in the organization must take personal risks in sharing information and revealing their own views, attitudes, and feelings.

There is growing evidence that individuals today wish to influence their work situation and to create a democratic environment at their place of work. People expect variety in their lifestyle in the organization as well as in their private lives. The flexibility and openness of the matrix design can accommodate these demands.

The degree to which people are committed to participating openly and fully in matrix organization effort can influence results. Murphy et al., in a study of over 650 projects including 200 variables, found that certain variables were associated with the perceived failure of projects. Lack of team participation in decision making and problem solving was one important variable associated with perceived project failure. In contrast, project team participation in setting schedules and budgets was significantly related to project success.[12]

[11] "Texas Instruments Shows U.S. Business How to Survive in the 1980's," *Business Week,* Sept. 18, 1978.

[12] D. C. Murphy, Bruce N. Baker, and Dalmar Fisher, "Determinants of Project Success," National Technical Information Services, Accession No. N-74-30392, 1974, Springfield, Va., 22151, p. 60669.

Increased Human Problems

Reeser[13] conducted research to examine the question of whether project organization might not have a built-in capacity of causing some real human problems of its own. This research was conducted at several aerospace companies. Reeser's research findings suggested that insecurity about possible unemployment, career retardation, and personal development is felt by subordinates in project organizations to be significantly more of a problem than by subordinates in functional organizations. Reeser notes that project subordinates can easily be frustrated by "make work" assignments, and by ambiguity and conflict in the work environment. Project subordinates tend to have less loyalty to the organization. There are frustrations because of having more than one boss. The central implication of Reeser's findings is that although there may be persuasive justifications for using a matrix design, relief from human problems is not one of them.

Even with formal definition of organizational roles, the shifting of people between the projects does have some noticeable effects. For example, people may feel insecure if they are not provided with ongoing career counseling. In addition, the shifting of people from project to project may interfere with some of the basic training of employees and the executive development of salaried personnel. This neglect can hinder the growth and development of people in their respective fields.

Consensus Decision Making

Many people are involved in the making of decisions. Members of the matrix team actively contribute in defining the question or problem as well as in designing courses of action to resolve problems in the management of the effort. Professionals who become members of a matrix team gain added influence in the organization as they become associated with important decisions supporting an effort. They tend to become more closely associated with the decision makers both within the organization and outside it. Perceptive professionals readily recognize how their professional lives are broadened.

A series of documents which describe the formal authority and responsibility for decision making of key project clientele should be developed for the organization. If a manager is used to a clear line of authority to make unilateral decisions, the participation of team members in the project decision process makes management more complex. However, the result is worth the effort, since the decisions tend to be of a higher quality. Also, by participating in the decision process, people have a high degree of commitment toward carrying out the decision in an effective manner.

[13] Clayton Reeser: "Some Potential Human Problems of the Project Form of Organization," *Academy of Management Journal,* vol. 12, December 1979.

Objective Merit Evaluation

This is an important area of concern to the individuals in the matrix design. If individuals find themselves working for two bosses (the functional manager or work-package manager and the project manager), chances are good that both will evaluate their merit and promotion fitness. Usually the functional manager initiates the evaluation; then the project manager will concur in the evaluation with a suitable endorsement. If the two evaluators are unable to agree on the evaluation, it can be referred to a third party for resolution. For the most part, individuals who are so rated favor such a procedure because it reinforces their membership on the project team, as well as ensuring that an equitable evalution is given. Project team members who have been assigned to the project team from a functional organization may find themselves away from the daily supervision of their functional supervisors. Under such circumstances, a fair and objective evaluation might not be feasible. By having the project manager participate in the evaluation, objectivity and equity are maintained.

New Criteria for Wage and Salary Classification

The executive rank and salary classification of project managers will vary depending on the requirements of their positions, the importance of a project to the company, etc. Most organizations adopt a policy of paying competitive salaries. However, the typical salary classification schema is based on the number and grade of managers and professionals that the executive supervises. In the management of a project, although the project manager may supervise only two or three people on the staff, this manager is still responsible for bringing the project in on time, at the budgeted cost, and in so doing, for managing the efforts of many others who do not report to him or her in the traditional sense. Therefore, new criteria for determining the salary level of a project manager are required. Organizations with a successful salary classification schema for a project manager's salary have utilized criteria such as:

Duration of project
Importance of project to company
Importance of customer
Annual project dollars
Payroll and level of people who report directly to project manager (staff)
Payroll and level of people with whom project manager must interface (e.g., functional managers)
Complexity of project requirements
Complexity of project
Complexity of project interfaces
Payroll and rank of individual to whom project manager reports
Potential payoff of project
Pressures project manager is expected to face

In many companies, the use of project management is still in its adolescence and suitable salary criteria have not been determined. In such cases, it is not uncommon to find individuals designated as project managers who are not coded as managerial personnel in the salary classification and executive rank criteria. Word of this will get around, and the individual's authority may be compromised. We have usually found this situation arising because of a failure of the wage/salary staff specialists to develop suitable criteria for adjusting the salary grade of the project managers. This problem is not as significant in industries where project management is a way of life, such as in aerospace and construction.

New Career Paths

Aspiring individuals typically have two career paths open to them: to remain as managers in their technical field or to seek general-manager positions. Or they may prefer to remain professionals in their field and become senior staff, e.g., senior engineers. Project management opens up a third career field in management. The individual who is motivated to enter management ranks can seek a position as a project manager of a small project and use this as a stepping stone to higher-level management positions. It is an excellent way to learn the job of a general manager, since the job of project manager is much like that of a general manager except that the project manager usually does not have the general manager's formal legal authority. This should not deter project managers; it should motivate them to develop their persuasive and other interpersonal and negotiation skills—necessary skills for success in general management.

Acceptable Adversary Roles

An adversary role emerges in project management as the primary project clientele find that participation in the key decisions and problems is socially acceptable. An adversary role may be assumed by any of the project clientele who sense that something is wrong in the management of the project. Such an adversary role questions goals, strategies, and objectives and asks the tough questions that have to be asked. Such a spontaneous adversary role provides a valuable check to guard against decisions which are unrealistic or overly optimistic. A socially acceptable adversary role facilitates the rigorous and objective development of data bases on which decisions are made. But the prevailing culture in an organization may discourage individuals from playing the adversary role that will help management to comprehend the reality of a situation. This situation is possible in all institutions of an hierarchical character.

An adversary role presumes that communication of ideas and concerns upward are encouraged. As people actively participate in the project deliberations, they are quick to suggest innovative ideas for improving the project or to sound the alarm when things do not seem to be going as they should. If the adversary role is not present, perhaps because its emergence is inhibited by the management style of the principal managers, information concerning potential organizational failures will not surface.

An example of the stifling of an adversary role is found in the case of a company in the management of an urban transportation project. In the late 1960s, this company attempted to grow from a $250-million-a-year subcontractor in the aerospace industry into a producer of ground transit equipment. In pursuing this strategy, it won prime contracts for two large urban rail systems. Heavy losses in its rail programs put the company into financial difficulties. What went wrong?

The company got into difficulty in part because the chief executive dominated the other company executives even though he was unable to face overriding practical considerations. When major projects in the rail-systems business were in difficulty, the unrealistic optimism demonstrated by the chief executive prevented any other executive from doing much about it. In the daily staff meetings, the executives quickly learned that any negative or pessimistic report on a project would provoke open and sharp criticism from the chief executive. Project managers quickly learned that, in the existing cultural ambience, revealing bad news would not be tolerated. Consequently, they glossed over problem areas and emphasized the positive in order to please him.

On one of their large contracts, they submitted a bid that was 23 percent below the customer's own estimate and $11 million under the next lowest competitive bid. The project manager had felt that this estimate was too low but had not argued against it because "I didn't want to express a sorehead minority view when I was in charge of the project." The cultural ambience within this company during this period might be summarized as follows: Don't tell the boss any bad news, only report good news; for if you bring bad news, you run the risk of being sharply criticized.

Members of a project team need to feel free to ask tough questions during the life of the project. When the members of a team can play an adversary role, a valuable check-and-balance mechanism exists to guard against unrealistic decisions. Within Texas Instruments, a cultural ambience exists where an adversary role can emerge. Consequently, "it is impossible to bury a mistake in this company. The grass roots of the corporation are visible from the top . . . the people work in teams and that results in a lot of peer pressure and peer recognition."[14]

Organizational Flexibility

The lines of authority and responsibility defining the structure tend to be flexible in the matrix organization. There is much give-and-take across these lines, with people assuming an organizational role that the situation warrants rather than what the position description says should be done. Authority in such an organizational context gravitates to the person who has the best credentials to make the required judgment.

The matrix design provides a vehicle for maximum organizational flexibility; no one has "tenure" on a matrix team. There are variable tasks that people

[14] "Texas Instruments Shows U.S. Business How to Survive in the 1980's," pp. 66–92.

perform, a change in the types of situations they may be working on, and an ebb and flow of work loads as the work of the organization fluctuates. When an individual's skills are no longer needed on a team, they can be assigned back to their permanent functional home.

There are some inherent problems in the flexibility of the matrix organization. The need for staffing tends to be more variable. Both the quantity of personnel and the quality needed are difficult to estimate because of the various projects that are going on in the organization. For example, a structural-design group may have a surplus of design engineers at a particular time who are not assigned to any one project. The personnel estimates for oncoming projects, however, may indicate that in several months' time these professionals will be needed for project assignments. A functional structural-design manager has the decision of whether to release the engineers and reduce overhead or to assign them to "make work" for the period and avoid the future costs of recruitment, selection, and training. The same manager may anticipate assigning these professionals to an emerging project; yet if the emerging project is delayed or even canceled, the project manager may not need these people for some time.

As the work effort nears its end, and perceptive individuals begin to look for other jobs, there can be a reduction in their output level. This reduction can damage the efficiency with which the project is being managed. Paradoxically, although morale takes on added significance in the matrix team, the design itself may result in lowering it.

The organizational flexibility of project management does, therefore, create some problems as well as benefits.

Improvement in Productivity

Texas Instruments attributes productivity improvement in the company to the use of project teams. Its productivity improvement over the past years has slightly more than offset the combined impact of its wage and benefit increases (average 9.2 percent annually) and its price decreases (averaging 6.4 percent per unit).[15]

At Texas Instruments, more than 83 percent of all employees are organized into "people involvement teams" seeking ways to improve their own productivity. The company views its people as interchangeable—"kind of like auto parts." The culture there is much like the Japanese: a strong spirit of belonging, a strong work ethic, competitive zeal, company loyalty, and rational decision making. The culture of Texas Instruments ". . . has its roots buried deep in a soil of Texas' pioneer work ethic, dedication, toughness and tenacity—it [the culture] is a religion. The climate polarizes people—either you are incorporated into the culture or rejected."[16]

The experience of Litton Industries in its Microwave Cooking Division shows that the use of project teams in the manufacturing function has increased

[15]Ibid.
[16]Ibid.

productivity. Since the manufacturing organization was grouped into team units, production increased fourfold in 15 months. Product quality has increased, 1,000 new production workers have been added to a base of 400 people, and unit production costs have declined 10 to 15 percent.[17] Some other claims of productivity increases are as follows:[18]

A steel company chief executive states: "Properly applied, 'matrix' management improves profitability because it allows progress to be made on a broader front; with a given staff size, i.e., more programs and projects simultaneously pursued (including those concerning productivity)."

The chief executive of a company in the microprocessor industry declares that the company's success (15 percent of the microprocessor market, $1.8 million in revenue, 18.1 percent ROI) would not be possible without matrix management.

A chemical company president claims: "Matrix management improves people productivity."

The experiences of these companies suggest that project-management techniques can assist in raising productivity.

Increased Innovation

In the private sector, in those industries where a fast-changing state-of-the-art exists, product innovation is critical for survival. There is evidence that the use of project teams has helped to further innovation within such organizations. For example, the teams are successfully used in the aerospace industry, where the ability to innovate is essential, particularly in the development and production of sophisticated weaponry.

Why does the project team seem to foster innovation in organizations? Innovation comes about because an individual has a technological or market idea and gathers some people who believe in the idea and are committed to it. A small team of people is formed, who become advocates and missionaries for the idea. The team of people represents a diverse set of disciplines who view the idea differently. It is difficult to hide anything in such an environment. The openness, the freedom of expression, the need to demonstrate personal effectiveness, all seem to be conducive to the creativity necessary to innovate. Within such organizations, decision making tends to be of a consensus type. A significant esprit de corps exists.

The objective in such undertakings is to create small teams that can operate using the advantages of their small size while still having access to the large resources of the parent organization. These teams can operate rather independently while still relating to the various functions and taking advantage of the resources and expertise that they have to offer. The team operates as a close-knit group relatively free from the bureaucratic constraints of the overall organiza-

[17]William W. George, "Task Teams for Rapid Growth," *Harvard Business Review,* March–April 1977, p. 71.

[18]These are productivity claims cited in personal correspondence.

tion. The net effect can be to create an organization in which entrepreneurs—innovators—can flourish by making available to them the technical resources needed to do the job.

Project teams, used effectively, can take advantage of the scale economies of large organizations, and through their team nature, the flexibility of a small innovative organization is realized. An early research effort in the use of program (project) organizations noted that such organizations seemed to have been more successful in developing and introducing new products than businesses without program organizations.[19]

L. W. Ellis, Director of Research, International Telephone and Telegraph Corporation, claims that temporary groups (project teams) that are well organized and have controlled autonomy can stimulate innovation by overcoming resistance to change. Cross-functional and diagonal communication within the project team and with outside interested parties helps to reduce resistance to change.[20]

Jermakowicz found that the matrix design was most effective of three major organizational forms he studied in ensuring the implementation or introduction of new projects, while a "pure" project organization produced the most creative solutions.[21]

Kolodny, reporting on a study of his own and citing some other studies as well, comments on the effect that matrix organizations have on new product innovation.[22] Kolodny cites Davis and Lawrence, who point to an apparent high correlation between matrix organization designs and very high rates of new product innovation.[23] In his summary, Kolodny notes that there is an apparent relationship between high rates of new product innovation, as measured by the successful introduction of new products, and matrix organizational designs.[24]

Realignment of Supporting Systems

As the use of project management grows in an organization, it soon becomes apparent that many of the systems that have been organized on a traditional hierarchical basis need to be realigned to support the project team. What initially appears to be only an organizational change soon becomes something larger. Effective project management requires timely and relevant information on the project; accordingly, the information systems have to be modified to

[19]E. R. Corey and S. H. Star, *Organization Strategy: A Marketing Approach,* Division of Research, Harvard Business School, Boston, Mass., 1970, Chap. 6.

[20]L. W. Ellis, "Effective Use of Temporary Groups for New Product Development," *Research Management,* January 1979, pp. 31–34.

[21]Wladyslaw Jermakowicz, "Organizational Structures in the R&D Sphere," *R&D Management,* no. 8, special issue, 1978, pp. 107–113v, as cited in Kolodny, below.

[22]Harvey F. Kolodny, "Matrix Organization Designs and New Product Success," *Research Management,* September 1980, pp. 29–33.

[23]Stanley M. Davis and Paul R. Lawrence, *Matrix,* Addison-Wesley Publishing Company, Inc., Reading, Mass., 1977.

[24]Kolodny, op. cit., p. 32.

accommodate the project manager's needs. Financial and accounting systems, project planning and control techniques, personnel evaluations, and other supporting systems require adjustment to meet project management needs.

Development of General-Manager Attitudes

An organizational culture is in a sense the aggregate of individual values, attitudes, beliefs, prejudices, and social standards. A change for these individuals means cultural changes. The matrix design, when properly applied, tends to provide more opportunity to more people to act in a general-manager mode. With this kind of general-manager thinking, the individual is able to contribute more to organizational decision making and information processing.

The matrix design, with its openness and demands for persuasive skills, provides an especially good environment for managers-to-be to test their ability to make things happen by the strength of persuasive and negotiative powers. Perceptive general managers know that there is little they accomplish solely by virtue of their hierarchical position; so much depends on their ability to persuade others to their way of thinking. Exposure to the workings and ambience of the matrix culture brings this point home clearly and succinctly.

Effective collaboration on a project team requires plenty of a needed ingredient: trust. To develop this trust, individuals must be prepared to take personal risks in sharing resources, information, views, prejudices, attitudes, and feelings to act in a democratic mode. Not everyone can do that, yet executives in successful companies are able to do so. For example, in the Digital Equipment Corporation, where a matrix environment prevails, the ambience is described as "incredibly democratic" and not for everyone. Many technical people cannot bear the lack of structure and indefiniteness. In such an organization, bargaining skills are essential to survival.[25] The matrix design is permanent—the deployment of people is changing constantly. In such a transitional situation, the only thing that prevents breakdown is the personal relationships as conflicts are resolved and personnel assignments are changed. Communication is continually needed to maintain the interpersonal relationships and to stimulate people to contribute to the project team efforts.

SUMMARY

The notion of a culture that is conducive to effective planning and implementation is imprecise. Yet there is widespread agreement on its importance.

Moreover, while the cultures that lead to effective planning and implementation are often discussed in different terms, they are more similar than they are different. Perhaps this is best illustrated and summarized by considering a firm that has developed such a culture—Intel Corporation.

[25]Harold Seneker, "If You Gotta Borrow Money, You Gotta," *Forbes,* Apr. 28, 1980, pp. 116–120.

Intel is described[26] as an open organization where people are willing to identify and discuss problems in a nonconfrontational way. Division managers work as an executive staff on both line and staff problems. Business planning is done by breaking the company into strategic business segments. A team of people is composed of middle managers and personnel below that level and a few senior managers who develop strategies and resource requirements for supporting future operations. Thus, strategic planning is embedded in the organization as an inherent responsibility of the line managers.

An important part of the culture at Intel is the measurement of "absolutely everything" that can be evaluated in terms of performance. Information concerning performance against objectives (the company is committed all the way in an MBO program) is open enough so that each manager's performance is examined in detail every six months by peers. The major organizational elements (divisions) of the company are heavily dependent on other divisions to get their job done. The result: an interactive disciplined company that gets things done and where innovation flourishes.

Robert Noyce, cofounder of Intel Corp., describes a key element of the culture at the company in this manner:

> What we've tried to do is to put people together in ways so that they make contributions to a wider range of decisions and do things that would be thwarted by a structured, line-type organization.[27]

Thus, in Intel, a variety of cultural factors impact on both the planning and implementation functions of management. We end the book on this point—that despite the taxonomies, such as the planning and implementation dichotomy that is used here, which are essential for the thorough study of ideas, the world is not so compartmentalized. Thus, the culture that is developed affects *everything* —planning, implementation, and all else. Conveniently, we discover that it is possible to identify common cultural features that will positively influence both planning and implementation. But such is not always the case in a world of complex systems, and we must be on guard, as noted in the beginning of the book, that "everything is related to everything else."

DISCUSSION QUESTIONS

1 How is culture defined? How can the concept of a culture be related to a wide variety of organizational contexts?

2 What are some of the cultural characteristics of contemporary business organizations known to the reader? What effect does the cultural ambience have on the management of the organizations?

[26]Quoted in *The New York Times Magazine,* Jan. 4, 1981, p. 17.

[27]Quoted in *The New York Times Magazine,* Jan. 4, 1981, p. 17, Copyright © 1981 by the New York Times Company. Reprinted by permission. See also "Creativity by the Numbers" (An interview with Robert N. Noyce), *Harvard Business Review,* May–June 1980, pp. 122–132.

3 How would you describe the nature of a business organization's culture in terms of the product or services that the organization offers?

4 One of the cultural characteristics of American business is a short-term orientation in the management of organizations. What effect does this orientation have on U.S. competitiveness with foreign multinational companies?

5 What are the eight characteristics of the well-run companies described in the McKinsey & Company report? Are these characteristics really important? Did these characteristics likely emerge as the result of deliberate planning for their development, or are they the result of how the companies are managed?

6 What are some of the factors in a corporation that can inhibit the development of a strategic planning culture? A project-management culture?

7 What are the key premises necessary in developing a future sense of purpose for an organization?

8 What are the key cultural characteristics of the project-driven matrix organization?

9 What is meant by participative management? Can the project manager benefit from using participative management techniques? How?

10 What are some of the human problems that the project manager should expect among the members of the project team?

11 Why is a consensus decision-making approach useful for the project manager? What are some of the dangers in making project decisions using the consensus approach?

12 Why do project managers have an excellent opportunity to compete for promotion to a general-manager position in the organization? Does the project manager have a better chance for promotion to general management than the functional manager?

13 Why is an adversary role important in the management of a project? How can the cultural ambience be developed in the project to ensure that such an adversary role is socially acceptable? Is social acceptance of such a role important?

14 Project management provides many opportunities to facilitate organizational flexibility. Describe how project management can be used to give the general manager more flexibility in managing the organization.

15 There is some evidence to suggest that the use of the project-driven matrix organization can improve productivity. Why? Is there a lesson here for contemporary managers who are concerned about productivity in their organizations?

RECOMMENDED READINGS

Banks, Robert L., and Steven C. Wheelwright: "Operations vs. Strategy: Trading Tomorrow for Today," *Harvard Business Review,* May–June 1979.

Bennis, Warren: "The Doppelganger Effect," *Newsweek,* September 17, 1973.

Cathey, Paul: "Matrix Method Builds New Management Muscle at ESB," *Iron Age,* June 20, 1977.

Cleland, David I., and William R. King: "Organization for Long-Range Planning," *Business Horizons,* June 1974.

———: "Developing a Planning Culture for More Effective Strategic Planning," *Long Range Planning,* September 1974.

Collins, John J.: *Anthropology: Culture, Society, and Evolution,* Prentice-Hall, Inc., Englewood Cliffs, N.J., 1975.

Corey, E. R., and S. H. Starr: *Organization Strategy: A Marketing Approach,* Division of Research, Harvard Business School, Boston, Mass., 1970.

Davis, Stanley M., and Paul R. Lawrence: *Matrix,* Addison-Wesley Publishing Company, Inc., Reading, Mass., 1977.

Ellis, L. W.: "Effective Use of Temporary Groups for New Product Development," *Research Management,* January 1979.

Gabriel, R. A.: "What the Army Learned from Business," *The New York Times,* April 15, 1979.

George, William W.: "Task Teams for Rapid Growth," *Harvard Business Review,* March–April 1977.

Hollingsworth, A. T.; Bruce M. Meglino, and Michael C. Shaner: "Copies with Team Trauma," *Management Review,* August 1979.

Jermakowicz, Wladyslaw: "Organizational Structures in the R&D Sphere," *R&D Management,* no. 8, special issue, 1978.

Kolodny, Harvey F.: "Matrix Organization Designs and New Product Success," *Research Management,* September 1980.

Main, Jeremy: "The Battle for Quality Begins," *Fortune,* December 29, 1980.

Mee, John F.: "Matrix Organization," *Business Horizons,* Summer 1964.

Murphy, D. C.; Bruce N. Baker, and Dalmar Fisher: "Determinants of Project Success," National Technical Information Services, Accession No. N-74-30392, 1974, Springfield, Va., 22151, p. 60669.

Paolillo, Joseph G., and Warren B. Brown: "How Organizational Factors Affect R&D Innovation," *Research Management,* March 1978.

Peters, Thomas J.: "Beyond the Matrix Organization," *Business Horizons,* October 1979.

Reeser, Clayton: "Some Potential Human Problems of the Project Form of Organization," *Academy of Management Journal,* vol. 12, December 1979.

Seneker, Harold: "If You Gotta Borrow Money, You Gotta" *Forbes,* April 28, 1980.

Walton, Richard E.: "Work Innovations in the United States," *Harvard Business Review,* July–August 1979.

"Corporate Culture," *Business Week,* October 27, 1980.

"Creativity by the Numbers" (An interview with Robert N. Noyce), *Harvard Business Review,* May–June 1980.

"Overhauling America's Business Management," *The New York Times Magazine,* January 4, 1981.

"Putting Excellence into Management," *Business Week,* July 21, 1980.

"Texas Instruments Shows U.S. Business How to Survive in the 1980's," *Business Week,* September 18, 1978.

CASE 17-1: Effect of Matrix Management on the Individual

Situation You are the general manager of a profit center in a corporation that has just been reorganized into matrix management. Your managers and key professionals have expressed serious misgivings about this idea of "matrix" and are deeply concerned about how this new organizational approach will affect them. Talk and concern about this issue have reached a point where it is necessary for you to do something!

Your Task Outline an agenda to include topic material that you would present for discussion on this matter at a forthcoming meeting of your key managers and professionals. Try to be prepared to field all questions that these people might raise about how matrix management will impact on them as individuals as well as on the organization. Have your team leader present your findings.

APPENDIX ONE

THE PROJECT PLAN

SPECIFICATIONS OF THE PROJECT PLAN

A project plan, or some such document outlining applicable plans and planning tasks, is an essential tool of project management. It provides the necessary guidance for all the participants in developing the project as it gains in maturity, and it forms the basis for the project operations. The project package plan will be described in detail in the following sections.

Part 1 Project summary
Part 2 Project schedules
Part 3 Project management
Part 4 Market intelligence
Part 5 Operational concept
Part 6 Acquisition
Part 7 Facility support
Part 8 Logistic requirements
Part 9 Manpower and organization
Part 10 Executive development and personnel training
Part 11 Financial support
Part 12 Project requirements
Part 13 General information
Part 14 Proprietary information

Part 1 Project Summary The project manager originates this section. It will be short in length and is prepared primarily with sufficient information to ensure understanding by top-level organizational officials who are interested in the key features of the project and what the project is intended to accomplish.

It should provide a brief description of the project and of the management structure, a summary of the guidance or constraints applicable to the project, the master project phasing chart, and the overall requirements of the project. The master phasing chart (Figure A-1) will contain a summary of the major milestones or key events contemplated in the project.

Part 2 Project Schedules The project manager prepares these schedules with the assistance of participating organizations. It is important that the project manager be party to the preparation of this section, since he or she is responsible for the overall compatibility and consistency of the participating organizations' roles and schedule requirements.

The project-schedules section should be prepared on a form similar to that shown in Figure A-2, using symbols illustrated in Figure A-3, and should provide a generalized picture of the major milestones, key events, or critical actions which the project manager deems vital to the execution of the project. Only the information required by participating organizations or agencies to determine the time periods or important dates applicable to their functional actions should be reflected. If detailed scheduling is required, event logic network techniques or individual-action Gantt charts can be used. Judgment is required in deciding whether or not to put all these networks in section 2 of the package plan, since they tend to become voluminous.

Part 3 Project Management The main purpose of this section is to provide the participating organizations with a summary of the management structure and philosophy applicable to the project. This section includes:

Details of how the project will be managed in terms of what is to be done, how it is to be done, who will do it, and when it is to be done

Specific identification of the participants in the project and a specification of their roles in terms of authority and responsibility

Identification of the advisory groups and committees that are required to support the project manager—their roles, functions, scope of activity, and relationships (authority and responsibility) to the other project participants

Copies of the memoranda of agreement that have been negotiated to support the project

Identification of all the parties that are participating on the project on a contractual basis, and the role these contractors play in managing some subsystem of the project

Specification of the management reports (the management information system) that are to be provided by the participants and used by the project manager in tracking the project

Organizational locations and functional breakdown of the project participants

A summary of measures to be taken for the protection of proprietary information

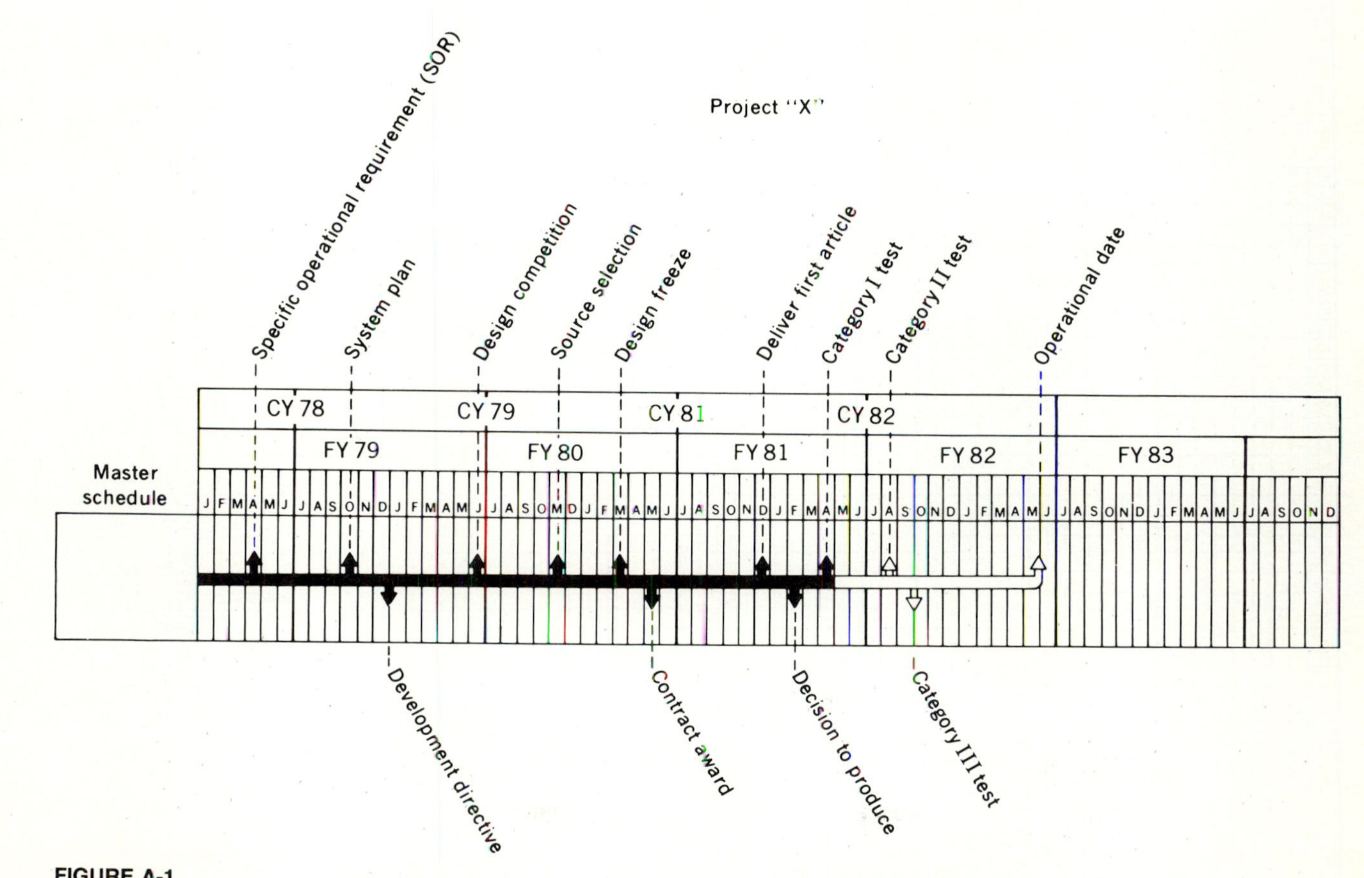

FIGURE A-1
Project master phasing chart for a weapons system.

PROGRAM SCHEDULE	SYSTEM (*Project*) NUMBER	SUBSYSTEM	TYPE OF SCHEDULE MASTER SUMMARY SCHEDULE	AS OF DATE

LINE		PRIOR SCHEDULE DATES	FY 1981 / CY 1981 (J F M A M J J A S O N D)	FY 19FY 1982 / FY 1983 / CY 1982 (J F M A M J J A S Q N D)	1984 (1 2 3 4)	1985 (1 2 3 4)	1986 (1 2 3 4)	COMPLETION DATES	LINE
1	Special operational requirement No. 2								1
2	PSPP approved								2
3	System program directive No. 3								3
4	Contract award								4
5	Preliminary model specification								5
6	Development engineering inspections								6
7	Make-up inspection								7
8	Contractor technique compliance inspection								8
9	First significant achievement								9
10	Start category I testing								10
11	End category I testing								11
12	Start category II testing								12
13	Start category III testing								13
14	Operational data								14
15									15
16									16
17									17
18									18
19									19
20									20
21									21
22									22
23									23
24									24
25									25
26									26
27									27
28									28
29									29
30									30
31									31
32									32
33									33
34									34
35									35
36									36
AUTHENTICATION			CY 1981 (J F M A M J J A S O N D)	CY 1982 (J F M A M J J A S O N O)	1983 (1 2 3 4)	1984 (1 2 3 4)	1985 (1 2 3 4)		

FIGURE A-2
Project schedules.

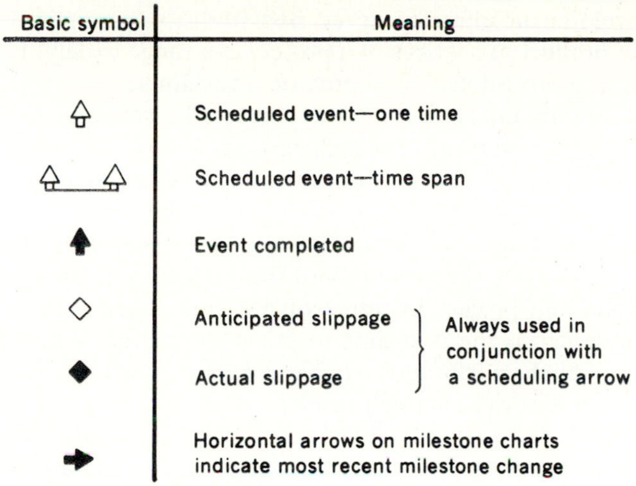

Basic symbol	Meaning
△	Scheduled event—one time
△——△	Scheduled event—time span
▲	Event completed
◇	Anticipated slippage } Always used in conjunction with a scheduling arrow
◆	Actual slippage } Always used in conjunction with a scheduling arrow
➡	Horizontal arrows on milestone charts indicate most recent milestone change

Figure A-3
Project milestone symbols.

Any unusual management agreements or conditions which are required to sustain the project

Specification of the control policy and procedures to be used in delivering the project on time, within cost limits, and with the desired performance characteristics

A charter for any special task teams that are to be organized for the project

Citation of management philosophy to be followed for the contingent or "withdrawal" conditions in the project activities

Procedures to provide for a periodic review of the project by all the participants used in the project

The project-management section is probably the most important section since it is the focus of the project actions. The section can propose the modus operandi to be followed in the project and define the interdependent roles of all the participants. It will be the most difficult to write; a considerable amount of negotiating is required to develop an understanding of the reciprocal roles that the participants have in the endeavor. While the technical aspects of the project can be demonstrated, the patterns of interpersonal relationships between the human sources can be critical; a well-thought-out plan to create the proper environment for the people can do much to alleviate the conflict that can arise when people from different parent units are working together on a common objective.

Part 4 Market Intelligence Traditionally, a market is defined as the place where buyers and sellers exchange goods. In our view of project management we must take certain liberties with that definition in order to make it meaningful and compatible with the definition of a project. Since we have held that a project can range from a product development effort to an ad hoc study of a corporate merger, the market for the product or services of a project can range widely. In other words, the objective of a project can be to provide a certain service to an organization. Presumably, the organization that is to receive the service has a need—or, perhaps better stated, a *demand*—for such service. If this is the case, we can proceed to reason as follows: *A market for the project is that environment where the product or service of the project is demanded.* Thus, the demand for specific recommendations regarding a corporate merger may stimulate the formation of a project group to study the feasibility of the merger. The project group that is created studies the project in terms of the cost (financial and opportunity), the schedule (when is the best time to effect the merger and on what types of time phasing), and the performance (what corporate performance, profitability, organizational continuity, etc., will result).

If we accept this definition, the market-intelligence section takes on added significance because the project manager and team must be as knowledgeable as possible about the environment in which the project is being conducted.

This section is an analysis of the environmental conditions which led to the project requirement, together with any special intelligence which would affect the project design and operation. In preparing this section, proprietary or sensitive information will be used, and it should be safeguarded carefully. It serves as a means of identifying the need to revive or alter the requirements for the project as the environment changes.

This section will include a current analysis of the competitive situation in the market (e.g., the competition to be faced by another company's product) and an estimate of all such potential capabilities to be encountered in the future. The basic forecasts which try to portray the environment at some future period of time will be contained in this section. The section on market intelligence is of prime importance to the project. It provides the master information file about the expected conditions under which the project is expected to survive.

Part 5 Operational Concept The operational concept will be originated by the project manager and contains in summary form the objective and a clear identification of the project or capability that it will replace or enhance. To be included are conceptual statements covering:

Objective
Limitations
Expected use
Readings
Support
Organizational Structure
Work force
Personnel
Training and education
Facilities
Availability dates

Part 6 Acquisition Section 6 portrays the development-test-production plan to achieve the project objectives as described in section 1. This includes:

Project description and performance, reliability, and maintainability. Describes the utility capability, anticipated performance, and design features and gives an illustration of the entire system.

Subsystem description. Essentially the same information will be shown here as for the above system.

Personnel area subsystem. A detailed explanation of the human factors of the project and how those factors integrate into the system.

Background, difficulties, and approach.

Test and evaluation data. Include test planning factors, test objectives, detailed test schedules, resources, test management, and test participants.

Production data such as manufacturing methods, processes, industrial facilities, tooling, special test equipment, quality-control procedures, and plant layout.

Part 7 Facility Support This section gives a description of the real facilities (plant, access roads, easements) required to support the project. If new construction is required, this section should contain a detailed plan of action to acquire the property, through lease, purchase, sale-leaseback, or other arrangements.

Part 8 Logistic Requirements This section will be originated by the organizational logistician and encompasses a comprehensive summary of the logistic support required for the project. It will include:

Logistic concepts, principles, and requirements in the areas of materials, supplies, spare parts, repairs, engineering, transportation, materials handling, quality control, test support, data processing equipment, support facilities (e.g., terminal facilities), and medical services

Material and supply control procedure

Peculiar or unusual circumstances that affect material and supply support requirements

Part 9 Work Force and Organization This section is concerned with the degree of work-force allocations for the project. It will become the basis for the recruitment action of the personnel department. This section will contain:

Assumptions and factors on which the work-force requirements are established

Projections of work-force requirements by type (salaried, wage-rated) and the skills required

Strategy for locating qualified people, e.g., through unsolicited applications, employment agencies, educational institutions, or professional meetings

Policies related to recruitment

Part 10 Executive Development and Personnel Training This section will be originated by the education and training office and is designed to provide a comprehensive summary of the personnel development and training required to support the project. It should be cross-referenced to other sections to reflect related actions and authorizations. The section contains:

Trained personnel requirements with specific identification of those peculiar to the project

Type, locations, and key dates of special courses that can be offered to increase the skills of the personnel

Specialized training equipment required

Planned management development program to include:

- An analysis of manager requirement
- An inventory of manager talent
- A determination of individual needs
- An appraisal of individual progress
- A means for program evaluation

Part 11 Financial Support This section, although compiled by the project manager, is a joint responsibility of all the project participants, and will be an integral part of the project package plan. The project manager, in collaboration with the finance officer, will be responsible for developing implementing instructions, ensuring overall consistency of data, and preparing and collating the section.

This section is designed to:

Provide the basis for the accounting period for the presentation of the estimates of *direct costs* of the project. Such estimates of direct costs will serve as backup data for the organizational budget.

Provide estimates of *total costs* (direct and indirect) for the development, production, and (if required) operation of the project (or product).

Cost categories will include:

Development: All resources (effort, material, facilities) for applied research, development, test, and evaluation directly associated with the project endeavor

Investment: Additional one-time resources (capital and operating) which are directly related to the establishment or buildup of a project, after an assessment of the usefulness of those assets which can be inherited from other organizations or activities

Recurring: Those resources (capital and operating) required to maintain the project during its projected operational phase, period by period

Part 12 Project Requirements This part will consist of a consolidation of the organizational requirements which generated the requirement for the project. Included would be such documents as a resolution of the board of directors, specific letters of authorization, approval of higher-echelon plans which this plan supports, unsolicited proposals, and bid packages. This section can be used as the central source of the overall documentation which justifies and authorizes the project.

Part 13 General Information There will be information which is not suitable for enclosure elsewhere in the project plan, but yet is vital to the plan documentation. This section should include a summary of the alternatives, such as trade-offs between costs and schedules considered in meeting the objective. A discussion of the variation in performance, schedule, and cost associated with each alternative should be included.

Part 14 Proprietary Information General matters relating to the security classification of the project should be included here. For example:

Physical security requirements and personnel background clearance (particularly appropriate when working on projects for the government)

A schedule to change the proprietary classification of the project as appropriate

Policy regarding the public release of information about the project

Instructions concerning any special handling of the equipment or documentation of the project

APPENDIX **TWO**

MASTER PROJECT MANUAL

1.0 General Information
- **1.1** Purpose
- **1.2** Name
- **1.3** Scope
- **1.4** Objectives
- **1.5** Statement of Work
  - X Company
  - Y Company
- **1.6** Production Specifications
- **1.7** Master Schedule
- **1.8** Configuration Status
- **1.9** Technology Summary
- **1.10** Cost Summary

2.0 Management and Organization
- **2.1** Project Management Concept
- **2.2** Project Charter
- **2.3** Role of Project Manager
- **2.4** Role of Functional Manager
- **2.5** Role of Manager of Projects
- **2.6** Project-Functional Relationship
- **2.7** Resolving Project-Functional Differences

2.8 Organization Chart
2.9 Key Project Personnel
2.10 Linear Responsibility Charts
2.11 Project-Management Systems
2.12 Project Implementation Procedures
2.12.1 Proposal Stage
2.12.2 Post-Proposal Stage
2.12.3 Contract Negotiations
2.12.4 Contract Award
2.12.5 Coded Cost Account Structure

3.0 Engineering Management
3.1 Systems Engineering
3.1.1 Introduction
3.1.2 Definitions
3.1.3 Policy
3.1.4 Organization
3.1.5 Systems-Engineering Management Tasks
3.1.6 Technical-Performance Management
3.1.7 Engineering-Test Management
3.1.8 Design Engineering
3.1.9 Product Engineering
3.1.10 Advanced Engineering

4.0 Project Planning and Administration
4.1 Definitions
4.2 Policy
4.3 Cost and Schedule Management Tasks
4.4 Project-Management Tasks
4.5 Project Control
4.6 Project Plans and Schedules
4.7 Financial Administration

5.0 Contracts Management
5.1 Contracts-Management Tasks
5.2 Post-Proposal Support
5.3 Contract Documentation
5.4 Contracts Management

6.0 Data Management
6.1 Data-Management Tasks
6.2 Data Requirements
6.3 Data-Management Plan

7.0 Financial Management
7.1 Policy
7.2 Organization
7.3 Financial-Management Project Tasks
7.4 Financial-Management Department Tasks
7.5 Financial-Management Interfaces
7.6 Project Profit Planning

- **7.7** Project Billing Tasks
- **7.8** Customer Cost Reporting
- **7.9** Cost Estimating Procedures
- **7.10** Internal Cost Reporting

8.0 Marketing Management
- **8.1** Organization
- **8.2** Marketing Services
- **8.3** Market Intelligence
- **8.4** Project Marketing
- **8.5** Proposals
- **8.6** Budgetary Estimates

9.0 Communications Policy
- **9.1** External
- **9.2** Internal

10.0 Reports
- **10.1** Trip Reports
- **10.1** Meeting Reports
- **10.3** Telephone Reports
- **10.4** Summary of Major Reports

11.0 Technical Summary
- **11.1** System X
- **11.2** Subsystem XI, etc.

12.0 Project Security
- **12.1** Proprietary Data
- **12.2** Work Classification
- **12.3** Visitations
- **12.4** Clearance Lists

13.0 Programming
- **13.1** Initial Reports
- **13.2** Monthly Report

14.0 Project Checklist
- **14.1** Systems
- **14.2** Management Systems

15.0 Meetings
- **15.1** Weekly Project Meetings
- **15.2** Internal Meetings
- **15.3** Customer Meetings

16.0 Quality Assurance
- **16.1** Requirements
- **16.2** Procedure

17.0 Reliability, Maintainability, Supportability
- **17.1** Requirements
- **17.2** Procedures

18.0 Field Service and Engineering
- **18.1** Requirements
- **18.2** Policy and Procedures

APPENDIX **THREE**

PROJECT-MANAGEMENT REVIEW CHECKLIST

Authority and Responsibility of the Project Manager

Are there any limitations on the project manager's executive authority? If so, identify and explain them.

Does the project manager exercise control over the allocation and utilization of all resources (people, money, material) approved in the project? If not, describe where control is lacking.

Are there any limitations upon the project manager's authority to make technical and business management decisions required by the project and authorized by the charter?

Does the project manager furnish information and requirements necessary for effective procurement planning and contract negotiation? Does the project manager approve all proposed contractual actions for the project consistent with the charter?

Does the project manager approve the scope and schedules of the project effort? Does the project manager approve the plans for accomplishing the objectives? If not, what limitations are placed upon him or her? Who imposes them, and by what authority? Who does approve the scope of the project, the schedule, and the plans?

How does the project manager report on the progress of the project?

Did the initial project manager have a primary role in the selection of key subordinates? Does the current project manager exercise this responsibility?

Does the project manager have a primary role in controlling the tenure of the key subordinates?

Did the initial project manager have a primary role in the determination of the organizational structure? Does the current project manager determine the organizational structure?

Does the project manager have a primary role in the assignment of tasks to the organization? Does the project manager, in turn, control assignments to others?

Project Charter

Does the project manager have a current and adequate charter approved by either the head of the organization having cognizance of the project or the head of the organization having dominant interest?

Is the project manager designated by name in the charter?

Does the charter designate the project elements or parts thereof for which the project manager will be responsible?

Does the charter define the interface relationships and the communication channels and identify the organizations which support the project manager in the following areas?

1 Production
2 Finance
3 Marketing
4 Contract administration
5 Customer communication

Does the charter provide for the project manager to control the allocation and utilization of all resources approved for the financial program?

Does the charter indicate the organizational and physical location of the project office and the organizations to provide administrative support? If not, what is lacking?

Does the charter delineate any special delegation of authority or exemptions from corporate policy?

Is the charter approved and signed by the chief executive officer? If not, by whom? What is the date of the charter? Is it current?

Does the charter clearly define the scope of the project?

Does the project charter provide that the project manager will do the following?

Organize, plan, and administer the project-management office
Make authorized technical and business management decisions
Establish initial and long-range objectives
Accomplish experimental test, engineering, and analytical studies
Delineate operational requirements, design specifications, performance specifications, technical approaches, etc.
Prepare a project master plan
Prepare, submit, and justify initial and long-range funding requirements
Exercise financial management controls over all allocated project funds

Define work efforts; approve plan of execution, scope, and schedule of work; and approve costs of work

Furnish information and requirements for contract negotiations

Approve, consistent with corporate policy, all proposed contractual actions

Establish and promulgate design interface specifications to ensure project integration

Respond to requirements of other project managers and head of functional activities in resolution of interface problems

Negotiate working agreements with organizations outside as appropriate

Develop and maintain the integrated logistic support plan for the project

Establish methods and procedures for project configuration control

Ensure that the quality assurance reliability, maintainability, and value engineering programs for the project are adequate

Ensure that technical documentation is prepared and available for concurrent delivery with hardware

Analyze project performance in relation to required performance specifications

Maintain a complete chronological history (significant events and decisions)

Institute appropriate management-control techniques (required by higher authority or selected) to provide status, progress, and forecasts

Report the current status and progress of the project to the appropriate people

Prepare a budget for and justify travel funds

Execute efficiency ratings for personnel

Does the project charter define the interface and operating relationships between (1) project manager and other designated projects, (2) project manager and functional groups, (3) project-manager and other agencies, etc.?

Does the project charter identify supporting organizations to participate in work in support of the project?

Does the project charter specify adequately personnel to staff the project-management office?

Does the project charter provide a staffing schedule?

Does the project charter identify personnel in liaison offices and in field organizations tied exclusively to, and under the management control of, the project manager?

Does the charter identify the organization responsible for "public information"?

Does the charter provide for a review for project disestablishment or a date for project disestablishment?

Is the charter in the proper format?

Introduction

Mission

Scope of project

Specific authority and responsibility of project manager

Specific interface and operating relationships of project manager
Personnel staffing for project-management office
Resources assigned to project
Project administrative support
Public information
Project disestablishment

Project Priority

Does the project have a priority? If so, what is the priority?

Project Complexity

What is the project manager's opinion of project objectives; specifically, do they have a significant effect on the organization's fortunes?

Does the project manager manage a group of projects which are conducted substantially on a concurrent basis with each having significant technical problems?

Does the project involve unusual organizational complexity or technological advancement?

Does the project require extensive interdepartmental, national, or international coordination or support?

Does the project present unusual difficulties which need expeditious handling to satisfy an urgent requirement?

Historical Data

Does the project manager maintain historical files?

Project Visibility

Is there any evidence that subcontractors have counterpart "managers" designated specifically and solely to manage their contractual efforts?

Project Manager's Rank

Does the project manager have sufficient executive rank to be accepted as the agent of the parent organization when dealing with outside organizations?

Project Manager's Staff

Is there evidence that the project-management staff is composed of persons with a high degree of technical and business managerial competence?

Is there evidence of recent experience in project management among the key subordinates of the project manager's staff?

What evidence is there of formal training in the special requirements of project management among the project manager's key subordinates?

Will any of the key subordinates of the project manager's staff (that he or she desires to keep) be available for the duration of the project?

Are all members of the project-management office assigned to the office on a full-time basis?

Communication Channels

Does the project manager have direct two-way communication between the office and key participants involved in support [contractors and/or bureaus, etc.] of the project which ensures timely and effective direction and interchange of information?

Reporting

Does the project manager provide formal written and/or oral briefings to top management on the status and progress of the project, including the identification of the problems?

Does the project manager attend formal briefings held by other company project managers?

Project Reviews and Evaluations

What procedure does the project manager use to identify problems and review the status of the project?

Personal contact with key subordinates?
Conference?
Formal, scheduled briefings by key subordinates?
Review of outgoing progress reports?
Combination of the above?
Other?

How frequently does the project manager review the status and progress of the project? How?

Do the project manager's procedures for program reviews and evaluations provide coverage for schedule accomplishment, technical performance, cost and logistic support, etc.?

Management Information Systems

Has the project manager applied management control techniques and developed information systems for effective control?

Financial Management

Does the project manager assess and document the effect of proposals to increase or decrease the resources authorized for the execution of the project upon cost, schedule, and performance objectives? How? Does the project manager reassess requirements? How?

Planning

Does the project manager have a project master plan? What is its station? Does the project master plan include the following?

Project summary
Project schedules
Management and organization plan
Market intelligence
Operational concept
Acquisition procedures
Facility support requirements
Logistics requirements
Work force requirements
Executive development and personnel training requirements
Financial support strategy
Policy for protection of proprietary data

Technical Direction

How does the project manager ensure the integration of schedule, performance, and cost considerations to manage the project?

Can the project manager issue technical instructions directly to prime contractors? To subcontractors? To others? If not, what are the limitations upon this authority?

Who in the project manager's organization exercises configuration change control?

How does the project manager ensure the adequacy of the following?

Space equipment
Training facilities and equipment
Documentation
Test equipment
Containers
Safety
Security (technical)
Failure analysis
Calibration of test equipment
Cost effectiveness
Reliability and maintainability

General

How has the project manager provided for ensuring an adequate implementation of the following?

Value engineering
Subcontractor information system(s)
Subcontractor performance evaluation

Does the project manager attend top-level policy meetings with the customers?

Has the project manager been overruled by seniors under customer pressure? If so, why?

Has the project been given sufficient publicity in the company paper?

Does the project manager encourage the primary project contributors to attend technical meetings or symposiums on related topics? Has the project manager arranged a visit by officials of the customer's organization?

Have procedures been established whereby outstanding contributions to the project by some of the participants can be recognized?

What assurance does the project manager have that the project contributors have developed a full understanding of the problem (as through the possession of a detailed statement of work)?

Have administrative procedures been set up in the project whereby creative groups are free from administrative paper work?

Have schedules been set up to conduct design reviews to assure in-process design adequacy?

INDEX

INDEX